ORGANIZING DATA FOR CIM APPLICATIONS

MECHANICAL ENGINEERING

A Series of Textbooks and Reference Books

1. Spring Designer's Handbook, *by Harold Carlson*
2. Computer-Aided Graphics and Design, *by Daniel L. Ryan*
3. Lubrication Fundamentals, *by J. George Wills*
4. Solar Engineering for Domestic Buildings, *by William A. Himmelman*
5. Applied Engineering Mechanics: Statics and Dynamics, *by G. Boothroyd and C. Poli*
6. Centrifugal Pump Clinic, *by Igor J. Karassik*
7. Computer-Aided Kinetics for Machine Design, *by Daniel L. Ryan*
8. Plastics Products Design Handbook, Part A: Materials and Components; Part B: Processes and Design for Processes, *edited by Edward Miller*
9. Turbomachinery: Basic Theory and Applications, *by Earl Logan, Jr.*
10. Vibrations of Shells and Plates, *by Werner Soedel*
11. Flat and Corrugated Diaphragm Design Handbook, *by Mario Di Giovanni*
12. Practical Stress Analysis in Engineering Design, *by Alexander Blake*
13. An Introduction to the Design and Behavior of Bolted Joints, *by John H. Bickford*
14. Optimal Engineering Design: Principles and Applications, *by James N. Siddall*
15. Spring Manufacturing Handbook, *by Harold Carlson*
16. Industrial Noise Control: Fundamentals and Applications, *edited by Lewis H. Bell*
17. Gears and Their Vibration: A Basic Approach to Understanding Gear Noise, *by J. Derek Smith*

18. Chains for Power Transmission and Material Handling: Design and Applications Handbook, *by the American Chain Association*
19. Corrosion and Corrosion Protection Handbook, *edited by Philip A. Schweitzer*
20. Gear Drive Systems: Design and Application, *by Peter Lynwander*
21. Controlling In-Plant Airborne Contaminants: Systems Design and Calculations, *by John D. Constance*
22. CAD/CAM Systems Planning and Implementation, *by Charles S. Knox*
23. Probabilistic Engineering Design: Principles and Applications, *by James N. Siddall*
24. Traction Drives: Selection and Application, *by Frederick W. Heilich III and Eugene E. Shube*
25. Finite Element Methods: An Introduction, *by Ronald L. Huston and Chris E. Passerello*
26. Mechanical Fastening of Plastics: An Engineering Handbook, *by Brayton Lincoln, Kenneth J. Gomes, and James F. Braden*
27. Lubrication in Practice, Second Edition, *edited by W. S. Robertson*
28. Principles of Automated Drafting, *by Daniel L. Ryan*
29. Practical Seal Design, *edited by Leonard J. Martini*
30. Engineering Documentation for CAD/CAM Applications, *by Charles S. Knox*
31. Design Dimensioning with Computer Graphics Applications, *by Jerome C. Lange*
32. Mechanism Analysis: Simplified Graphical and Analytical Techniques, *by Lyndon O. Barton*
33. CAD/CAM Systems: Justification, Implementation, Productivity Measurement, *by Edward J. Preston, George W. Crawford, and Mark E. Coticchia*
34. Steam Plant Calculations Manual, *by V. Ganapathy*
35. Design Assurance for Engineers and Managers, *by John A. Burgess*
36. Heat Transfer Fluids and Systems for Process and Energy Applications, *by Jasbir Singh*
37. Potential Flows: Computer Graphic Solutions, *by Robert H. Kirchhoff*
38. Computer-Aided Graphics and Design, Second Edition, *by Daniel L. Ryan*
39. Electronically Controlled Proportional Valves: Selection and Application, *by Michael J. Tonyan, edited by Tobi Goldoftas*
40. Pressure Gauge Handbook, *by AMETEK, U.S. Gauge Division, edited by Philip W. Harland*
41. Fabric Filtration for Combustion Sources: Fundamentals and Basic Technology, *by R. P. Donovan*
42. Design of Mechanical Joints, *by Alexander Blake*
43. CAD/CAM Dictionary, *by Edward J. Preston, George W. Crawford, and Mark E. Coticchia*

44. Machinery Adhesives for Locking, Retaining, and Sealing, *by Girard S. Haviland*
45. Couplings and Joints: Design, Selection, and Application, *by Jon R. Mancuso*
46. Shaft Alignment Handbook, *by John Piotrowski*
47. BASIC Programs for Steam Plant Engineers: Boilers, Combustion, Fluid Flow, and Heat Transfer, *by V. Ganapathy*
48. Solving Mechanical Design Problems with Computer Graphics, *by Jerome C. Lange*
49. Plastics Gearing: Selection and Application, *by Clifford E. Adams*
50. Clutches and Brakes: Design and Selection, *by William C. Orthwein*
51. Transducers in Mechanical and Electronic Design, *by Harry L. Trietley*
52. Metallurgical Applications of Shock-Wave and High-Strain-Rate Phenomena, *edited by Lawrence E. Murr, Karl P. Staudhammer, and Marc A. Meyers*
53. Magnesium Products Design, *by Robert S. Busk*
54. How to Integrate CAD/CAM Systems: Management and Technology, *by William D. Engelke*
55. Cam Design and Manufacture, Second Edition; with cam design software for the IBM PC and compatibles, disk included, *by Preben W. Jensen*
56. Solid-State AC Motor Controls: Selection and Application, *by Sylvester J. Campbell*
57. Fundamentals of Robotics, *by David D. Ardayfio*
58. Belt Selection and Application for Engineers, *edited by Wallace D. Erickson*
59. Developing Three-Dimensional CAD Software with the IBM PC: A Practical Guide to Microcomputer-Aided Drafting and Design, *by C. Stan Wei*
60. Organizing Data for CIM Applications, *by Charles S. Knox with contributions by Thomas C. Boos, Ross S. Culverhouse, and Paul F. Muchnicki*

Mechanical Engineering Software

Spring Design with an IBM PC, *by Al Dietrich*

Mechanical Design Failure Analysis: With Failure Analysis System Software for the IBM PC, *by David G. Ullman*

ORGANIZING DATA FOR CIM APPLICATIONS

Charles S. Knox
Knox Consulting Services
St. Paul, Minnesota

with contributions by
Thomas C. Boos
Ross S. Culverhouse
Paul F. Muchnicki

MARCEL DEKKER, INC. New York and Basel

LIBRARY OF CONGRESS CATALOGING-IN-PUBLICATION DATA

Knox, Charles S.
Organizing data for CIM applications.

Includes index.
1. Computer integrated manufacturing. 2. CAD/CAM systems. I. Title.
TS155.6.K675 1987 670.42'7 87-15740
ISBN 0-8247-7768-9

MARCEL DEKKER, INC.
270 Madison Avenue, New York, New York 10016

Current printing (last digit):
10 9 8 7 6 5 4 3 2 1

PRINTED IN THE UNITED STATES OF AMERICA

I dedicate this book to the following companies, who hired me and taught me so much

Iowa Manufacturing Company, Cedar Rapids, Iowa
Caterpillar Tractor Company, Peoria, Illinois
Allis Chalmers Corporation, Milwaukee, Wisconsin
Control Data Corporation, Minneapolis, Minnesota
Minnesota Mining and Manufacturing, St. Paul, Minnesota

I also dedicate this book to the following companies, who courageously hired me to assist them as a consultant

American Hoist & Derrick, St. Paul, Minnesota
Anaren Microwave, Inc., Syracuse, New York
Babcock & Wilcox, Barberton, Ohio
Bally Manufacturing Company, Chicago, Illinois
Bose Corporation, Boston, Massachusetts
Buhler-Miag, Minneapolis, Minnesota
Chemical Industry, Bucharest, Romania
Communication Techniques Company, Budapest, Hungary
F. L. Smidth Company, Copenhagen, Denmark
Domain Industries, New Richmond, Wisconsin
Hazelett Strip-Casting Corp., Colchester, Vermont
Ingersoll-Rand Co., Turbo Products Division, Phillipsburg, New Jersey
Ingersoll-Rand Co., Cameron Pump Division, Phillipsburg, New Jersey
Mitsubishi, Hiroshima, Japan
Morgan Construction Co., Worcester, Massachusetts
New York Air Brake Corporation, Watertown, New York
North American Turbine, Houston, Texas
Orion Electric, Budapest, Hungary
Pacific Engineering Co., Henderson, Nevada
Seat (Fiat) Automotive, Barcelona, Spain
Standard Electric (ITT), Rio de Janeiro, Brazil
Sulzer Brothers, Winterthur, Switzerland
Verolme Shipyard, Rotterdam, Holland

Finally, I dedicate this book to the many fine men and women in these companies with whom I have worked for their personal dedication to improvement of manufacturing through modern techniques and effective creativity.

PREFACE

During the past ten years hundreds of seminars have been held to discuss data base formats, types, and theories. It seems that everyone in the United States should have attended one by now. In these presentations such terminology as *schema* and *subschema* are used with abandon and are combined with *"flat" files*, *hierarchical data model*, *network structures*, and *relational data model*. Unfortunately, the data increments in these data bases, their interrelationships, and the systems which use them are not always considered as assiduously.

This book does not pretend to cover the technical aspects of data base types and their optimum utilization. Instead, it is dedicated to describing the types of data and data relationships necessary for an integrated engineering manufacturing system, including items normally identified with computer-aided design/computer-aided manufacturing (CAD/CAM). The general data contents described here may be considered those substantially necessary to implement computer-integrated manufacturing (CIM).

A large, monolithic, data base is not mandatory for CIM. In fact, the main illustration used in this book implies several data base *segments*, each designed both to be used as a separate entity and to work smoothly with other segments. This concept, with careful design and wise use of computers and data bases, will forever free us from the tyranny of large computers and monolithic data bases. This does not mean large computers will be useless--on the contrary. Once data base system segments are established, those segments requiring large quantities of data can be placed on a large computer without damaging the integration principles. The goal has much merit. Implementation of the concept will take time and effort.

The following explanation of the chapters may help the reader to utilize this book fully.

1. The word *integration* has been greatly misused in business systems. Lip service to this important concept has been universal. The reality behind it probably has not been examined as thoroughly as it must be if CIM is to be implemented in its full meaning. Let me cite just one example. Suppose a rule is made which states that, once an engineer designs a part and the geometry is entered into a computer, the rest of the disciplines in engineering and manufacturing must use this data with no alteration. Won't this frighten a lot of people? Won't someone say, "I don't trust the data an engineer creates?" The Introduction of this book examines some of the islands of data presently available and questions how we must change to make CIM a reality. It also asks the question, "How will we find such data after it is created?"

2. Chapter 1 describes the fundamentals of a data base system with simple examples to explain data associations.

3. Chapter 2 describes and identifies the key "input" data necessary for CIM. The data is often described and pictured as though paper forms are filled in and a person enters it into a computer data base. The future implies that much data will be entered into the computer on line or computed using design analysis programs or preestablished menus or formats from which the data can be selected.

The 12 data inputs shown on the chart are in the general sequence in which they are used. While the data on such input may be equal to several hundred data increments, only a general description of such data has been prepared. For a rigorous description of these data increments, classical data processing or CAD/CAM graphics systems need to be studied and such increments extracted. While the use, inclusion, or omission of specific data increments is a highly important consideration of an effective CIM system, describing each would entail such detailed explanations that the original concept for developing an overall picture of a CIM system would be lost.

Not all input needed for a specific industry or manufacturing plant may be itemized. Special requirements may need to be added for customizing a particular engineering or manufacturing facility or function.

4. Chapter 3 expands the data requirements for CIM. Resident information may be considered part of the data base contents. It may also be considered in another way as the "working" data base, or those increments of data required to operate the engineering manufacturing system. Until such data is resident in a data base, an effective operating system is probably impossible. Where it resides, the manner in which it is stored, and the data increment relationships and redundancies, if any, are planned in a data base analysis of the entire system requirements.

5. Chapter 4 explains the segments of data needed for CIM. As emphasized previously, these data base segments are not named or necessarily organized in a classical data processing manner. In some cases, such as part geometry and perishable tools, considering this data as part of the integrated whole has only been established recently. If it is assumed that the resident information described is resident in these data bases, it is apparent that care must be taken to establish concise and logical arrays of data and data association to make the system operate effectively.

6. Chapter 5 has been designed to present a more detailed description of the CAD/CAM portion of CIM. Since the use of a geometric data base and associated designs are still going through a significant transition period, the exciting ideas presented in this concept of feature-based design using entities familiar to the engineer and the use of solid geometry in the concept are worth such detail.

7. Chapter 6 is defined as unstructured data. The unstructured output information can be considered operating data or analysis data available as a result of the data increments available in the data base or as the results of calculations using the data base and dynamic requirements, such as the master schedule. Only a small portion of data combinations have been shown here. For inventory control alone there may be several different on-line screens or reports with various data increments used to operate the system.

8. Chapter 7 is defined as structured data. The structured outputs described here are classical since most of them have existed in manual as well as mechanized systems previously. It is important to emphasize these critical outputs to indicate the broad relationships of data in an integrated CIM system.

9. Chapter 8 is another chapter devoted to expanding on one ingredient of CIM. It is singularly important to retrieve old data for reuse and/or to avoid generating redundant data. The impact of a design retrieval system on CIM is emphasized.

10. Chapter 9 is also intended to expand on one ingredient of CIM. It is common knowledge that significant economic benefits will be gained in the use of CIM within the manufacturing environment. For many years, the use of industrial engineering has not been emphasized sufficiently, so it is hoped that this chapter will help one recognize this basic and important component of a manufacturing company.

11. Chapter 10 challenges management to change. The best computer, the best data base manager, the best data, and the best of CIM ingredients will fail us until management becomes aware of the needs for their active participation. One of the authors remembers a day in the 1960s when a particularly advanced data collection system was turned down by management of a company as "too advanced" and "dangerous." A colleague remarked, "We are fighting

World War II with a Spad." Modern systems require modern management.

12. Appendix A is a glossary of terms describing each of the items found on the chart used throughout the book.

13. Appendix B is a summarized set of requirements for a successful CIM system. Extensive references are given. They have been included to help the reader if such basic requirements need review and implementation.

We hope this book will jog the memory of any and all who are embarked on the implementation of CIM. The successful use and the future of computer, data base, and business system relationships rest on those who conceive, design, and implement them. They need all the help they can get. If this book helps in any way, we have met our goal.

CHARLES S. KNOX

CIM DATA BASE
HOW TO READ THE CHART

The accompanying chart (p. xiii) has been devised to illustrate the interrelationship of data, activities, and data base in a computer-integrated manufacturing system. Since it represents a comprehensive view of the main segments and actions of a CIM system, it has been compressed so that a general picture of the relationships may be illustrated.

The chart is divided into 5 segments for clarity:

12 Input data
19 Resident information
17 Data base segments
20 Output information
10 Structured outputs

INPUT DATA

For anyone intimately familiar with an integrated engineering manufacturing system, the 12 inputs designated and described in a separate section may seem grossly inadequate. Endless numbers of different input forms or on-line input screens actually exist in such a system. An attempt has been made here to express the major and critical types. Various means to update inventory and other files, variations to the different inputs, and many others, are not indicated. It is believed, however, that those which are itemized will cover a substantial amount of activity. The 12 data inputs are:

Part Number Log
Release Notice
Parts List
Engineering Changes

Generic Bill of Material
Geometry Features
Part Design
Assembly Design
Manufacturing Geometry
Process and Routing
Master Schedule
Inventory-Labor Transactions

RESIDENT INFORMATION

In one sense, resident information can be considered input data also. However, the difference is that such data is developed and maintained as a series of closely integrated but often separate sets of data which are stored in various data bases and are either the result of a "mass" input, or the result of activities in the use of the system.

The 19 resident information segments are:

Design Standards
Drafting Standards
Design Analysis
Finite-Element Meshes
Design Catalog
Family Geometry
Parametric Drawings
Simulation Models
Configuration Management
Producibility Tips
Traceability Structure
Microfilm Microfiche
Process Standards
Family Processes
Work Center Parameters
Process Polycodes
Transit Times
NC Tool Paths
Labor Material Costs

DATA BASES

The data bases designated on the chart have not been segregated in a classical manner, but are shown to emphasize both new types of data which must be recognized for a CIM system as well as to emphasize segments of the data base which have, in the past, been considered of less importance. The segregation shown should in no way be considered rigorous in that, through analysis, segments of these data bases may be combined, added to, or made subordinate one to another. Thus the data bases shown here are more for purposes of explanation of data and its relationship to CIM rather than a description of the optimum means for storage and manipulation of data.

The 17 data base segments are:

Design Retrieval
Family Processes
Part Family Data
Process Polycodes
Item Masters
Perishable Tools
Product Structures
Jigs and Fixtures
Engineering Changes
Production Planning
Part Geometry
Inventory Control
Assembly Geometry
Group Technology Cells
Generic Bill of Material
Raw Material Inventory
Electronic Mail

Please notice in previous and subsequent sections that a table showing each of these data base segments has been prepared with a "O" or an "X" tabulated beside each of them for the subjects being discussed. This table is an extraction from the chart and shows whether there is a relationship between the item being discussed and that segment of the data base.

OUTPUT INFORMATION

Output information should be considered that which may be examined, analyzed, or interrogated on an informal basis or used as a master output file of data for back-up. The data may, however, never appear in the form of hard copy but be available on-line as required. For example, the Item Master represents a complete record of all parts and associated data of the Item Master data base. Not only could this be considered a back-up document it may also be considered a data base which is on-line for interrogation.

The 20 sets of output information are:

Item Master
Parts Catalog
Old Drawings
Indented Bill of Material
Level by Level Bill of Material
Summarized Bill of Material
Where Used Bill of Material
Parametric Drawings
Job Shop Bill of Material
Layouts Design Models Simulations
Group Technology Planning
ABC Inventory Analysis
Purchase Order Status
Material Requirements
Production Schedule
Shop Floor Control Data
Inventory Control Data
Purchasing Data
Costed Bill of Material
Quality Assurance Data

STRUCTURED OUTPUT

Structured output may be considered formal documents used either in the course of business and/or as a formal company document. For example, the detached parts list is actually an extension of the assembly drawing and must be considered so.

The 10 structured outputs are:

Microfilm Microfiche
Part Geometry
Part Drawing
Assembly Geometry
Assembly Drawing
Detached Parts List
Picking List
Effectivity
Routings
Processes

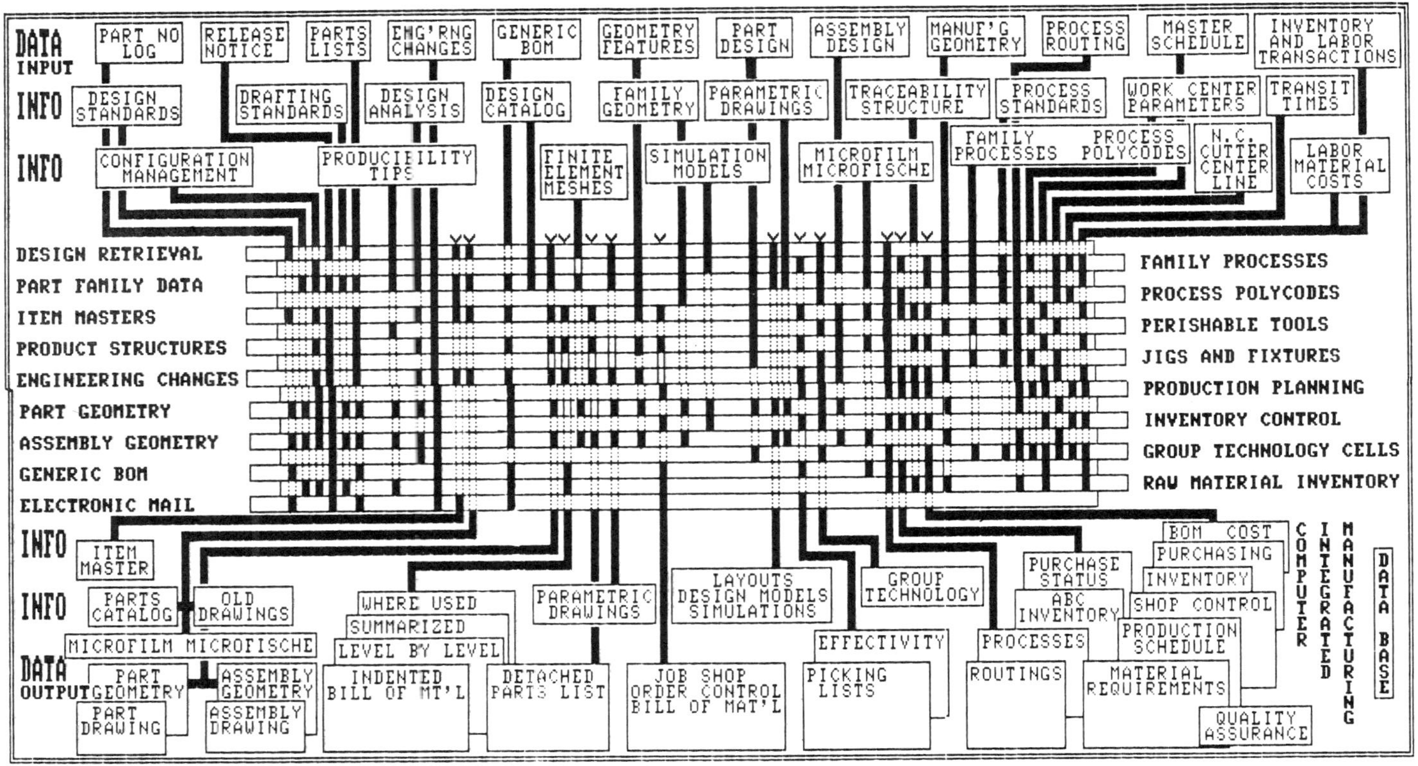
DATA INPUT
PART NO LOG
RELEASE NOTICE
PARTS LISTS
ENG'RNG CHANGES
GENERIC BOM
GEOMETRY FEATURES
PART DESIGN
ASSEMBLY DESIGN
MANUF'G GEOMETRY
PROCESS ROUTING
MASTER SCHEDULE
INVENTORY AND LABOR TRANSACTIONS
INFO
DESIGN STANDARDS
DRAFTING STANDARDS
DESIGN ANALYSIS
DESIGN CATALOG
FAMILY GEOMETRY
PARAMETRIC DRAWINGS
TRACEABILITY STRUCTURE
PROCESS STANDARDS
WORK CENTER PARAMETERS
TRANSIT TIMES
INFO
CONFIGURATION MANAGEMENT
PRODUCIBILITY TIPS
FINITE ELEMENT MESHES
SIMULATION MODELS
MICROFILM MICROFISCHE
FAMILY PROCESSES
PROCESS POLYCODES
N.C. CUTTER CENTER LINE
LABOR MATERIAL COSTS
DESIGN RETRIEVAL
PART FAMILY DATA
ITEM MASTERS
PRODUCT STRUCTURES
ENGINEERING CHANGES
PART GEOMETRY
ASSEMBLY GEOMETRY
GENERIC BOM
ELECTRONIC MAIL
FAMILY PROCESSES
PROCESS POLYCODES
PERISHABLE TOOLS
JIGS AND FIXTURES
PRODUCTION PLANNING
INVENTORY CONTROL
GROUP TECHNOLOGY CELLS
RAW MATERIAL INVENTORY
INFO
ITEM MASTER
INFO
PARTS CATALOG
OLD DRAWINGS
MICROFILM MICROFISCHE
DATA OUTPUT
PART GEOMETRY
PART DRAWING
ASSEMBLY GEOMETRY
ASSEMBLY DRAWING
WHERE USED
SUMMARIZED
LEVEL BY LEVEL
INDENTED BILL OF MT'L
PARAMETRIC DRAWINGS
DETACHED PARTS LIST
LAYOUTS DESIGN MODELS SIMULATIONS
JOB SHOP ORDER CONTROL BILL OF MAT'L
GROUP TECHNOLOGY
EFFECTIVITY
PICKING LISTS
PURCHASE STATUS
ABC INVENTORY
PROCESSES
ROUTINGS
BOM COST
PURCHASING
INVENTORY
SHOP CONTROL
PRODUCTION SCHEDULE
MATERIAL REQUIREMENTS
QUALITY ASSURANCE
COMPUTER INTEGRATED MANUFACTURING
DATA BASE

CONTRIBUTORS

Thomas C. Boos, General Manager, Architectural Design and Control, Control Data Corporation, Arden Hills, Minnesota

Ross S. Culverhouse, Principal Consultant, Architectural Design and Control, Control Data Corporation, Arden Hills, Minnesota

Charles S. Knox, Knox Consulting Services, St. Paul, Minnesota

Paul F. Muchnicki, CAM Consultant, Professional Services, Control Data Corporation, Waltham, Massachusetts

CONTENTS

ORGANIZING DATA FOR CIM APPLICATIONS

INTRODUCTION: WHAT IS INTEGRATION?

THE PROBLEM

Manufacturing companies are embarked upon an ambitious program to fuse their individual versions of CIM into an integrated system which will affect both the engineering and manufacturing systems within their company. The "total" system concept is being studied. Inherent within this plan is the need for an integrated database and a data base manager.

How does a manufacturing company receive this idea? How well can it absorb, implement, and utilize it? Unfortunately, the integrated system philosophy is being planned by manufacturing companies who have never had a truly integrated manual system either within a department or between departments. As managers of engineering and manufacturing become more aware of the implications of CIM, they perceive the need for a larger computer with a commensurate high cost or a combination of low-cost turnkey systems.

The turnkey system, even though limited in application, is packaged and priced low enough that a chief engineer or manufacturing manager can authorize purchase of the system without preparing a formal presentation for the board of directors. The cost of CIM often seems to eliminate this course of action. Since the interested management is not clear as to what the "total" system includes, and does not understand how an integrated data base will increase CIM productivity, it is difficult for them to picture how to develop a justification for presentation.

There is also uncertainty as to the contents of the database and how to obtain, maintain, or retrieve data from it. Since the present manual database of designs, drawings, and bills of material are used sporadically, concern about how future data will be used more effectively is also expressed. These uncertainties about the unknown make the potential customer extremely wary in considering CIM.

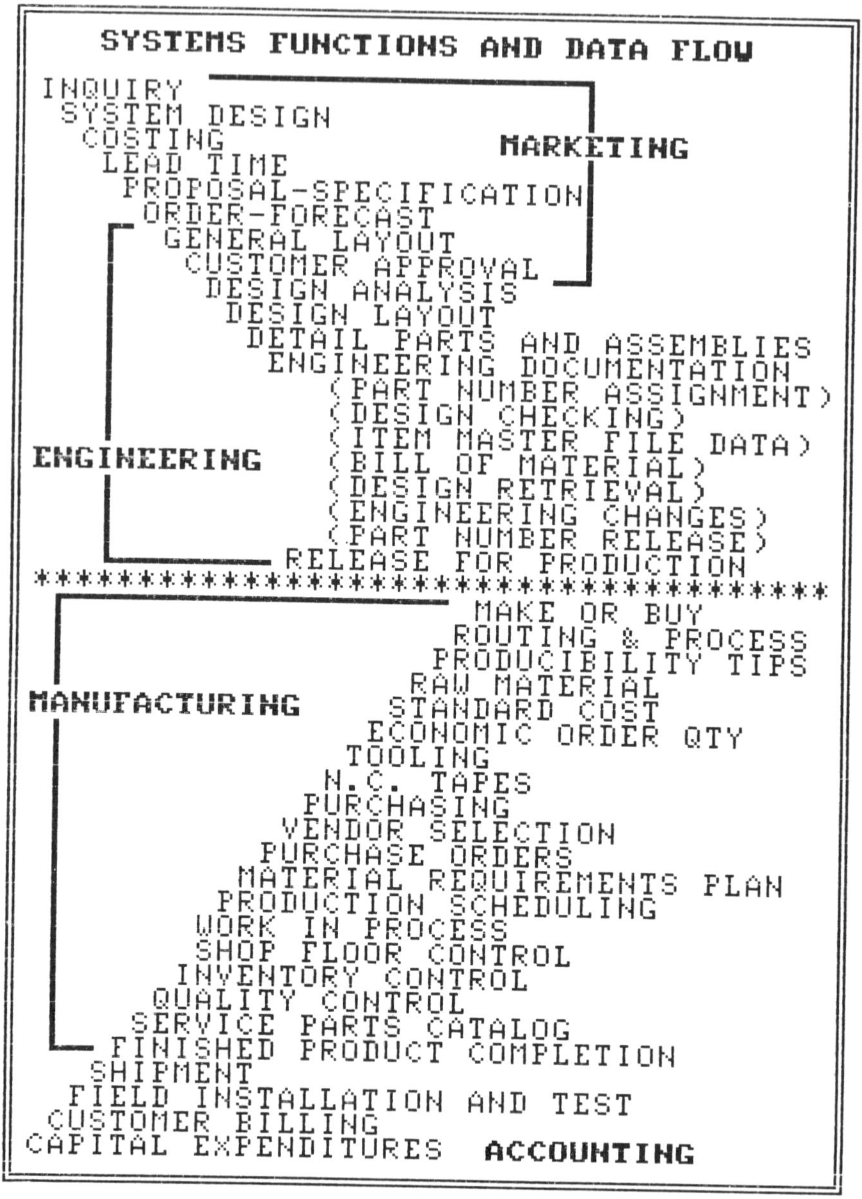

FIGURE 1 Organization of a database must accommodate the flow of data through these functions and activities.

A paradox is found also when describing CIM, the CIM data base, and flow of data. Figure 1 shows the general system ingredients of a manufacturing company. These activities are listed in the general sequence of events or flow of data. The general illustration for this book shows no flow, only data input, storage segments, and outputs. These two pictures seem to represent different concepts. In fact, each illustration represents one of potentially many views of a manufacturing company. Figure 1 shows a general sequence of events while the general illustration shows the data requirements to achieve them. In fact, in the future most sequential flow charts will need to show more interaction, in and out, from databases in conjunction with event sequences to completely explain CIM concepts.

GEOMETRY-DRIVEN CIM

Let us consider the new types of data available from a CIM system and consider the geometry of the designs as being source data. This geometry "drives" the rest of the system. The approach presumes that the geometry of a design, once developed, is "pushed down" from engineering to all other departments who need it. Unfortunately, and in all innocence, this does not describe the present segmentation of engineering and manufacturing. Both engineering and manufacturing are divided into a series of compartments, each performing individual activities. The same data is being continually regenerated. This causes each compartment to be loaded with redundancies and inefficiencies. Thus care must be taken in explaining the concept and developing means to implement the concept effectively.

SEGMENTATION

Since engineering departments do not have an integrated system today, even the integrated data base concept may be meaningless and/or unusable. An illustration of the activities performed in each compartment of a manufacturing company will assist in clarifying this statement. The segmentation of both engineering and manufacturing are pictured in Figure 2.

Marketing

Marketing is the first compartment which might use the system. Today they may require a new product or a variation of an old one. A sketchy, incomplete, and hurried specification with time constraints may be prepared and passed on to engineering. The database concept is first ignored here. Perhaps a design variation meeting the

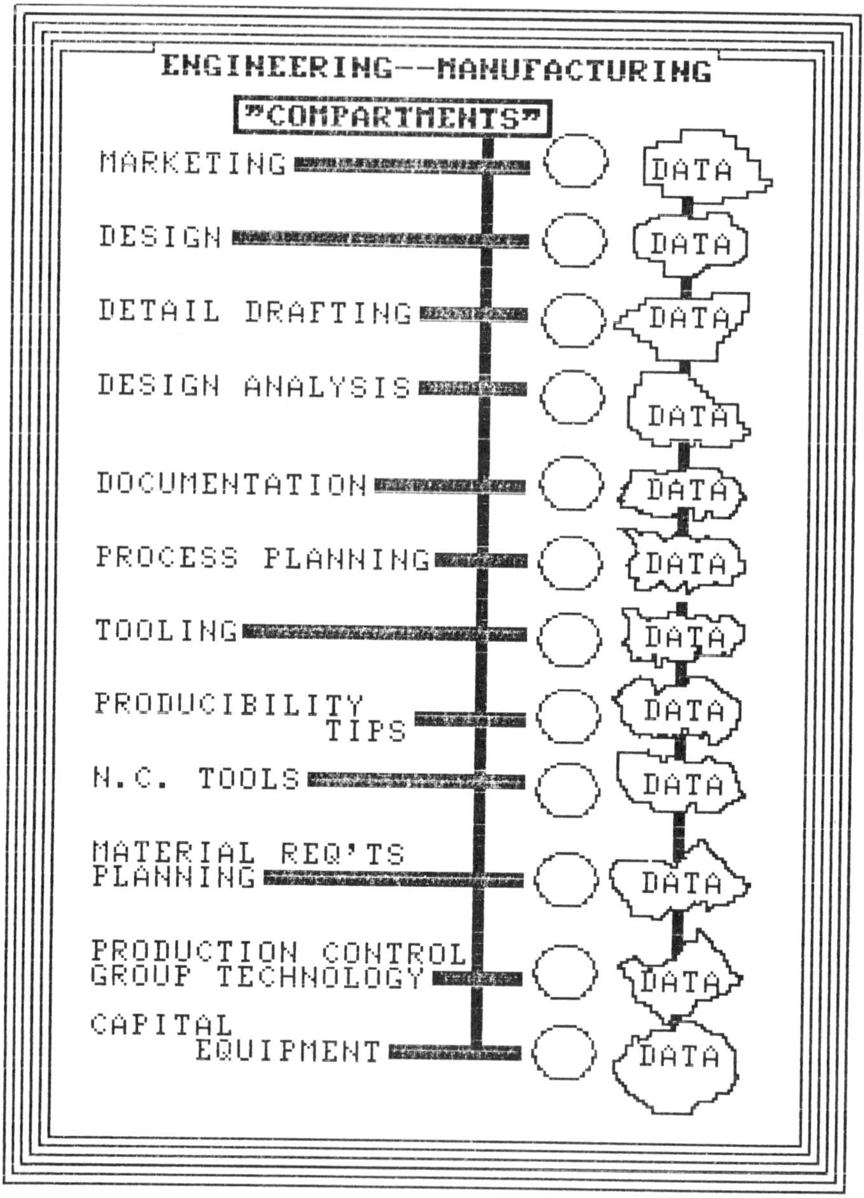

FIGURE 2 The present engineering manufacturing compartments are not conducive to either an integrated system or an integrated database.

exact marketing requirement has already been engineered. However, so many variations have been built in the past, and described and filed so poorly, that marketing leaves it up to engineering to see if one exists. (It should not be thought that this and following activities are based only on fast turnaround requirements. The same actions will be found in a rational environment also.)

Design

Since time constraints are significant, the new project is assigned to a senior designer. The word *senior* has no direct correlation to ability. It does mean length of service and thus a larger recall of previous designs. Once again, the total system will be ignored. Perhaps there are 10 senior designers with comparable lengths of service. Each has a wealth of knowledge about various products which have been designed in the past. However, there is no simple or logical means for the senior designer assigned the new project to interrogate all of them, or find their past designs. Thus he will rely on his own knowledge and memory only, thereby ignoring and eliminating a significant amount of information.

The senior designer will try to remember what he has designed previously that meets the new requirement. Suddenly he remembers a project he developed 10 years ago. Referring to the customer file, he finds the proper customer order number, obtains a bill of material, and pulls all the design layout drawings from the drawing file. Once again the database has been ignored and lies unused. Never mind that new design concepts, new design standards, new methods for performing the same function have been successfully prepared as a product during the past 10 years. The designer has found a familiar, if antiquated, design, and he is content.

The designer does not next obtain a fresh piece of drawing paper. Rather, he makes a blue-line or sepia of his old design, and marks it up to show the new design variations. Where a complete new set of assemblies are required, he does not refer to any old assemblies in the drawing file—there is rarely a method for him to find them—so he ingeniously has them drawn from the beginning, making new assembly drawings as required. It doesn't matter that potentially hundreds of assemblies and parts (with better and more economical design features) have been prepared. He has developed new ones. The database has again been avoided.

Detail Drafting

When the designer has completed his set of layouts, he turns them over to another compartment of engineering—drafting. The drafter surveys the design layouts, and begins to break them down into piece parts. Although hundreds, perhaps thousands, of piece part designs are already lying in a drawing file, he knows better than to search for the series of his 20 or 30 new parts needed among the 60,000 drawings on file. He would be fired for doing so—a waste of time. Consequently he creates them. With time constraints, he will care very little whether they are designed for economical manufacturing. He will, instead, breathe a sigh of relief that he has ended his share of the project on time when he releases 30 new part numbers and their drawings.

Design Analysis

While the previous design steps are in process, it may also be necessary to analyze the design for structural integrity. The senior designer may follow one of two courses. Since he is familiar with his old design, and if the new specification upon which he is working is not too radically different, he may avoid any analysis at all, and increase the thickness of a plate, the size of a bearing, or the diameter of a shaft based on experience or "gut" feeling. He disregards the facts that the factor of safety of the new design is now 40, the weight is twice that necessary, the cost is 50% higher than necessary—it works. In the noncompetitive days when anything sold, this lack of sound and economical design was lost in the congratulations for a project finished on time to meet the customer's delivery data. (However,, many of these same products became popular and are now under heavy price competition from designers in other countries and anyone else who designs them correctly, more efficiently, and more economically.)

The senior designer may take the other course, however, and submit the design for computer analysis. This calculation is performed in still another compartment of engineering, jealously guarded and segregated from the others. Once again, the database will be avoided. The computer analyst, skilled in preparing a mesh for finite element analysis (FEM), may have never seen this type of design before. His neighboring analyst may have just finished one, but there is no way to find it, or know he did so. Consequently, the analyst prepares a completely new mesh from the beginning. With no previous guidelines to go by, many consultations must be scheduled with the senior designer and many calculations performed to make certain the results are correct. If this sequence of events doesn't make the project late, it may make it impossible to redraw the design to utilize the optimum materials. Therefore, if the design analysis proves the design is adequate or more than adequate, chances are it will not be changed to take advantage of the economies discovered.

nade nonstandard, requiring a complete new stock of
. Dimensions on fabricated parts may require pur-
ndard material and thus create additional raw material
art may be dimensioned, so it must be burned from
ther than sheared from standard bar stock. Tight
require grinding instead of turning. Hundreds of
variations may exist. The planner has two choices.
is to call the engineer, persuade him the changes
nd worthy of a drawing change, and then wait as
ks for the paperwork.
choice (which usually becomes the first choice) is to
s as designed and place the burden and cost of manu-
the shop. Some planners may be new to their position.
ecognize these design aberrations. Thus proliferation
oling and manufacturing costs will continue. Neither
d drafter nor the process planner has a data base
tips which can be used to eliminate, at the design
oblems affecting cost. No database is available. (If
tinues to call the senior designer weekly and remind
gn problems he has generated, the planner will not
r in engineering. He most certainly won't be invited
nas party.)

ntrolled Machine Tools

plan is prepared, certain, or all, operations may be
numerically controlled (NC) machine tool. The new
s transmitted to a parts programmer, who faithfully,
eproduces the portion of the part geometry he requires.
res the proper tool paths, inserts specific machine tool
makes a paper tape or other medium for the machine
e part can be manufactured. He has previously pre-
geometry for other parts, but it is difficult to find
tter center lines, so the new part data is prepared
ning.
t programmer prepares the cutter center line, he
adii, grooves, and drilling and tapping operations that
ified if minor dimensional changes were made. Past
s him these changes probably aren't significant to the
uality control. He makes them to simplify the machine
. Thus an unrecorded design has been prepared. (If
during operation of the product, marketing and en-
spend weeks creating a host of new designs to allevi-
m without ever finding the source of the real problem.)

Documentation

Once the drafter has completed his portion of the project, he may be saddled with the paperwork for preparing computer data for the release notice or item master, and the "parts list" or bill of material. Never mind that when the project was first introduced to the senior designer, it was evident that 60% of the old design was reusable. The designer did not enter this data into the computer at the beginning, nor was each new design entered as it was completed, allowing manufacturing the opportunity to receive data in an orderly flow. The entire project is entered in a "gush." Some of the parts required have been noted on the layouts. Some have been noted on the assemblies. Someone must now find all this miscellaneous data, make sure none is overlooked, and enter it into the bill-of-material system accurately. This creates a flood of paperwork because the bill-of-material system isn't on line to the computer. Thus, another compartment of the company, engineering services (and/or data processing), must reinterpret the data from the paperwork, keypunch, verify, confirm, reconfirm, correct, and resubmit the data, all until final approval is given by the drafter or designer.

Retrieval

What has this explanation of the various departments described? The answer is simply that an "integrated" database probably doesn't exist today, and if it did, no one would or could use it easily. Why? Several reasons exist:

1. Rarely is there a logical means of data or design retrieval.
2. There is no rapid means of retrieval.
3. There is a natural suspicion of using someone else's design data.
4. Some of the key data, such as FEM meshes, has no means for being filed for easy reference.
5. There is rarely means for finding data other than one's own.
6. There is no easy means for storing combinations of data.
7. There is no human memory large enough to remember all that has gone before.
8. There has been no continuous major effort in engineering to cut costs of new designs.
9. There is no widespread incentive for engineers to work together.

When a manufacturing company begins to understand the present state of affairs, the manager of engineering immediately thinks either consciously or unconsciously of many questions:

1. If the present manual system, containing designs, assemblies, specifications, part drawings, and other data, is not being utilized effectively today, how will an integrated database be more cost effective and useful?
2. In many complex designs, there may be several iterations through design concept, analysis, layout, detailing, and back again. How will this be changed and improved?
3. What should be stored in the database?
4. If geometry is stored in the computer by part number, won't the file eventually get so large it will resemble our present large drawing file where we can't find anything now?
5. Where and how will we store all the meshes we generate for FEM? Should we?
6. How will we persuade those senior designers to enter item master and bill-of-material data as it becomes available? Should we? How will they enter it? Will it assist them in their design responsibilities?
7. Would it be better to prepare new design layouts of old projects for later variations, or continue to mark up old drawings?
8. Should we find some way for marketing to find old designs quickly so they won't bother engineering so much?
9. How will we persuade one designer to use the designs of another?
10. If we find a design has been over designed, how will we correct and change it easily if we have a project completion deadline?
11. Can we economically justify looking for old piece part designs, or should we continue their proliferation?
12. If designer uses a design that is 10 years old to make modifications for a new one, how will the database concept change that?

Let us move along with our description of the various compartments, continuing with manufacturing.

Process Planning—Parts

Just as engineering releases the new project in a gush, so does manufacturing receive it. However, they may receive gushes from 12 senior designers, not one. Each part must be studied to make a process plan and routing. Once again the database will be ignored. Thousands of parts have passed through the hands of process planners. Perhaps just yesterday an identical part passed across the desk of one of them. It is almost impossible to find. The planner thus starts from the beginning.

His problem is worse th
parts each. Thus 360 part
process planners' desks. N
parts with only his memory
of parts requiring different
bushing, a plate, a shaft, a
requires reconsideration of
His cohorts are working und
planners may be assigned sp
some of this problem, the sh
to consider a wide variety o
their memories.

Process Planning—Assemblies

One of the planners is charg
assemblies. He soon finds ou
assemblies incorrectly for man
parts to match the way the s
this means recomposing the p
meet the change configuration
back to engineering. In some
enough to require new assemb
not agree to make them. If t
forced to assemble the parts
compartment in manufacturing
them. A new and separate pa
also. The database has been

Tooling

During the planning process,
tooling, jigs, or fixtures. Me
old tools which might be used
to meet this need. The databa
mentation is usually in worse c
thousands of closely similar and
in some dark storage area, or ru
lead time for tools may be six mo
ject on time may be threatened.
be repeated here to describe the

Producibility Tips

As the process planner prepares
may see a variety of noneconomic

may have been
perishable tools
chase of nonsta
inventory. A p
a large plate ra
tolerances may
these potential
The first choice
are economical
long as six wee
The second
release the par
facturing onto
They will not r
of expensive to
the designer an
of producibility
stage, these pr
the planner co
him of the desi
be very popula
to their Christ

Numerically Co

As the process
planned for a
part drawing i
if tediously, r
He then prepa
commands, and
operator so th
pared similar
these other cu
from the begin
As the pa
sees certain r
could be simp
experience tel
designer or q
operator's job
this part fails
gineering may
ate the proble

Material Requirements Planning

The paperwork finished, the part data is released to the manufacturing system. Materials are scheduled for delivery by purchasing, based on the material requirement planning system. If nonstandard material is required by the designer, material lead time may extend final assembly of the finished product beyond delivery data. The new and individual material requirement will create a purchase order for a small and therefore expensive quantity of material.

Production Scheduling—Group Technology

Once the material is received, the part may be manufactured. No database is available which can be scanned to see whether other similar parts are required for manufacture at the same time. Thus a set of different parts cannot be combined as one batch to reduce the cost of set-up time. The result is that the total cost for designing a new part may have to be absorbed on its first and only use.

Capital Equipment

The new tooling, the shelf for the tool, the storage space for the new material, all increase the capital investment requirements of the company.

We have now covered many of the significant compartments of manufacturing. If, as with engineering, an integrated data base doesn't exist in manufacturing, the presentation of CIM as a total system concept to manufacturing management will pose some serious questions for them:

1. Can I rely on the geometry that engineering develops? Can I make changes to it, without the permission of engineering, to make the part more economically?
2. How will producibility tips be agreed to between engineering and manufacturing, and how will they be entered and maintained in the data base? Is it cost effective to enter them on a part-by-part basis? How can I persuade engineering to create bills of material that truly reflect how a product is assembled?
3. If I keep the geometry and tool paths for parts and store them in the computer, won't this be expensive and futile if I can't find them?
4. How will process planning be improved if I can't find old similar processes and compare them to new processes required?

5. What data that we use should reside in the business systems database, and what data should be in the CAD/CAM database? Can I go back and forth between them easily and economically, and/or combine data from the two systems on line or in reports?
6. Who is responsible for each data increment now that we are combining forces with engineering? Who will be authorized to make changes?
7. Who will be responsible for the computer hardware? Who will decide how to merge and operate with two databases and their corresponding database managers?
8. After the many years we have developed NC tool part programs, how will we convert them to the new database? When and how will we change from our present system to another? Will conversion to another system create only a marginal improvement to the present methods, or will the improvement be significant? How?
9. Will changes in organization be needed?
10. How will management of data processing be affected?
11. How will I incorporate tool design?

All management within a manufacturing company will have global questions:

1. What is the savings differential, not in hardware, but in my operations between a turnkey system and a total integrated system. Don't all turnkey systems have a database and a database manager? What is the penalty we might pay if we do not make use of a total system concept? How will we use the system?
2. How can we take our engineering and manufacturing process and measure the before and after costs and savings so we can convince our board of directors that an integrated CIM system is superior to a smaller turnkey system? What is the return on investment we should expect?
3. Can we really believe it is possible to implement CIM as an integrated system between engineering and manufacturing when we have been unable to create a manual integrated system? If we are not using efficiently the data we have already generated, how will an integrated database make it possible?
4. Can we find some other company that has implemented a total system concept?

A manufacturing company must be willing and able to answer these questions. If they are not, they will find themselves with an inadequate, costly system. Thus the concept of system requirements must be defined carefully, including the process flow and the data base contents.

REQUIREMENTS

Understanding Users' Needs

The customers, the users, are confused. A concise description of the system must be prepared, including not only what it is, but how it will work, how it interfaces to existing systems, what it will do, what it will not do, how it should be implemented, the database increments required, and the persons responsible for their preparation, update, and maintenance.

Figure 3 is a picture which contrasts the present compartmentalization of a company to the integrated database concept. Once a company understands the basic differences, much progress can be made.

Some of the ingredients which must be considered in a CIM system are as follows:

Specific database contents
Database integration
Bills of material
Design retrieval
Producibility tips
Design and drafting standards
NC parts programming and database
Group technology
Tooling design
Material requirements planning
Production scheduling
Implementation schedules
Measurement of cost and savings

Bills of Material

A manufacturing company may be confused between the systems required for CAD/CAM and the business systems, and consider them as separate entities. Many arguments occur about how the bill-of-material system should interrelate in CIM. Often, in the past, the mechanized bill of material was not created in the engineering department. Thus this concept may create a host of questions by the user. Nevertheless, the parts list and corresponding bill of material are natural results of preparing designs on a CIM system. The emphasis to integrate the bill of material system into CIM is both logical and

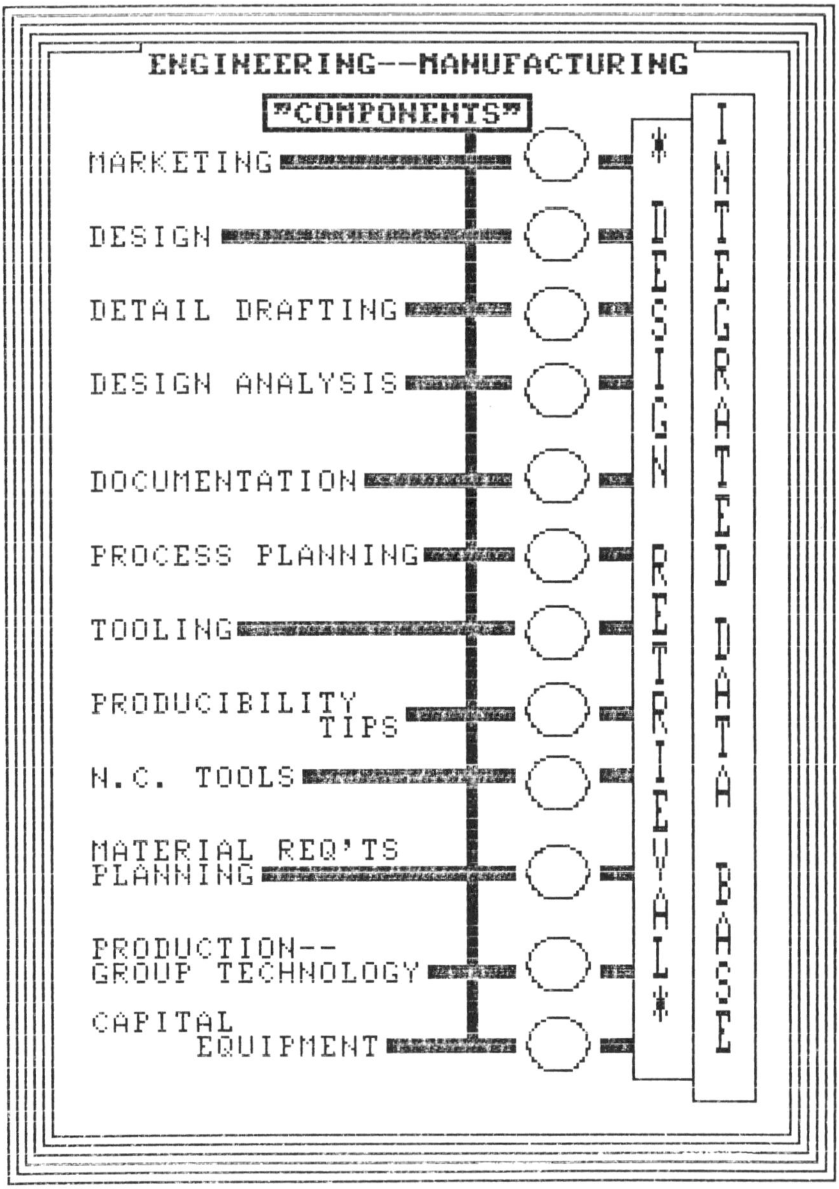

FIGURE 3 The engineering manufacturing compartments will be integrated and flow much more smoothly with a design retrieval system and an integrated database.

economical. The data thus derived from the system can be passed on to other segments of the business system without causing a major problem in the operation. (It is being performed manually in some fashion now.) This can become the natural break between design data and other segments of the system. The bill of material system planned, however, must be as sophisticated as any now used by manufacturing so that it will not cause two separate bill-of-material databases to be generated.

Producibility Tips

Design engineers and process planners have been feuding since the dawn of the industrial revolution. A designer designs, but the planner must find a way to manufacture it. It is symbolic of the process that design faults, errors, omissions, and more will be found by the process planner. If CIM systems planning can assist these two departments to find ways to work together cooperatively, both will breathe sighs of relief. A means to implement a series of producibility tips to assist both persons will alleviate the feud and create more economical designs as a byproduct.

Process Planning

As a special section of manufacturing systems tied to CIM, the need for explaining the relationship of process planning is important. That it should become a portion of the CIM database rather than of business systems is evident when one recognizes the need for viewing and/or using the geometry of the part as the plan is prepared. Much closer relationships between portions of engineering and some of the separate manufacturing disciplines will show the economies of merging some compartments into a single entity. Thus the data for such a set of merged personnel will become obvious. The process plan is to manufacturing what a bill of material is to engineering.

NC Machine Tool Plan

Ever since the numerically controlled machine tool was invented, confusion on the varieties of computer programs for generating paper tapes has existed. The lack of standards in the machine tool industry, the significant varieties of machine tools, and the large differences in the post processor requirements to customize tool paths have contributed to a Babel of different solutions.

With a combination of tablet menus, the database of a CIM system can be used rapidly and economically when it is integrated into the total system.

Tooling Applications

CIM can be considered an application which will be highly effective in preparing tool designs. However, the state of the art in tool design, documentation, and retrieval is by no means up to the level of documentation that is found in product design. Thus a study of this segment of manufacturing must be made to determine what must be changed to make tool design economical on CIM.

Group Technology

For over 40 years, group technology has existed in Europe. It was developed by a Russian named Mitrafanov. The Brisch Consulting firm expanded upon it in the 1940s and developed several systems in Europe. While a considerable amount of talk has been generated in the United States, the reality is that few companies have used the technique. One of the main reasons for not implementing group technology techniques is the lack of a compatible classification system and a means to gather and store the necessary data in some database. All potential customers can now take advantage of group technology. Many excellent algorithms to optimize the use of group technology exist. They, too, must be studied to determine how best to use them within the company.

Classification and Design Retrieval

The key feature for combining a host of disciplines and ingredients of a database will rely on a means for retrieving data. In our description of the compartments of both engineering and manufacturing, it becomes obvious that no database, no integrated system can be effective if the data generated within it is not easily found so that it may be reused. The concept is so important to the success of CIM that a section of this book has been set aside to develop the concepts and various alternatives of a classification system.

A manufacturing company is not fully aware of the tremendous amount of data it will generate once it obtains a CIM system. The classification system will be the major solution to the problem of retrieval of this data.

Decision Trees Automation

Another important portion of a design retrieval system is automation of the decision tree logic used to find a family of parts. This has been achieved as a separate entity by professional classification consulting firms. This retrieval logic will become an important part of the system to increase the speed of retrieval.

1
DATA BASE SYSTEMS—WHAT? WHY? HOW? TWO PERSPECTIVES

THOMAS C. BOOS

PERSPECTIVE

Over the years the term *database* has come to mean different things to different people. *Database* and *database management system* are often used interchangeably among technical people when discussing the application of this technology. More important than the terminology, however, is the perspective of the individual referring to the database or database management system. If the individual's perspective is that of an end user, a database management system is seen as a repository of operational data that the individual can draw upon to perform a specific task. If the end user is a production manager or inventory clerk, this database could be a filing cabinet in which there is a record (file folder) for each part this company uses (Figure 1). In its computerized form, all the information in this filing cabinet would reside in one of the company's computer memory units, very much like a music selection resides on a phonograph record. The records would be accessed by the clerk typing in the part name on a terminal screen. To this end user the database management system is a "part" information system. The fact that it resides in a computer rather than in a filing cabinet is of little consequence to the clerk.

The clerk needs to access the part records, which are the same whether they are in a file cabinet or in the computer's memory. What the clerk does see is a significant difference in speed of access and the amount of space it takes to store the records. This perception of a database management system holds for most end users. These end users can be hospital employees using the database to access patient data or manufacturing employees using the database to access part data, or engineers using the database to access drawings and specifications.

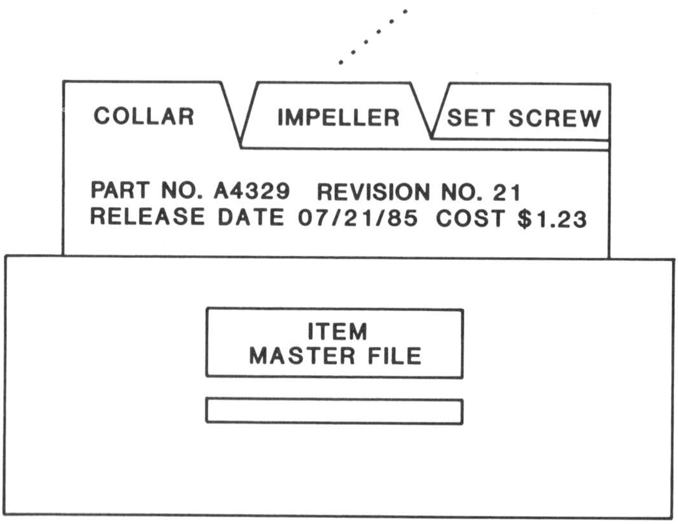

FIGURE 1 Databases originated with end user's need to store operational data about his/her business.

On the other hand, looking at a database management system from the perspective of the computer scientist or the computer professional is somewhat different. Computer professionals generally think of database management systems as the underlying container into which the end user places operational data, such as part records and product information, for safekeeping and convenient access.

EVOLUTION OF DATABASE SYSTEMS

History

Computerized data management systems of the mid to late 1960s were conceptually quite simple. These early data management systems looked very much like the filing cabinets they were intended to replace. So in the early days the analogy we have set forth so far held quite nicely. The end user looked at his database management system as a filing cabinet. It was accessed by a program that generated a report or a inquiry from a terminal, rather than opening a specific drawer of a filing cabinet and flipping through paper records until the sought-after document was retrieved. The challenge to the computer scientists, on the other hand, was and is to utilize the hardware and software of the computer to design an electronic filing cabinet into which end users can put their data and the reporting mechanisms to access that data when needed.

Prior to the advent of computerized systems, end users had devised clever ways to access individual records within these filing systems using more than one criterion. An example of this is keeping a filing system of parts which is maintained primarily on the part's number rather than the part's name. This technique is commonly used because part names are often similar to other part names, and in some cases identical. To uniquely identify a part that has a name which is identical to some other part can be difficult, if not impossible. Therefore, the practice of assigning unique part numbers to parts is very widespread. The world being what it is however, a substantial number of requests for information about parts is not accompanied by a part number. For example, if this were a purchased part, the vendor's part number might be kept in the record in addition to the part number assigned by the purchasing company. If you wanted to see what you paid for that part the last time you ordered it, it would be convenient to look for the record using the vendor part number. If the database were ordered on the part number assigned by the purchasing company, this request would be an exhaustive search of the database. Every record would have to be examined until the right one was found.

Prior to computerized information systems a technique referred to as *cross referencing* was in common use (see Figure 2). Cross-referencing simply meant that separate lists were maintained which were ordered either alphabetically or by numeric value and cross-referenced a part record by name or by number, or by any other criteria required to run the business.

These lists, of course, were quite clumsy to maintain. Whenever a new part was placed in the item master filing cabinet sequenced on part number, several separate lists had to be updated in order to reflect this new entry. One of those lists would be the alphabetized list of all part names (descriptions). This list (Figure 3) would be sequenced on the part's name and contain the part number so that whenever one had to locate the record by name, one could first find it in the alphabetized list, thereby locating the part's number. One could then go to the filing cabinet and pull out the record associated with that part number.

Immediately after updating the part name cross-reference list, the clerk would have to update the vendor part number cross-reference list so that if a request for information from a vendor came based on vendor part number the record could be located. From these examples it's not difficult to see how an organization with relatively few parts could operate quite nicely using a file drawer

PART NUMBER–PART NAME CROSS REFERENCE LIST

PART NO.	PART NAME
A4329	COLLAR
B2362	SHAFT
H9641	IMPELLER
→ Y1947	SET SCREW
.	.
.	.
.	.
.	.

FIGURE 2 Cross reference lists were developed to speed the search for information.

and paper database management system. However, as the number of parts in an organization grows and the number of assemblies and products increases, maintaining accurate and up-to-date operational data in a file cabinet and paper database management environment soon becomes an intractable problem. The simple request, "What parts do I buy from vendor X?" in a company that has several thousand types of parts could cause clerks to work for days looking

PART NAME/NUMBER CROSS REFERENCE LIST

PART NAME	PART NO.
COLLAR	A4329
IMPELLER	H9641
SET SCREW	Y1947
SHAFT	B2362
.	.
.	.

FIGURE 3 Cross reference list to locate part number (record) if name is known.

through every part record to check the vendor name associated with each part.

Of course, if the request to know "what parts do I buy from vendor X" had been anticipated at the time the file cabinet database management system had been designed, the database designer would have stipulated that a vendor cross-reference document be maintained (Figure 4), which then would have led to another cross-reference index that some clerk would be required to update every time a new part type was purchased or a new vendor was used. From these relatively simple examples it is not difficult to see how what appeared to be a rather simple and straightforward recordkeeping application can soon turn into a complicated, time-consuming, and error-prone process, which if allowed to continue could at some point strangle an otherwise successful company's ability to do business.

Computerized Cross-Referencing

Computer scientists in the early days of data processing dealt with cross-referencing requirements in what we now refer to as "sort and report" file management systems. These systems really were the forerunners of today's modern database management systems. These sort and report systems were, and are today, simply a computerization of the file drawer and paper database management systems. Records, whether they be of employees, parts, subassemblies, or products, are most often kept in some numerical order based on part number or employee number. The records on each individual item contain such information as part number, part name (description), revision level, cost, and classification code.

VENDOR/PART LIST

ABC COMPANY	VENDOR NO.	PART NO.
COLLAR	JJ66	A4329
SHAFT	KL22	B2362
XY2 COMPANY		
SET SCREW	17369	Y1947
IMPELLER	77629	H9641

FIGURE 4 Other lists were maintained so managers could quickly determine which companies parts were purchased from.

Because those early computers could work so much faster than humans, and because we really didn't know how to adapt cross-referencing to computer technology, the early systems relied heavily on sorting to give management the information it required for day-to-day operations. So even though the individual records were in the computer ordered on part number, it was quite acceptable to run a reporting program which would read the entire database file and sort it so that parts were grouped by vendor ID, and thus management could determine what parts they bought from whom in a much more timely manner than they could in the old file drawer and paper system.

This evolution in information management greatly increased an organization's ability to make decisions based on fairly recent and accurate operational data. However, reports were expensive to acquire because they had to be tediously programmed by highly skilled professionals. And there were so many other demands on a company's computer system that companies set up procedures so that reports were run either once a month or once every two weeks, or if it was a really important report, maybe once a week. It doesn't take a very perceptive individual to see that working with a monthly report means working with data that is at least two days old and on the average two weeks old or more.

Of course, the timeliness of reports is only one aspect of database management. Another important aspect is security of the data. As soon as all that very sensitive data about part or products moved from the locked offices and locked filing cabinets of the manufacturing or marketing department to the computer room, the issue of data security took on new meaning. Who can change records in the database; who can add new records to the database; who can retrieve information from the database; who can see the reports generated from the database—these have all combined to provide a significant challenge to the computer scientists whose job it is to develop a "container," a database management system (DBMS), into which end users can place their operational data. For the end user who keeps detailed engineering drawings and specifications of piece parts in a database, tight control on who can change those records could mean the difference between whether or not a final product will function properly.

In addition to effective and timely presentation of data (e.g., reports) to end users and maintaining the security of the records, the computer scientists have many other challenges to meet in the design of a DBMS container. One of these deals with volumes of data. The volume of data encountered in many applications is often hundreds of millions of records. These volumes preclude the use of "sort and report" techniques which were discussed earlier. Sort and report techniques required that all the records in a file be examined by a program, and then those which qualify to some criteria, for example, purchased from a certain vendor, be sorted in some order and reported.

Examination of every record by a computer could take hours when dealing with large numbers of records. This was unacceptable because end users were demanding faster turnaround time. Moreover, the advent of the computer terminal was creating demand for instantaneous response to requests for information. Therefore, it made sense for computer scientists to develop a technique to include cross-reference list capabilities when designing a DBMS which is expected to accommodate a very large number of records. The computerized technique is very similar to that used in the old file drawer and papers systems (Figure 3). However, the speed of a modern computer and flexibility of some DBMSs relieve the end user from the burdensome task of manually updating each of those cross-reference lists. These lists are now updated automatically by the DBMS.

Protecting the Integrity of Data

Another of the challenges faced by the computer scientist who is designing the DBMS (container) is guarding the integrity of the data which is stored in the DBMS. We've all heard the saying "garbage-in-garbage-out." This well known precept has been uttered many times since bad information caused the defeat of Rommel at El Alamein, and I'm sure many times before that.

The designers of some DBMSs have gone to great lengths to protect the integrity of data. Integrity differs from security in that integrity deals primarily with human error while security deals primarily with protection from willful tampering. These designers are painfully aware that people make mistakes. It doesn't matter whether these people are data input clerks, computer programmers, or managers.

These mistakes can take many forms. An example might be a clerk entering the incorrect part number or vendor number when a new shipment of parts is placed in stock. Worse yet, the clerk may input the incorrect quantity. If the program does not detect the error, this could cause a production line to be shut down (when you unexpectedly run out of parts) with the resultant loss of product shipments and revenue. The more advanced DMBSs go to great lengths to protect the database from becoming contaminated with misinformation (garbage). There is very little disagreement that misinformation is worse than no information. We were never able to have this protection with file drawer and paper systems. Today the more advanced DBMSs provide data integrity checking (Figure 5) as an integral part of the system. This can be accomplished if the end user's management

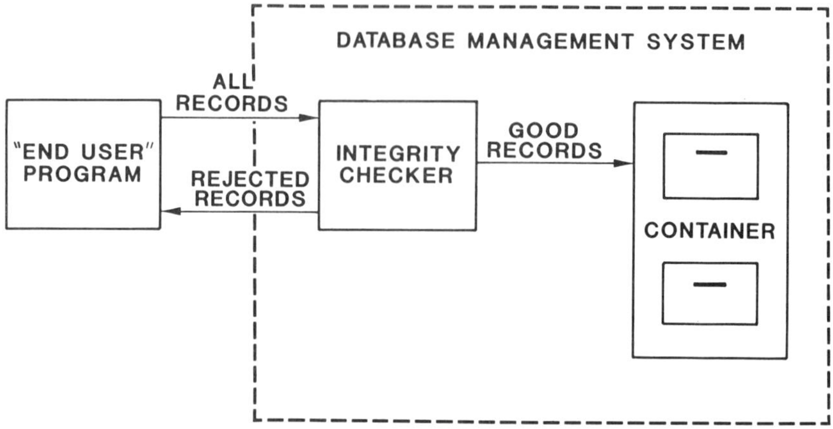

FIGURE 5 The most technically advanced DMBS products have built-in integrity checking.

appoints a senior person, usually called the database administrator (DBA). These DBAs can specify rules, commonly called integrity assertions, which are automatically checked by the DMBS every time a record is added to, changed, or read out of the DMBS. This integrity checking is done completely separate from any programs used to access the database, so the rules can be changed at any time without reprogramming—and no program can be written to get around the rules. An example of a rule which might be placed in the DMBS integrity checker is: The unit cost of parts (U) must equal total cost (T) divided by quantity (Q): ($U = T/Q$). This simple check would make it much less likely that an incorrect part quantity, unit cost, or total cost would be entered into the DBMS.

Record Relationships

Up to now, for the most part, we have been discussing records of data as if they existed on their own with no relationship to other records of data which might be stored in other containers. Examples of this might be the relationship between a record which contains information about a vendor from whom we buy parts and the individual piece parts that we buy from that vendor. Another relationship might be records we keep about assemblies of a finished product which contain such information as cost, number on hand, the parts that compose those assemblies, and the engineering drawings of those parts. The computer scientist refers to that type of relationship as a subset relationship. For the purpose of this example, that means that the parts supplied by a specific vendor are a subset of the total amount of information that we maintain about that vendor. Using our other example, a computer scientist looks at the list of parts that compose an assembly as a subset of the total amount of data maintained for that assembly (Figure 6). This type of relationship is very often also referred to as hierarchical. The concept of a hierarchy is very easy for an end user to grasp because all our lives we have dealt with hierarchies.

In the schools we went to we were able to see hierarchical charts depicting the principal of the school at the top of the chart. Below were listed all the teachers that worked for the principal. Then below all the teachers, all the students were listed. This method of representing data could be referred to as a trilevel hierarchy. Because most of the records we deal with in the engineering and commercial world are in some way related to other records, computerized DBMSs began to support record organizations that allowed the end user to relate his operational data in a more convenient format. These new ways of organization were quite impractical to do in a file drawer and paper DBMS. No one would argue that hierarchical systems could not be maintained in a file drawer and paper system, but the complexity and duplication of information that would be required to do so would make it extremely difficult.

With the speed and accuracy of the computer, many of these problems were overcome. Although designing the DBMS to efficiently manage these hierarchies was a very complicated task, it could be accomplished by a computer scientist, and the end user could be masked from the inherent problems. Once the system was implemented, a computer programmer could locate a particular vendor in the database and then, with considerable ease, access the list of all parts

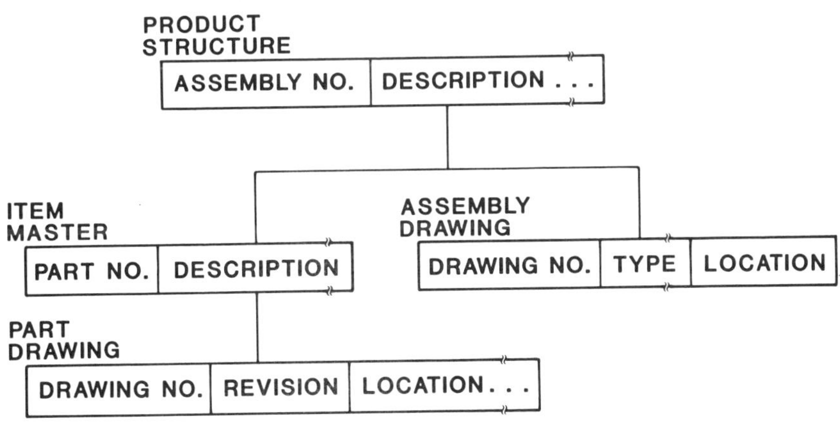

FIGURE 6 "End user" model a of hierarchical database organization.

and products supplied by that vendor. Just as easily, a programmer could locate a product structure, and then, using the part path, directly access the records of the parts that compose that assembly (Figure 7). This technique made information processing and the business it supported much more efficient and cost effective than single file systems.

Of course, as with most things in life, the added benefits of this hierarchical access approach brought along with them some severe drawbacks. True, an engineer could locate a product structure record and then locate all the parts used in that product, but what would happen if the same part were used in more than one product? Now a list is needed of all parts used in these two products. This implies that the same part record would have to be maintained in two hierarchies, the part list for product A and the part list for Product B. One way to do this would be to have duplicate copies of the part record in the database. This means redundant data, and worse yet, trying to identify which products are affected if a certain part has to be changed would require a long, expensive search of the whole database.

Another way to do this would be to develop a means by which a single part record could be included in more than one list. In other words, certain records must participate in more than one subset (hierarchical) relationship. Being faced with this real world

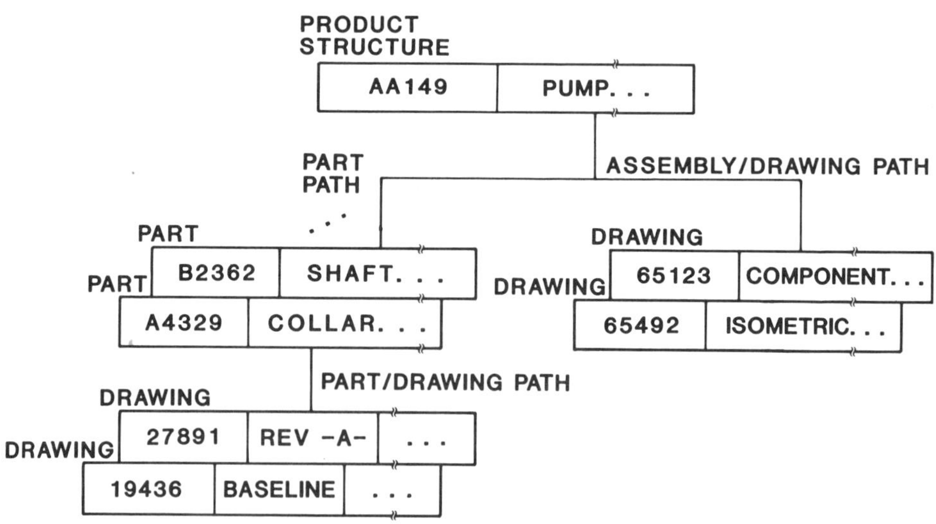

FIGURE 7 Operational data in the DBMS (container) modeled in Fig. 6.

requirement, computer scientists developed DBMSs which support network relationships.

DBMSs which support the network model allow the end user's DBA to configure the DBMS for an organization's operational data in such a way that a record can exist in multiple hierarchies. This capability is very useful in computer-aided manufacturing applications where a certain part may be a component of numerous assemblies or a certain drawing may be in a list of drawings which comprise a specific assembly and may also be in a list of drawings to be maintained by a certain project.

Relational Database

Hierarchical and network DBMSs are very common in today's data processing industry. With these systems we are able to organize data in such a way that highly repetitive requests for records can be handled quite efficiently. The term *highly repetitive* refers to those applications where the end user organization can predict with a fair amount of certainty the relationships that will be required to sustain the business. For an engineering application this might be predetermining that whenever an end user wants project information the end user will also require information regarding the associated engineering drawings. For a payroll application it may be predetermining that each time employee information is required, the employee's salary history or educational history will be required.

Predetermining the most likely accesses and then physically linking the related records together so that programmers can navigate through the hierarchy or network to get results is really just a method of applying a computer's speed to a further refinement of the file drawer and paper database management system. This analogy can be drawn because the end users programmers are still locating and examining individual records to determine if the record is of interest for a particular request.

Wouldn't it be much easier if we could just tell the DBMS what we wanted—for example, "Tell me what I buy from vendor X"—rather than require a programmer to instruct the computer how to locate each vendor X product record? In addition to this, because hierarchical and network DBMSs require relationships between records (subsets) to be predefined, any change to these relationships can require hours of programmer and computer time to accomplish. Because of these and other problems with hierarchical and network systems, a fundamental new direction was sought.

Several years ago, in order to try to understand the most fundamental concepts of data storage and end user access requirements, a group of computer scientists stepped back to take a look at how data should be organized and structured. Their objective was to minimize data redundancy and the programming effort required to access and maintain large databases. These scientists wanted to approach the problem from a sound mathematical theoretical foundation rather than further refine the file drawer and paper DBMS approach. The direction taken by these computer scientists is called the relational approach and is based on the mathematical theory of relations. Although, at first glance, this approach seems somewhat more complicated than the hierarchical or network approach, it is primarily the terminology used in relational mathematics that makes it appear so. Of the three approaches, the relational approach is as easy to conceptualize as the hierarchical or network approach and is much simpler for the end user to deal with.

The relational approach deals with the concept that most users prefer to deal with their information in a tabular form. By *tabular* we mean a form similar to that which one finds in a spreadsheet, or in one of the popular microprocessor-based products, such as Visicalc or Lotus 1-2-3. There are many advantages to the relational approach to DBMSs, but one of the most easily demonstrable and powerful is the method in which data can be retrieved. The reader might recall that in the network or hierarchical approach, the method of retrieval if one wanted to know all of the drawings associated with a particular engineering project was to have a programmer write a program which first located the project in question and then read down a particular path and through the subset of associated drawings. This programmer would then write this information to a file which could be later viewed or printed.

Because of the data organization in the relational model, it is possible to issue a single command which says FIND DRAWINGS WHERE PROJECT NAME EQUALS "TURBINE BLADE." This single command would locate all the drawings associated with the turbine blade project. No program has to be written. No explicit navigation is required. The command can be issued from a terminal (or from a program), and the results are returned to that terminal within moments.

The relational model is based on the theory that records about employees or parts or drawings are best represented as a row of related data elements. This row of related data elements is very similar in concept to the way records are represented in the hierarchical or network data model. There are, however, some very significant differences as to how the relational model is organized within the computer. These structural differences are beyond the scope of this text, but many papers and books have been written analyzing these differences in great detail. Because of the relational model's flexibility and ease of use, hierarchical and/or network record relationships are easily composed from the operational data residing in the DBMS (Figure 8). The relational DBMS also makes it much easier

PRODUCT STRUCTURE RELATION

ASSEMBLY NO.	DESCRIPTION	$ COST
AA149	PUMP	326.00
JF776	WING	1769.00
KL694	CLUSTER	2764.00
22934	BRAKE	14731.00

MATCHING "LIKE" FIELDS

PART RELATION

PART NO.	ASSEMBLY NO.	DESCRIPTION
A4329	AA149	COLLAR
B2362	AA149	SHAFT
H9641	AA149	IMPELLER
Y1947	AA149	SET SCREW

DRAWING RELATION

DRAWING NO.	PART NO.	LOCATION	REV
D137	H9641	FILE BR23	A
D169	B2362	FILE H196	
D583	H9641	FILE K371	B
D998	Y1947	FILE F193	J

FIGURE 8 Operational data in "relational" DBMS format. Relationships are maintained by matching "like" fields. The pump assembly consists of all parts with assembly number AA149 in their record.

to change the database as requirements change, and recover from mistakes in design when they are discovered. Most computer scientists agree that using a good relational DBMS increases programmer productivity by at least 40%.

Bibliographical/Text Databases

Up to now we have been discussing the organization and maintenance of records about specific people or things. We have looked at how employee records and part records might be organized in a DBMS. In the engineering and scientific community there is also a very substantial requirement to maintain and index free-form textural data. Nearly all research and development programs, scientific investigations, engineering analysis, and other activities result in a substantial amount of documented information. Often the value of this information could be greatly extended if it could be made available to others engaged in similar activities. Historically one of the problems with managing these large volumes of technical information has been our inability to catalog and classify it so that interested parties could easily locate all the papers which were written about their subject of interest.

An example of this might be that a large corporation had several divisions doing research on how heat and stress affected various chromium alloys. Each of these research activities would, of course, have resulted in a paper being done by the research team. It is not hard to imagine in a busy corporation, that two years after that research had been completed, very few people, if any, would remember that it had been undertaken. Furthermore, the paper documenting the results of that research would be very difficult to find.

If, however, the research paper were entered into a DBMS that supported documents and "key work in context" indexing techniques, the results of that study would be available and accessible for years to come. Accessing that study and others like it would be a very simple matter. A scientist or engineer could ask the DBMS to tell him or her how many papers were in the database that referenced the term chromium alloy. Finding numerous papers, the scientist or engineer could then review each document from a terminal or could ask the DBMS for a more fine-grained search. To accomplish this the user could issue the following command to the DBMS: LIST ALL DOCUMENTS THAT REFERENCE "CHROMIUM ALLOY" AND "HEAT STRESS." This would produce a listing which detailed how many documents in the database referenced both chromium alloy and heat stress and the catalog number of each. Another simple command, PRINT, followed by the catalog number, would let the user view the documents on the computer terminal.

DBMSs which support this type of text management have become very sophisticated. They can search on key words, multiple key words, portions of key words, and many other criteria. DBMSs which support text management in addition to normal record handling provide much more value to organizations requiring complete and flexible management information.

Fourth Generation Languages

Fourth generation languages are very close relatives to modern DMBSs. Very often they are referred to as 4GLS or Fourth Generation Systems (4GS). It is nearly impossible to pick up any data-processing-related paper or magazine which does not in some manner reference fourth generation languages. Fourth generation languages have been developed in response to the growing requirement that computers must be made easier to program. Although the number of computer programmers has increased rapidly over the last two decades, the cost to employ these programmers and the number of applications to be programmed is moving at an even faster pace. In an attempt to match the number of people who could program a computer with the amount of work to be done, computer scientists have been attempting for many years to develop computer languages which are easier to learn and can instruct a computer what to do with less painstaking detail.

This text will not attempt to describe the first and second generation languages, but we will make brief mention of the two most popular third generation languages. These languages are, of course, FORTRAN and COBOL. FORTRAN was designed to allow the engineer or scientist to communicate with a computer in a form comfortable to his background and training. FORTRAN made it relatively easy to deal with (describe to the computer) formulas, arrays and matrixes, scientific notation, and other constructs with which they were familiar. COBOL was designed to be the computer language of the commercial world. It dealt more with records of data similar to those we discussed in the earlier sections. It was designed to keep count of inventory items and/or maintain payroll records rather than perform stress analysis or calculate a rocket's trajectory.

These languages and many other third generation languages were and are very valuable tools in today's data processing repertoire. These tools do, however, require a great deal of tedious thought and effort. Much of the tedious effort they require is the repetitious defining of the records that they must access from the database. Third generation languages also require a great deal of effort in programming all the requests to bring data into the computer's memory, checking it for correctness and writing results to the database.

Fourth generation systems are designed to minimize the amount of instruction that a programmer must give to a computer. Although there are many different fourth generation systems, all claim to improve programmer productivity from 10 to 50 times over third generation systems. In reality the amount of programmer productivity gained by use of these systems is highly dependent upon the complexity of the problem which is being programmed. There is little doubt that if the problem being programmed is relatively simple, such as increasing the cost of a specific part in a database or generating a report on items which have been in inventory for over six months, the use of a fourth generation language could easily decrease the amount of time it would take a programmer to accomplish those goals with a third generation language by a thousand percent.

If, on the other hand, the requirement was to only update a specific record if certain criteria were met, and that criteria could only be determined by accessing and making computations against data items contained in other database records, the productivity gain of a fourth generation language over a third generation language would be considerably less. Fourth generation languages gain much of their power from their ability to perform implicit input/output (I/O). This means that with many fourth generation languages, the programmer can simply specify the types of records to be accessed, the action to be taken on those records, and the name of the input file or terminal screen that states specifically which individual operational records are to be acted upon.

Figure 9 is an example of a program written in a fourth generation language. This program is designed to change the quantity on hand (Qty-on-Hand) field of part records (invent) in a manufacturing database, and to add the records, detailing which parts and how many were received from each vendor to the RECV section of the database. The input records come from TRAN-SOURCE, which could be a terminal or a file of transaction records. Implicit I/O is accomplished because the key word CHANGE tells the computer that part records must be read into the computer memory, their quantity-on-hand field changed by adding the quantity in the TRAN-SOURCE record to the existing quantity, and the record with the changed field replaced in the database. The addition of the detail records is accomplished by the ADD command followed by the section of the database (RECV) to which it should be added.

These 15 lines of fourth generation code will generate approximately 500 lines of third generation COBOL code. Applying the U.S. government formula that each line of code written by a programmer costs $12.00, it is easy to calculate the cost savings (not to mention the time savings) in this approach.

```
100 DEFINITION SECTION.
120    TRANS-SOURCE IS FILE TRANSF.
140    TRANSACTION,ADD-RECV,RECV.
160    REJECTS TO REJFILE.
200 PROCESS SECTION.
220    BEGIN-TRANSACTION ADD-RECV.
240       CHANGE INVENT USING PART-NO EQ RECV.PART-NO.
260            COMPUTE QTY-ON-HAND = QTY-ON-HAND + QTY-RECVD.
280          ON-FAILURE.
300             IGNORE TRANSACTION.
320       END-CHANGE.
340       ADD RECV.
360          ON-FAILURE.
380       END-ADD.
400    END-TRANSACTION.
```

INVENT – INVENTORY RECORD TYPE IN WHICH IS STORED CURRENT QUANTITY ON HAND FOR EACH PART;

RECV – RECEIVING RECORD TYPE WHICH CONTAINS A COPY OF EACH RECEIVING RECORD THAT IS PROCESSED.

FIGURE 9 4th Generation language program example. These 15 statements generate a third generation program of about 500 statements.

Almost every software vendor that markets data management products claims to have fourth generation capabilities in those products. Because the term *fourth generation language* is difficult to define, there are no firm criteria on which to measure a particular vendor's claim to fourth generation capabilities. Fourth generation languages are, however, a very definite and substantial plus to any DBMS environment. For this reason any organizations contemplating the acquisition of a new DBMS should investigate thoroughly its fourth generation language capabilities.

Structured Design/Information Analysis

The only reason for DBMS (paper or computerized) is for the storage and protection of the end user's operational data. End users invest large sums of time and money in the hardware and software required to manage their operational data. Therefore, it is of extreme importance that the data in the database be the data that is required to support their business. Who determines what data should be kept in a DBMS? How are the integrity rules established? These and many other questions cannot be answered by a computer scientist or programmer. They can only be answered by the subject matter experts—the people who know what information they require to run the business. Of course, these people who know what information is required to run the business have no idea how to develop a safe efficient container (DBMS) in which to place their data. Nor should they have to know. Their job is to run the business. The problem with all this is that the people who understand the business don't understand the intricacies of the computer and the DBMS. The computer scientists, on the other hand, don't understand the information requirements of the business. What we have is a communications problem between end users (subject matter experts) and the DBA (computer scientist).

One approach to minimizing this problem is to use a well-structured communications vehicle. This vehicle must be easily understood by both subject matter experts and computer scientists. This vehicle must have enough structure to support the precise definition of what data is required to support a business or enterprise. A vehicle such as this can also be a great aid to productivity. This is accomplished if the vehicle consists of a language/methodology in which the data requirements are written down in a format which can be used directly by the DMBS. One such vehicle is "Information Analysis," an information requirement analysis methodology developed and marketed by Control Data Corporation. Information Analysis allows a subject matter expert (after a few days of training) to specify his or her information requirements in a very precise and unambiguous manner.

Information analysis deals with "objects of interest." For a manager, administrator, or clerk involved in product development or manufacturing, these objects might be products, assemblies, parts, etc. Figure 10 is an Information Analysis diagram that might have been done by a manufacturing employee. This employee is dealing with objects called parts. In this example, the object, part, can take on more than one definition. The first definition is as an item master part. To define this object, a circle is drawn and labeled item master part. Then, the attributes of this object are specified (e.g., part number, description, manufacture data, or release data). The arrows above the boxes precisely document identification and location attributes. For instance, examine the item master part/part number relation.

The arrow over the left side of the box indicates that for each item master part, there is only one part number. The arrow over

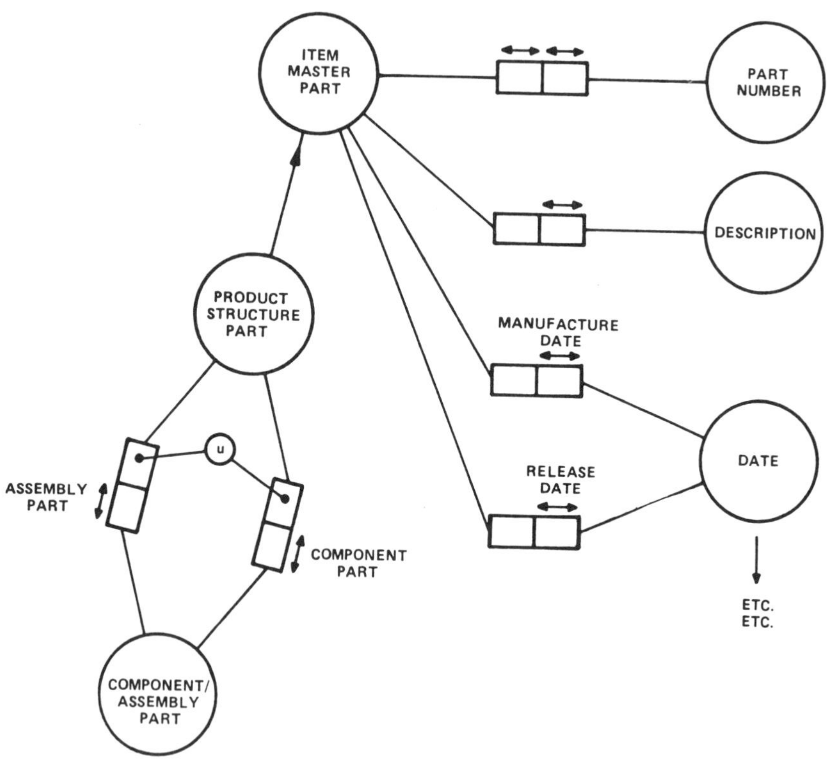

FIGURE 10 Information analysis diagram.

the right side of the box indicates that part number is a record identifier, so if part number is known, the part can be located. The other attributes defined for the item master part have an arrow over the right side of the box only. This location indicates that for each part object there is only one description, one manufacture date, and one release date. It also indicates that for each value of these three attributes, there could be more than one item master part.

The line to the item master part object from the product structure part in Figure 10 precisely defines the fact that all attributes of the item master part are also required for the product structure part. This is called inherited attributes in Information Analysis. In either case, this form of notation saves the subject matter expert from having to define the same attributes more than once. It also serves the purpose of stating this fact of inheritance to the computer scientists so she or he can tell the DBMS how to form records.

The object linked to the product structure part object is representative of assembly and components objects that comprise the end product. For the purpose of explanation, this is kept very simple and, therefore, lacks the accuracy and complexity that an actual diagram of this design would have. From this portion of Figure 10, it is important to make two points. The first is the ability of the Information Analysis methodology to represent constraints. The circled U that connects the assembly part and the component part boxes indicates a unique constraint. In the example, it precisely specifies that no product can have a component part and an assembly part of the same part number. In addition to the unique constraint, it is very important that a design or analysis methodology support other constraints commonly found in the real world (e.g., subset and exclusion). Another point that can be made from the component/assembly portion of this diagram is the position of the arrow over the connecting boxes. The position of these arrows indicates that the subject matter expert wanted the computer scientists to build this database so that if a component or assembly part number were known, the DBMS could locate all the products that component was contained in.

Structured design and analysis is an important component of any large-scale application development. One of the most difficult obstacles to overcome in any large undertaking is the communications barrier between the people who know what they need to run the business and the computer experts who must organize the data and put in a DBMS. Information Analysis was developed to minimize that barrier by using natural language techniques, such as objects, attributes that describe the objects, and constraints that must be placed on these objects as they are stored in and retrieved from the DBMS.

Artificial Intelligence

The next great challenge for DBMS is in the area of artificial intelligence. There is a great deal of activity in the information management community directed at this emerging technology. The result of this activity is that today we are finding ways to represent knowledge in database systems in such a way that computers programmed to simulate human reasoning power can make recommendations based on that knowledge. Systems of this type will allow us to debrief persons who have expert knowledge on some subject and place it in a knowledge base. Knowledge is much more than just facts. Knowledge is a line of reasoning which can draw conclusions based on incomplete data. An example might be that if a car won't start and there is no sound when the starter is engaged, there is a 85% chance that the battery is dead. Systems such as this can be called upon when an expert is not available to recommend a solution or direction to be taken. This category of AI technology, which we call expert systems, has already proven itself extremely valuable in areas of material requirement planning, medical diagnosis, mineral exploration, financial analysis, mechanical design and many other areas.

Expert systems and other forms of artificial intelligence are being applied in the many areas where experts are in short supply. There is little doubt that wherever human experts are now required to make decisions or give recommendations, expert systems will be of great value.

SUMMARY

Information is an extremely valuable organization resource. Finding better ways to make that information available to those who require it for day-to-day operation of their business and for long-range planning is a challenge for the computer scientists of today and the future. In this chapter we have looked at how database systems began and how they have evolved since the advent of the electronic computer. Database management technology is continuing to evolve at a rapid pace. Such phenomena as bibliographic databases, graphical databases, and fourth generation languages were not even heard of a decade ago. The advent of the microcomputer and high speed reliable communications over telephone lines and other media are opening doors to still further enhancements in database management technology. Database management is commonplace today and becoming more efficient as technology improves.

To keep all this in perspective, it is important to remember that all this complicated technology has evolved so that the end user can have a safe, secure, and efficient container into which operational data can be placed and from which he or she can extract the timely information required to successfully operate an enterprise.

2
INPUT DATA

The interrelationship of data, activities, and a database in a computer-integrated manufacturing system (CIM) is illustrated in the main visual at the beginning of this volume. Since it represents a comprehensive view of the main segments and actions of a CIM system, it has been compressed so that a general picture of the relationships may be illustrated.

The visual is divided into five segments for clarity:

12 Input data
19 Resident information
17 Database segments
20 Output information
10 Structured outputs

INPUT DATA

For anyone intimately familiar with an integrated engineering manufacturing system, the 12 inputs designated and described in this chapter may seem grossly inadequate. Endless numbers of different input forms or on-line input screens actually exist in such a system. An attempt has been made here to express the major and critical types. Various means to update inventory and other files, variations to the different inputs, and many others, are not indicated. It is believed, however, that those which are itemized will cover a substantial amount of activity. For each input described, and accompanying table has been prepared as an extraction from the large illustration. An "X" represents a database segment influenced by the specific input.

Part number log
Release notice
Parts list
Engineering changes
Generic bill of material

Geometry features
Part design
Assembly design
Manufacturing geometry
Process and routing
Master schedule
Inventory-labor transactions

CIM INPUT PART NUMBER LOG

DESIGN RETRIEVAL--------------X
FAMILY PROCESSES------------O
PART FAMILY DATA----------- O
PROCESS POLYCODES----------O
ITEM MASTER--------------------X
PERISHABLE TOOLS------------O
PRODUCT STRUCTURES--------O
JIGS AND FIXTURES----------- O
ENGINEERING CHANGES--------O
PRODUCTION PLANNING------- O
PART GEOMETRY----------------X
INVENTORY CONTROL--------- O
ASSEMBLY GEOMETRY-----------X
GROUP TECHNOLOGY CELLS--- O
GENERIC BILL OF MATERIAL----X
RAW MATERIAL INVENTORY--- O
ELECTRONIC MAIL---------------X

The part number log should be considered an engineering tool to avoid extra paperwork. An on-line system, with or without CAD/CAM, should allow the designer to receive a part number automatically when a new design is released. As mentioned previously, significance in the part number and/or other restrictions may make this automatic assignment impossible.

The following actions occur as the result of a new part number issue:

1. *Design Retrieval*. A number (or numbers) from the log is assigned to a specific person or department and an abbreviated description of the new design documented. The new design may be prereleased to the design retrieval system.

2. *Item Master*. A new item master is initiated.

3. *Part Geometry*. An opening entry is made in the part geometry database in preparation for the design if it is a piece part design.

4. *Assembly Geometry*. An opening entry is made in the assembly geometry database if the new design is an assembly.

5. *Generic Bill of Material*. If a generic bill of material is being used, the new part number is entered correctly into it, replacing a generic part number.

6. *Electronic Mail*. The electronic mail system is used to inform all departments that a new design has been initiated and part numbers assigned or reserved.

CIM INPUT RELEASE NOTICE

DESIGN RETRIEVAL-------------- X
FAMILY PROCESSES-------------O
PART FAMILY DATA--------------X
PROCESS POLYCODES-----------O
ITEM MASTERS-------------------X
PERISHABLE TOOLS--------------X
PRODUCT STRUCTURES----------X
JIGS AND FIXTURES--------------X
ENGINEERING CHANGES----------X
PRODUCTION PLANNING----------X
PART GEOMETRY-----------------X
INVENTORY CONTROL------------X
ASSEMBLY GEOMETRY------------X
GROUP TECHNOLOGY CELLS------X
GENERIC BILL OF MATERIAL-----X
RAW MATERIAL INVENTORY------X
ELECTRONIC MAIL----------------X

The release notice is a key document prepared by engineering when a new design is created. Consequently it affects all portions of the database which is part number oriented. The release notice is closely associated with the item master database, which, in turn, is closely associated to the product structure file. The two, operating as a unit, are often considered the bill of material system.

The following actions occur as the result of a new release notice:

1. *Design Retrieval*. The new design is entered into the design retrieval system by cataloging the part with its own unique family and sector number. (It should be emphasized that the issue of a release notice for a design and its corresponding part number issued from the part number log presumes that care has been taken to previously search the design retrieval system for an identical or closely similar design.)

2. *Part Family Data*. Part family data is updated by entry of the new design and corresponding data into the appropriate family.

Data may include critical dimensions, chemistry of the material, revision level, and specific comments as needed.

3. *Item Master.* The new item master in the database is completed with all the necessary engineering data increments.

4. *Perishable Tool Inventory System.* Once the part is released, perishable tools may be entered onto the appropriate process and routing if the part is manufactured. A cross index or "where used" will then be available in the perishable tool inventory system.

5. *Product Structure.* If the new design is an assembly, a parts list will also be required. Introduction of this data will update the product structure.

6. *Jigs and Fixtures.* Often a new design will require jigs and fixtures to manufacture it. If the jigs and fixtures inventory is connected to the part numbers for which they are used, and the part numbers are also collected together as part families, initiation of a new design will allow the search for tools which may be adequate for the new design.

7. *Engineering Change Notice.* Once the new design has been released, documentation control will require an engineering change notice to be used to notify all concerned of any changes. Such a change will be reflected in a revision level change for the part number (if the change is interchangeable), or replacement of one part with another (if the change is not interchangeable). Revision level changes will be documented in all sections of the database which are part number oriented.

8. *Production Planning.* The production planning system requires a process and routing for the new design if it is to be manufactured. If the new design if it is to be manufactured. If the new design is to be purchased, it will require the purchasing department to obtain quotations, a possible new vendor code(s), lead time, and quantity price discounts.

9. *Part Geometry.* As the new design is created, part geometry is stored and retrieved by part number. During this design process, the part is not yet released. The release notice is therefore a formal notification that the geometry is available for use by other departments.

10. *Inventory Control.* Once the design enters the production control system and is manufactured or purchased, the part must be recorded in the inventory control system.

11. *Assembly Geometry.* As the new design is created, assembly geometry data is stored and retrieved by part number. As with part designs, it is not yet released. The release notice is the formal notification that the geometry is available for use by other departments.

12. *Group Technology Cell.* As the new part is designed, the appropriate design retrieval code and associated manufacturing

polycodes have been designated. Based on manufacturing criteria, the new design may automatically be regarded as a candidate for a group technology cell in manufacturing. Predetermining this criterion expedites release of the part and elimination of unnecessary tooling, processes, and routings.

13. *Generic Bill of Material.* Within a manufacturing company which is a "job shop" operation, a generic bill of material may have been issued prior to detailed design. The new design release will automatically replace a generic part number with the actual part number.

14. *Raw Material.* As previously mentioned, elimination of many product structures can be achieved by placing the raw material requirements for designed parts into the item master. Thus the release notice may become a multiuse document. Manufacturing engineering may use the release notice as the trigger to enter raw material requirements into the item master of the design. In turn, the raw material requirement may cause the purchase of a new type of raw material.

15. *Electronic Mail.* Excess time is often generated because the release of a new design causes serial, not parallel activities. Use of electronic mail will notify all departments at the same time that a new design has been released. In addition, now that the geometry of the design is also available from the database, this parallel release of data is even more powerful, appealing, and time saving.

CIM INPUT PARTS LIST

DESIGN RETRIEVAL-------------- X
FAMILY PROCESSES------------- O
PART FAMILY DATA--------------X
PROCESS POLYCODES----------- O
ITEM MASTERS------------------- X
PERISHABLE TOOLS-------------- X
PRODUCT STRUCTURES---------- X
JIGS AND FIXTURES--------------X
ENGINEERING CHANGES---------- X
PRODUCTION PLANNING-------- O
PART GEOMETRY-----------------X
INVENTORY CONTROL---------- O
ASSEMBLY GEOMETRY------------X
GROUP TECHNOLOGY CELLS---- O
GENERIC BILL OF MATERIAL-----X
RAW MATERIAL INVENTORY---- O
ELECTRONIC MAIL----------------X

The parts list is closely associated with the product structure file, which, in turn, is closely associated with the item master file and its corresponding release notice. The two, operating as a unit, are often considered to be the bill-of-material system. Every assembly requires both a release notice and a parts list.

In a mechanized system, the parts list input becomes an abbreviated document showing only the parent-component relationship of parts to the assembly. Additional data, such as find numbers, quantities, units of measure, and effectivity data (if changes have been made to the structure), are entered into the system at this time.

Since the parts list generated by the computer is considered a document of equal significance, and a part of the assembly drawing, both documents are usually released at the same time. The parent-component relationships are combined in the computer with data previously entered into the item master of each component to print the parts list with minimum redundancy of data entry. The parts list is often designated as detached, since in manual systems it was often found as a material list on the face of an assembly drawing. Today this has been removed and detached.

The following actions occur as the result of a parts list input:

1. *Design Retrieval*. The new design is entered into the design retrieval system by part number, assigning a corresponding family and sector number to the design.

2. *Part Family Data*. Additional data is cataloged in the part family data, including revision level, critical dimensions, and other associated data.

3. *Item Master*. The item master is a completed with all the necessary engineering data increments.

4. *Perishable Tools*. As the process and routing for the new assembly is prepared, the proper perishable tools will be designated. In the perishable tool system a cross index or where used index will be maintained.

5. *Product Structure*. The product structure database for the bill of material will be updated, including where used information.

6. *Jigs and Fixtures*. In a manner similar to that used with perishable tools, the jigs and fixtures database will be updated to show new ones required or old ones reused. A cross index will be maintained as a where used file.

7. *Engineering Change*. Once the parts list has been released with its corresponding assembly drawing, documentation control will require an engineering change when design changes are made. In addition, if one component on an assembly is replaced by another, disposition of the old parts in inventory and in process must be made. The difference between the time the design is changed in engineering and implemented in manufacturing is called effectivity

and is maintained both in the product structure and the engineering change database, controlled by revision level.

8. *Part Geometry*. The parent-component relationship will be utilized as pointers to maintain part geometry relationships including the associated assembly transforms. (It should be accentuated that a similar product structure or parts list may be used to record the geometrical feature relationships required for a piece part design.)

9. *Assembly Geometry*. The assembly geometry can be considered a database of special instructions and notes required for the finished assembly together with the product structure or parts list as an index to find the part geometry for the assembly, geometric features added to the assembly, and the geometric transforms needed for the part association.

10. *Generic Bill of Material*. For those companies using a generic bill of material, the parts list becomes a means for updating the database and/or a means to collect parts issued previously without a structure into a meaningful assembly. It is also a means to replace a generic assembly with a "real" one.

11. *Electronic Mail*. Excess time is often lost because the release of a new design causes serial, not parallel activities. Use of electronic mail will place the knowledge that a new design has been released in front of all departments which must contribute to its manufacture or purchase. Now that the geometry of the design is also available from a database, this parallel release of data is even more powerful, appealing, and time saving.

CIM INPUT ENGINEERING CHANGES

DESIGN RETRIEVAL-------------- X
FAMILY PROCESSES-------------- X
PART FAMILY DATA--------------X
PROCESS POLYCODES------------ X
ITEM MASTERS------------------- X
PERISHABLE TOOLS-------------- X
PRODUCT STRUCTURES---------- X
JIGS AND FIXTURES------------- X
ENGINEERING CHANGES---------- X
PRODUCTION PLANNING----------X
PART GEOMETRY-----------------X
INVENTORY CONTROL------------X
ASSEMBLY GEOMETRY------------X
GROUP TECHNOLOGY CELLS------X
GENERIC BILL OF MATERIAL-----X
RAW MATERIAL INVENTORY------X
ELECTRONIC MAIL----------------X

An engineering change is probably the most significant document used in the system since the data entered on it can affect every department in the company. Thus it represents a potential change to every database segment which is controlled by part number and its corresponding revision level.

The actions required due to an engineering change are based on the rules of interchangeability which are associated with configuration management rules.

It should be noted that only a specific set of data increments in a manufacturing system is controlled by an engineering change. Care must be taken to define which elements are in this set. An engineering change usually creates the need for one or more documentation changes, a revision level change, a part number change, and/or a change to an increment of engineering data or geometry data. A revision level change to a part number implies that the old and new designs are interchangeable. A part number change usually means that a new design will replace an old design and may occur at any level of the bill of material and corresponding parts list. Since the parts list is controlled by revision level, a part change in an assembly is reflected on the parts list and requires a revision level change to the assembly parts list. If the change of designs within an assembly makes the assembly noninterchangeable, a new assembly part number is required. This test of interchangeability continues upward through all levels of a bill of material until interchangeability occurs, at which time a revision level change only occurs. The often complex relationship of changes versus levels of a bill of material must be carefully studied before a change is made. Once this is completed, the following actions occur as a result of a revision level or part number change:

1. *Design Retrieval*. The part number revision level is changed in the design retrieval system.

2. *Family Process*. The family process must be analyzed to determine if a part of the process has been augmented or changed.

3. *Part Family Data*. Part family data must be analyzed to determine if a part of the general design has been augmented or changed.

4. *Process Polycodes*. The process polycodes must be analyzed to determine if any of them have been augmented or changed, or a new polycode is required.

5. *Item Master*. The item master for the design must be updated to show the latest revision level. In some systems it is possible several changes are in process at the same time for the same design. It may be necessary to record several manufacturing effective changes versus the latest engineering revision change.

6. *Perishable Tools*. The perishable tools for the design may require analysis to see if they are still appropriate.

7. *Product Structure*. If the change requires a revision level change to an assembly, the product structure of the assembly and

its associated assembly drawing may require changes in documentation. If a revision level change to an assembly is caused by the replacement of one part with another, effectivity of change will need analysis.

8. *Jigs and Fixtures.* The jigs and fixtures for the design may require analysis to see if they are still appropriate.

9. *Engineering Change.* The engineering change records must be updated with the new change. (It should be noted that the engineering change database could be used for change tracking and costing.)

10. *Production Planning.* Any engineering change which affects a design being manufactured may pose a significant problem. Parts in process may have already passed the operation in which the change could be made readily. In some cases a design change can be implemented on a later requirement, in others the design may require rework, even scrapping. Thus the timing and severity of the change makes production planning changes important as a consideration. In addition, the process and routing for the part may require change and a corresponding change to the process revision level.

11. *Part Geometry.* The part geometry may change, depending on the design change. In addition, the geometry feature used in the design may require change or creation of a new geometric feature.

12. *Parts in Inventory.* In theory, parts in inventory should not be affected by a design change which requires a revision level change only. By definition, designs with different revision levels are interchangeable and could be found mixed together in inventory with no harmful affects. If the design change is to an assembly, the same reasoning should apply also. Complex changes, however, requiring rework, disassembly, and reassembly, may cause parts or assemblies to be removed from stock, altered, and then placed back in inventory.

13. *Assembly Geometry.* Assembly geometry may change, depending on the design change. If such is the case, two cases of such change may occur. If the change is to the assembly geometry separate from the parts within the assembly, the change is probably relatively simple. If the change is to a part in the assembly or a part is replaced by a new part within the assembly, the changed geometry of the part may affect several assemblies on which it is used. If the change is in note form not geometry, no geometry change will probably be reflected at the assembly level. In any case, a systematic method must be established to determine how various types of engineering changes will affect both assembly and part geometry and how such changes will be recorded.

14. *Group Technology Cell.* If the part is manufactured in a group technology cell of machines, the manufacturing process must be analyzed to determine whether a machining practice must be changed or augmented.

15. *Generic Bill of Material.* If a company uses a generic bill of material, the bill of material must be altered to reflect the change in a manner similar to that described for the item master and the product structure for normal bills of material.

16. *Raw Material.* It is possible that the engineering change will create the need to alter the raw material required to manufacture the part. As a result analysis of a new raw material availability must be made and the product structure for the raw material-part relationship changed to reflect the difference.

17. *Electronic Mail.* Electronic mail will place the knowledge that a design has been changed in front of all departments which must contribute to its manufacture or purchase. Now that the geometry of the design is also available from a database, this parallel release of an engineering change becomes a powerful, appealing, and time-saving tool.

CIM INPUT GENERIC BILL OF MATERIAL

DESIGN RETRIEVAL-------------O
FAMILY PROCESSES-------------O
PART FAMILY DATA--------------X
PROCESS POLYCODES-----------O
ITEM MASTERS-------------------X
PERISHABLE TOOLS------------ O
PRODUCT STRUCTURES----------X
JIGS AND FIXTURES------------O
ENGINEERING CHANGES----------X
PRODUCTION PLANNING----------X
PART GEOMETRY-----------------X
INVENTORY CONTROL------------X
ASSEMBLY GEOMETRY------------X
GROUP TECHNOLOGY CELLS----O
GENERIC BILL OF MATERIAL-----X
RAW MATERIAL INVENTORY------X
ELECTRONIC MAIL--------------- X

A generic bill of material should be considered the tool for a hybrid type of manufacturing system. Primarily used for job shop material and labor scheduling, the generic bill of material represents an engineered product which is not yet fully designed. It may even be considered that the company that uses this technique will maintain separate generic bills of material for each individual customer order and control them individually.

The bill of material may contain a special set of generic part numbers which are used only to document and maintain the product structure. These numbers will be replaced with "real" part numbers for each part and assembly as the design progresses. However, all other systems must be closely tied to this sytem. The following actions occur once a generic bill of material is entered into the system:

1. *Part Family Data.* The generic set of part numbers may be associated within the classification system to part family data to ensure obtaining proper lead times, process times, and other general family data prior to complete design documentation.

2. *Item Master.* As each generic part number is replaced with the actual design part number, a new item master is generated for it as well as changing the generic bill of material documentation from the special part number to a real one.

3. *Product Structure.* As new assemblies are documented, a new product structure and corresponding parts list are used to update both the product structure file and the generic bill of material, once again replacing special part numbers with real ones.

4. *Engineering Changes.* All engineering changes are handled in the same manner as previously described. However, a tabulation of all engineering changes as they reflect changes to the specific generic bill of material is recorded also. (Generic part numbers do not require engineering change documentation.) This maintains a consolidated bill of material and related engineering changes as a unit.

5. *Production Planning.* The generic bill of material should be considered a replacement for the master schedule and thus is the major database for production planning in a job shop. A series of generic bills of material may then be exploded in a manner similar to the way a master schedule is used to generate material and work center requirements. The process may require a separate customer order file used to segregate customer orders and bills of material from standard engineering documentation and bills of material. The customer order file then replaces the master schedule.

6. *Part Geometry.* Part geometry will be created and maintained exactly as described previously for any new part design. The part number of the new design, however, will replace a generic part number.

7. *Inventory Control.* As a generic bill of material is updated with actual designs, some of them may have been previously released for other projects. When this occurs, the inventory control system should reflect additional usage for the design.

8. *Assembly Geometry.* Assembly geometry will be created and maintained exactly as described previously for any new design, using real part numbers and product structures for the new design.

9. *Generic Bill of Material.* The generic bill of material should be considered a special system and a separate database used in a job shop operation. Replacing or augmenting a master schedule, the generic bill of material is a tool to generate lead times, material requirements, and other data for a customer order before the complete design has been released. Material requirements planning has often

been felt incompatible with job shop manufacturing since the bill of material for a customer order is often generated from the bottom up as individual parts and subassemblies are designed. Since material requirements planning requires a complete bill of material which chains all parts and assemblies from the top level part number down through all piece parts and raw materials required, the bottom-to-top design methods of a job shop are incompatible with this requirement. A generic bill of material is designed to reflect the general product structure of the finished customer order using generic part numbers which reflect general, not specific, design requirements. These special part numbers maintain an active chain in the bill of material until they can be replaced with actual designs part numbers. In this manner a generic bill of material can be utilized in all systems which are normally considered to be effective only in mass production manufacturing.

10. *Raw Material.* Raw material is required for all manufactured parts in a generic bill of material and should be treated in a manner similar to that discussed for all new designs. It may be impractical to document raw material requirements for a generic bill of material using the special part number system, however.

11. *Electronic Mail.* Since a job shop manufacturing system is a continuous mixture of new releases, engineering changes, and special considerations for a specific customer, use of electronic mail to notify all departments of these activities will enhance communications and corresponding requirements for action.

CIM INPUT GEOMETRY FEATURES

DESIGN RETRIEVAL-------------- X
FAMILY PROCESSES-------------- X
PART FAMILY DATA--------------X
PROCESS POLYCODES------------ X
ITEM MASTERS------------------X
PERISHABLE TOOLS-------------O
PRODUCT STRUCTURES---------- X
JIGS AND FIXTURES------------ O
ENGINEERING CHANGES---------- X
PRODUCTION PLANNING-------- O
PART GEOMETRY----------------X
INVENTORY CONTROL---------- O
ASSEMBLY GEOMETRY------------X
GROUP TECHNOLOGY CELLS---- O
GENERIC BILL OF MATERIAL--- O
RAW MATERIAL INVENTORY---- O
ELECTRONIC MAIL-------------- O

Possibly the most powerful feature of an integrated CIM manufacturing system is the consideration of design features as individual entities which can be combined in various arrays to generate designs. Almost all designs can be divided into individual geometric features, by function, form, or shape. These features also represent a set of manufacturing processes. If each of them is represented by solid geometry, they may be combined in an almost endless array to meet any finished design requirement. Geometric features should be considered the tools to create a new design.

Once design features have been defined and described, they must be stored in a manner that makes them easy to retrieve and also easy to combine in necessary groups. A combination of the bill-of-material system and a classification system seems the most practical and powerful means to achieve this goal. If each geometric feature is considered a design entity, it can be assigned a part number, entered into a database as an item master, and collect one with another onto a parts list. The parts list, however, should represent a family of geometry which in turn represents a family of similar designs. Thus a collection of geometry features will represent individual aspects of a part family of designs which have also been classified by form, shape, or function. The power of this associative process makes design of new parts and assemblies economical, rapid, and effective.

Since each geometry feature as an entity has its own peculiar function, form, or shape, it may also be classified in a retrieval system.

The following actions occur once design features have been defined:

1. *Design Retrieval*. All design features are entered into the design retrieval system by cataloging the feature with its own unique family and sector number.

2. *Family Process*. All geometry required for a family process will be collected, including rough and finished geometry implied by the geometry feature. In some cases, the geometry features may be sequences in a logical manner so that generative processes may be created for new designs.

3. *Part Family Data*. Part (feature) family data is updated by entry of appropriate family. Data may include restrictions, critical dimensions, and so on.

4. *Process Polycodes*. Each geometry feature may be considered a manufacturing attribute, such as drilling, milling, or boring, and be classified with process polycodes in special operation categories.

5. *Product Structures*. A series of product structures, one for each family of parts and assemblies, will be generated. Thus when a new part design is required, for example, the designer selects the proper part family. As the part family is selected, a list or a

set of pictures, or the actual geometry features, can be displayed which will be used to create a new design in the family. This geometry will be used to create the new design by manipulating it and locating it in the proper position and/or entering the proper dimensions. Assembly designs will be created using the predesigned parts combined with any geometry features required for the assembly. Thus assembly families will have associated to them two different product structures, one the parts list and product structure of the piece parts to make the assembly, the other the parts list and product structure of the additional geometry features needed on the assembly after the parts have been assembled.

6. *Engineering Changes.* As a relatively new consideration, engineering changes may alter the geometry features for all parts or assemblies upon which they are used, or only some of these designs. It should be noted that the power of associativity is such that, should a geometry feature require a significant change due to design failure, or other major reason, the individual change can be chained to every part and assembly upon which it has been used.

7. *Assembly Geometry.* The classification of assemblies into families is more difficult than the classification of piece parts. Thus it may be impractical to consider a parts list of assembly geometry features by assembly families. However, any feature required for an assembly can be classified into three categories, "normal" features, such as holes and milling, which are already classified for piece parts, feature geometry peculiar to assemblies, such as welding, spot welding, and bolting, and processes, such as brazing and press fitting. Through this means, the designer of an assembly can find the tools necessary for additionl assembly design requirements.

CIM INPUT PART DESIGN

DESIGN RETRIEVAL-------------- X
FAMILY PROCESSES-------------- X
PART FAMILY DATA--------------X
PROCESS POLYCODES------------ X
ITEM MASTERS-------------------X
PERISHABLE TOOLS--------------X
PRODUCT STRUCTURES----------X
JIGS AND FIXTURES--------------X
ENGINEERING CHANGES----------X
PRODUCTION PLANNING-------- O
PART GEOMETRY-----------------X
INVENTORY CONTROL---------- O
ASSEMBLY GEOMETRY---------- O
GROUP TECHNOLOGY CELLS------X
GENERIC BILL OF MATERIAL--- O
RAW MATERIAL INVENTORY---- O
ELECTRONIC MAIL-------------- O

A significant change in manufacturing systems occurs with the new means by which a part design is prepared. A CIM system now has a new geometry database of parts which must be integrated into the system. Where, before, a design was represented by a piece of paper which was interpreted, the actual geometry of the part is now available. To eliminate interpreting engineering design after it is released to manufacturing, the designer must be more aware of manufacturing practices and limitations.

A part design should now reflect the best of all manufacturing practices for economy and machinability. Thus producibility tips become an important feature of design. New part requirements also are derived from several sources. Design of a new product and its corresponding layouts, alteration of an old product for economy and operation characteristics, and alteration of an old product to make it larger or smaller are some of these sources. In any case, the following actions occur when and after a new part is designed:

1. *Design Retrieval*. The new design part number is entered into the design retrieval system by cataloging the part with its own unique family and sector number. The proper parametric drawing and its corresponding geometry features are used in the preparation of the new design. It should be emphasized that the design retrieval system should have been used previously to make certain no identical or closely similar design is available.

2. *Producibility Tips*. As the new design is prepared, producibility tips will confirm it to be compatible with other parts in the family. Of considerable importance is the comparison of the geometry of the new part with the geometry displayed in the parametric program used to make it, particularly if the part will be machined on numerically coded (NC) machine tool. The cutter center-line data geometry should be equal in sequence and reference points with that of the family cutter center-line. In addition, the producibility tips for such manufacturing requirements as bevels, bends, and hole sizes must be considered.

3. *Family Data*. Once the design is complete, entry of family data should be accomplished. All tabulated data found in the family should be entered.

4. *Process Polycodes*. Manufacturing will enter the proper process polycodes for the new design.

5. *Item Master*. Documentation of the new design will be initiated for the item master.

6. *Perishable Tools*. As the part is designed, care should be taken that recommended hole sizes and other specific features are standardized where possible according to the producibility tips. Otherwise the part, if it is manufactured on an NC tool, will cause a tool package to be altered or exceeded for the machine tool. Care should be taken that standard perishable tools be utilized.

7. *Product Structure.* When the part design is complete, provision for the raw material requirements in the form of a product structure or as an addition to the item master must be made.

8. *Jigs and Fixtures.* Many machined parts require jigs and fixtures to manufacture them. Using the part family data, similar tooling can be retrieved and it can then be determined whether a slight alteration on the new design will make the old tool usable, and/or whether a slight alteration to the tool will make it useable on the new design. If a new tool is required, tool design should follow all the principles described for a manufacturing system, including part numbers, piece part drawings, and assemblies. It should be noted that implementation of geometry of the design into the CIM database will now allow jigs and fixtures to be designed around the actual geometry of the part rather than around a hole in space, or requiring an actual part to be made as a prototype.

9. *Engineering Change.* Any change to the geometry of a piece part will require an engineering change. Since, today, changes to engineering geometry can be made in manufacturing by interpreting the drawing, this is a radical change to present practices. The specific geometry prepared by engineering must be used to manufacture the part. In order to achieve this goal, one must meet two requirements, integrity of engineering geometry, and the recognition that manufacturing "adds" geometry to the design and doesn't alter it. Thus engineering must become acquainted with shop practices, and also work in cooperation with manufacturing to develop producibility tips which reflect these practices. On the other hand, manufacturing must regard the finished design as the "what" and add the necessary envelope of material to perform the "how" of manufacturing.

10. *Part Geometry.* When need for a new part design is first recognized, a file for the design is opened and assigned a part number from the part number log. This allows the part geometry to be stored immediately under its proper identification number. During the design process three stages of development can be predicted. During the first stage the design may be almost a sketch, certainly incomplete. In this stage no revision level should be assigned. Rather, some indication of its incompleteness should be designated to avoid use of the design by others. In the second stage the part may be prereleased. At this time the design may be considered unstable, but documented sufficiently to make a prototype part or for limited production. A revision level numbering system that differs from the standard is recommended to notify users of this status. (Purchased parts which are prereleased must have sufficient data to identify them, but such data may be one-vendor oriented.) In the third stage of development, all design requirements have

been met, and the design appears stable. The design may now be released to all other departments using the standard revision level system. As described previously, the part geometry may be stored as a complete entity, or stored as text data only with a geometry feature parts list which calls and "assembles" them as required.

11. *Group Technology Cells*. Part geometry is intimately associated with group technology cells. Each of them may be used to define the machine tool attribute necessary for their manufacture. Once a coding system for them has been implemented, the natural result will be to take those parts with the appropriate machining characteristics and "load" and route them to the appropriate cell.

CIM INPUT ASSEMBLY DESIGN

DESIGN RETRIEVAL-------------- X
FAMILY PROCESSES-------------- X
PART FAMILY DATA------------ O
PROCESS POLYCODES-----------O
ITEM MASTERS------------------- X
PERISHABLE TOOLS-------------- X
PRODUCT STRUCTURES---------- X
JIGS AND FIXTURES------------- X
ENGINEERING CHANGES---------- X
PRODUCTION PLANNING-------- O
PART GEOMETRY-----------------X
INVENTORY CONTROL---------- O
ASSEMBLY GEOMETRY------------X
GROUP TECHNOLOGY CELLS---- O
GENERIC BILL OF MATERIAL--- O
RAW MATERIAL INVENTORY---- O
ELECTRONIC MAIL-------------- O

Under the principles of feature-based design, an assembly is composed of geometry features associated to it and also individual piece parts which, in turn, are composed of geometry features. This data may be stored in two ways, at the assembly level in the form of a parts list which calls for the finished part designs, or at the geometry feature level. In the latter case, the assembly is redrawn completely when the geometry for it is required. Many of these activities have been explained under geometry features.

Since it is more difficult to classify assemblies into entities with similar shapes or form, we can presume such a classification will be an exception rather than the general rule. How then will the database affect assembly design?

A designer today generally follows two courses when he designs a new assembly. Either the new requirement is based on an old design, in which case he need only find a similar assembly design (through experience or by old order data and other often clumsy retrieval systems), or he must create a completely new assembly based on a product layout. If the design is "new," retrieval of piece part designs which might fit his new assembly may be the only effective retrieval system to achieve economies. In any case, the assembly will be prepared utilizing a series of piece parts.

If a new assembly is to be created, it must be extremely easy for the designer to go from one part of a database to another. He may first make a rough sketch of his conception of the finished assembly. He may next enter the retrieval system and look for the families of one or more of the design entities he thinks he will need. Perhaps he will find two or three parts he could use in one family of parts and five or six parts in another family. These must be saved and/or viewed at any time the designer wishes.

He may return to the original layout and place the parts he has found into it. He may return to the families of parts in the retrieval system again and again, and then view the use of parts he has found as they fit his layout. As part of the process, he may be unable to find a part meeting his requirements exactly, and pause while he prepares one, issues the item master for it, and stores the geometry of the part. When he is satisfied the assembly design is complete, he will store it under the part number he assigned to it at the beginning. As a byproduct of this entire activity, he has also created a parts list, which is automatically entered into the product structure and is used to prepare a detached parts list.

The following actions occur when and after an assembly is designed:

1. *Design Retrieval*. The new design part number is entered into the design retrieval system by cataloging the assembly, if a classification system exists. In many cases a parametric drawing to design an assembly will be practical also. These may be classified under a special retrieval system designed for parametric assembly drawings only.

2. *Family Process*. As the new assembly is prepared, a review of the family process should be made. Producibility tips for similar assemblies will be examined for compatibility of operations.

3. *Classification*. As the assembly design is prepared, all components, already designed, which might possibly be used to make the new assembly are examined through a classification system. If an assembly classification exists, part family data will be entered.

4. *Item Master*. Documentation of the new design will be initiated for, and contained in, the item master.

5. *Perishable Tools*. As in the case of designed parts, all assemblies should be designed to include standard hole sizes and other manufacturing operations. Producibility tips should be developed for assembly designs also. Through these procedures, standard perishable tools will be utilized.

6. *Product Structure*. As an assembly is created, the parts list for the components should be prepared automatically. Once released, the assembly design should be recorded in the product structure database as well as in the geometry design.

7. *Jigs and Fixtures*. Welded assemblies particularly require fixtures. These holding and trueing devices are expensive. Thus a means to retrieve them for similar assemblies is important. Similar assemblies may cause only a minor change to an existing fixture. Thus the cross-index of jigs and fixtures and assemblies upon which they are used is an important feature. Use of a CIM database will allow a new jig or fixture to be designed around the assembly geometry rather than a hole in space.

8. *Engineering Change*. An engineering change to an assembly is usually more complex than that of a piece part. As described previously, the rules of configuration management dictate the steps required to document the changes made to an assembly. The geometry of all parts of an assembly which might be altered because of an engineering change as well as additional assembly geometry are stored. Thus the integrity of the geometry data must be maintained through careful analysis of the engineering change requirements. It must be recognized also that an engineering change to one assembly may inadvertently change many other assemblies if care isn't taken to assure that the change is limited by changing the part number of a segment of the design, or analyzing where the portion of the changed design is used on other assemblies.

9. *Part Geometry*. As mentioned previously, an assembly design is composed of part geometry plus additional geometry features relating only to the finished assembly design. The parts list of the assembly should be considered the basic relationship used both to prepare a parts list for the assembly and also to record the parent-component geometry relationships.

10. *Assembly Geometry*. The assembly geometry should be considered an assemblage of unique part and geometry features combined with assembly transforms which locate and orient all piece part designs to their assembly location. This assemblage is controlled with a series of product structurers or parts lists which control both parent-component relationships and parent-geometry feature relationships. Creation of a new design prepares this relational data for the database.

CIM INPUT MANUFACTURING GEOMETRY

DESIGN RETRIEVAL-------------- X
FAMILY PROCESSES-------------- X
PART FAMILY DATA---------------X
PROCESS POLYCODES------------ X
ITEM MASTERS------------------O
PERISHABLE TOOLS--------------X
PRODUCT STRUCTURES---------O
JIGS AND FIXTURES--------------X
ENGINEERING CHANGES---------- X
PRODUCTION PLANNING-------- O
PART GEOMETRY-----------------X
INVENTORY CONTROL---------- O
ASSEMBLY GEOMETRY------------X
GROUP TECHNOLOGY CELLS---- O
GENERIC BILL OF MATERIAL--- O
RAW MATERIAL INVENTORY------X
ELECTRONIC MAIL-------------- O

Except for such parts as those made from plastic, molded to extremely close tolerances, most designed parts are made by removing excess material using one or more machining operations to meet the dimensional requirements of the designer. The excess material should be considered an envelope which is added to the engineering design. With this thesis, the original engineering design need not be altered. Thus no additional "copies" of the engineering design are needed, but provisions for the addition of material and its geometry are made. This principle has some economic benefits. For example, a rough casting or forging can be compared to the finished part and the kerf or excess material, seen by the designer. If it is excessive and therefore costly, alterations can be made to the design. The means by which the excess material is removed is also a function of the amount in terms of thickness, material machinability, and other factors. These variables can be made into rules governing the amount of manufacturing material required to machine a design.

The following actions occur as a result of creating manufacturing geometry for a new design:

1. *Design Retrieval.* The new design will be cataloged in a family of parts through a design retrieval system and can be compared to other similar parts.

2. *Family Process.* The family process for the part family of which the new design is a member can be analyzed and used to determine the best amount of excess material to generate for the process.

3. *Part Family Data.* Part family data can be used to compare one part to another including the new design to optimize material removal.

4. *Process Polycodes.* Specific process polycodes may confirm the manufacturing methods and procedures to be used and consequently the manufacturing geometry as well as the sequence of removal necessary for optimum manufacturing.

5. *Perishable Tools.* The perishable tools selected to manufacture the new design will depend on the material removal requirements and the corresponding manufacturing geometry.

6. *Jigs and Fixtures.* In a manner similar to that used to select perishable tools, so will the manufacturing geometry affect the design of jigs and fixtures. For example, sufficient material to hold the design during machining is required, and, in the case where several faces of a part need machining, a means for holding the part during each operation.

7. *Engineering Change.* Any engineering change can have direct effect on manufacturing geometry since design dimensions, including tolerances, may be altered. Thus care must be taken to equate all engineering dimensional changes with potential manufacturing geometry dimensional changes. When changes are made to part geometry, the corresponding manufacturing geometry must be analyzed. These changes could include notes, dimensions, tolerances, finishes, and more. If the engineering design of a part is changed and requires a revision level change only to the part number, such a change might have little affect on the manufacturing geometry. However, it is possible to change from a broad tolerance to a close tolerance on the turning of a part, for example. This might require additional machining operations on the same material, requiring a process change. Thus all changes should be analyzed carefully.

8. *Part Geometry.* Once the designer has completed and released a new design, the part geometry is available for use by manufacturing to create additional material requirements in the form of manufacturing geometry.

9. *Assembly Geometry.* Once the design of a new assembly is complete, the assembly geometry must be examined to determine what additional manufacturing geometry must be added to machine it. This added geometry must be completely separate from the piece part design geometry already recorded for parts in the assembly. The manufacturing geometry tied to an assembly geometry is analyzed and maintained at two levels, the piece part level and the assembly level. If assembly geometry, such as weld size, hole sizes, or machining of portions of the assembly, changes, these must be carefully distinguished between manufacturing changes to the parts within the assembly.

10. *Raw Material*. A new design which is manufactured requires some sort of raw material—plate, bar, sheet, rough forging, etc. in one sense, the manufacturing geometry required is represented in the raw material dimensions selected for the part under consideration. For example, the tolerances on a finished piece part design which is 3 in. in diameter may be such that 3 in. cold-finished bar stock may be purchased. On the other hand, the tolerance and finish requirement may be such that the purchase of 3 1/4 in. hot-rolled bar stock may be required, with several machining operations necessary. Thus manufacturing geometry has a significant and direct relationship to the raw materials selected.

PROCESS AND ROUTING

DESIGN RETRIEVAL-------------- X
FAMILY PROCESSES-------------- X
PART FAMILY DATA--------------X
PROCESS POLYCODES------------ X
ITEM MASTERS------------------ X
PERISHABLE TOOLS-------------- X
PRODUCT STRUCTURES---------- X
JIGS AND FIXTURES------------- X
ENGINEERING CHANGES---------- X
PRODUCTION PLANNING----------X
PART GEOMETRY-----------------X
INVENTORY CONTROL---------- O
ASSEMBLY GEOMETRY------------X
GROUP TECHNOLOGY CELLS------X
GENERIC BILL OF MATERIAL--- O
RAW MATERIAL INVENTORY------X
ELECTRONIC MAIL-------------- O

The process and routing database is as important to the industrial engineer as the bill of material is to the design engineer. The process for each new design is prepared after the engineer releases a new design. Material, methods, tools, and sequence of machining or manufacturing are described through use of the process and routing system. The process is enhanced with the use of standard time data, work center information, and the use of NC tools or robots as special kinds of work centers. In some cases the more economical manufacturing method is batch, or group, technology. In such a case the machine tools for various operations may be located together in one center. Those parts which are manufactured in batches may have no individual routings, only a family process. This may cause some problems with accounting procedures, which tend to require individual part and process costs.

The following actions occur as the result of preparing a process and routing for a new design:

1. *Design Retrieval.* The design retrieval system will show that not only has an additional part been released, but also the "make-buy" decision has been reached and a process prepared.

2. *Family Process.* The family process will be reflected in the process for the new part or assembly design. If new segments of geometry are required for the new part and the family process is updated to reflect this condition, a family process revision level will have changed. In any event, the family process has been used to prepare the process for the new design.

3. *Part Family Data.* The new design has previously been entered into the part family data. In addition, new tooling and the cross-index between them and the design family will be updated.

4. *Process Polycodes.* The process polycodes will have been utilized to develop the means by which the new part will be entered into the database in a manner such that the production schedule can bring together all parts with the same characteristics if and when such a combination is required.

5. *Item Master.* Data developed for the new design, such as make or buy is entered into the item master database.

6. *Perishable Tools.* The proper perishable tools have been selected as requirements for each operation on the process for the new design. If new types or sizes of perishable tools are required, they have been noted and ordered.

7. *Product Structure.* In some manufacturing companies, all parts for an assembly are not issued at the same time but, rather, by operation sequence. If the product structure of the assembly is used to portray this condition, each part on the assembly will be associated with the operation number found on the process plan. This synchronization will serve to coordinate manufacture of an assembly and issue of the parts within it.

8. *Jigs and Fixtures.* The proper jigs and fixtures have been selected or designed as requirements for each new operation on the process for the new design. In addition, for those designs which are manufactured partially or totally on an NC machine tool, proper paper tapes, or computer data, and/or hold-down clamps and other specialized tools for such equipment has been prepared. Adequate programs for the movement of robots have been prepared if necessary, also. Because the new design falls into a family of designs, a family cutter centerline for the NC tools utilized may be available and altered to meet the requirements of the new design.

9. *Engineering Changes*. All engineering changes must be examined to determine whether a design change has caused a change to the process and routing of a part. An engineering change may require taking parts already complete and reworking them to the latest change. In a complex assembly, several parts may require rework as a unit.

10. *Production Plan*. The production plan is directly affected by the process and routing. If too many designs are processed over the same machine tools or other work centers, some will be overloaded, and some will be underloaded. Production variables may require alternate routings to be provided. Unnecessary design restrictions, as close tolerances, may force a process to be performed on a specific machine tool, creating an artificial load. Simulation of a shop load utilizing the processes and routings of active designs can be used to forecast shop load and bottlenecks as well as to forecast new machine tool requirements. The process data is used as the basis for preparing the production plan also.

11. *Part Geometry*. In a relatively stable product with designs whose shape form and process are also relatively stable, the part geometry can be considered a set of sequential design features such that a generative process plan could be developed. Since the geometry can be considered a set of sequential operations, entry of the dimensional data combined with the operation sequence can create a process and routing. For the general condition, a family process should have been made based on part families. As the new part geometry is prepared in engineering, the dimensional data required to prepare the process and routing should be transferred directly from the design to the process. With tables of machinability and corresponding material chemistries, a variant process plan may be prepared semiautomatically as the design is created.

12. *Assembly Geometry*. In a manner similar to that developed for part geometry, process plans can be created utilizing the assembly geometry created for a new assembly design.

13. *Group Technology Cells*. Since each new design falls into a family of parts which also has a family process, several part families may be combined to be manufactured in a "cell" of machine tools or as a batch of composite parts routed as a unit through manufacturing. Each new design, once classified, will possibly fall into one of these group technology cells.

14. *Raw Material*. Since every new design, other than a purchased part, requires a raw material to manufacture it, a new process and routing is the place where such material is designated. Of course, as is the case in so many other examples, this is not the only place where such data is found, but it is a part of the basic data developed with the process and routing.

MASTER SCHEDULE

DESIGN RETRIEVAL-------------O
FAMILY PROCESSES-------------O
PART FAMILY DATA------------ O
PROCESS POLYCODES-----------O
ITEM MASTERS------------------- X
PERISHABLE TOOLS-------------O
PRODUCT STRUCTURES---------- X
JIGS AND FIXTURES------------ O
ENGINEERING CHANGES---------- X
PRODUCTION PLANNING----------X
PART GEOMETRY--------------- O
INVENTORY CONTROL------------X
ASSEMBLY GEOMETRY---------- O
GROUP TECHNOLOGY CELLS------X
GENERIC BILL OF MATERIAL-----X
RAW MATERIAL INVENTORY------X
ELECTRONIC MAIL-------------- O

The master schedule is the main driver of all manufacturing systems. It is an input media which creates action as opposed to planning data. The master schedule is the forecast, usually by dollars, by product, and by modules, of the goods which must be manufactured and/or purchased by the company. The master schedule may originally reflect finished goods requirements but then be converted into a shipping or a build schedule.

The master schedule, in conjunction with the engineering bill of material, including data obtained from the item master, is "exploded" to obtain the total requirements of the raw materials, parts, and assemblies which compose the forecasted products. If the forecast has been prepared by daily, weekly, or monthly time periods, all requirements will be computed in the same manner. By subtracting the parts and assemblies on order and on hand, the total new requirements will be determined.

The master schedule is not considered practical in a job shop since often no forecast is possible, the finished product is extremely customer oriented, or the company makes few but large pieces of equipment. Considering customer orders as entries into the master schedule and also utilizing generic bills of material to document highly customized engineered products will make it possible for a job shop to implement the powerful techniques which can be used in conjunction with a master schedule.

The following actions occur as the result of a master schedule:

1. *Item Master.* Data from the item master is used in the bill of material explosion to determine requirements. Lead time, make or buy, parts on hand and/or on order, low level code, and more, are data which assists in converting the master schedule into a series of raw material, part, and assembly requirements by time period.

2. *Product Structure.* The product structure is the most significant tool utilized with the master schedule to develop requirements. The master schedule represents the products or the modules and attachments of products which have been forecast to manufactur. Each product, module, or attachment has its specific product structure, which is composed of assemblies and piece parts. The master schedule manipulates data in what can be considered a five-dimensional array:

Product mix
Forecast time periods
Bill of material "levels"
Part and assembly lead times
A "low level" coding system

The product structure for each forecasted item on the master schedule is exploded to determine by time period and by part number the total requirements of each part and assembly represented under them. The results are tabulated using material requirements planning (MRP).

3. *Engineering Change.* An engineering change is significant in the computation of requirements based on a master schedule, since the change may create a new requirement while is also eliminates an old one. An engineering change may call for the replacement of one design with another at a period later than the present date, requiring an "effectivity" date to be computed. This type of change may affect future design requirements as well as those in process and those in inventory, and must be considered carefully.

4. *Production Planning.* Production planning is normally considered the total amount of facility requirements necessary to meet the manufacturing requirements represented by the master schedule explosion. (Purchased part requirements are also computed, but have no direct influence on shop loading and production planning except as they are required for assembly and/or for additional machining.) The production planning system provides general and detail schedules for the manufacturing operation by work center, department, and time period, often in priority sequence.

5. *Inventory Control System.* During the process of generating the material requirements data, the master schedule represents the

gross requirements for the finished product. Each level of the product structure may be found in the inventory control system. To accurately reflect net requirements, all parts and assemblies on order and in process must be subtracted from these gross requirements. The result is true net requirements.

6. *Group Technology Cell.* Once actual net requirements are generated, it is logical to summarize them by part family and/or by parts which are designated for manufacture in a group technology cell of machine tools.

7. *Generic Bill of Material.* In a job shop type of manufacture, a master schedule may well be impractical. On the other hand, the generic bill of material represents completed product still in the design stage. This bill of material, and all like them, may be considered a replacement for the master schedule and manipulated just like one. Use of the data will be similar to that of a company which mass produces a product and allows the manufacturer to take advantage of all the techniques and methods of such a company.

8. *Raw Material Inventory.* The lowest level of all bills of material is some sort of raw material or purchased part. Thus the bills of material which are exploded as a result of a master schedule will create raw material requirements for all manufactured parts. This raw material requirement will be reflected in the raw material inventory requirements.

CIM INPUT INVENTORY AND LABOR TRANSACTIONS

DESIGN RETRIEVAL-------------O
FAMILY PROCESSES-------------- X
PART FAMILY DATA--------------X
PROCESS POLYCODES-----------O
ITEM MASTERS------------------- X
PERISHABLE TOOLS-------------O
PRODUCT STRUCTURES---------- X
JIGS AND FIXTURES------------O
ENGINEERING CHANGES---------- X
PRODUCTION PLANNING----------X
PART GEOMETRY--------------- O
INVENTORY CONTROL------------X
ASSEMBLY GEOMETRY---------- O
GROUP TECHNOLOGY CELLS------X
GENERIC BILL OF MATERIAL-----X
RAW MATERIAL INVENTORY------X
ELECTRONIC MAIL-------------O

Labor Reporting

As products are manufactured and parts purchased, material and labor are reported and recorded by and for accounting systems. Generally, cost accounting is charged with the responsibility of both collecting costs and combining them into meaningful product costs. In more cases accounting systems are designed to utilize methods to determine more accurate forecasting of costs and to utilize standard costing techniques to "smooth" product costs as well as reflect inflation rates.

Two basic types of cost systems are used today, job cost and part cost. A job cost system requires that all manufacturing and purchasing costs be tied to a customer job number. If a job number becomes the main control mechanism, engineering documentation becomes the main control mechanism, engineering documentation becomes of secondary importance in the system. Job number manufacturing means it is possible to find an identical part duplicated and manufactured many times under different job numbers.

With the implementation of a bill of material system, a computer, and part number costing, the problems of a job order system become even more obvious. A bill of material system is almost diametrically opposed to job number accounting since the meticulous documentation of engineering is lost under the pseudopart numbering schema found with job numbers.

The following explanations and actions are based on a part number, not job number, costing system:

1. *Family Process*. As actual labor costs are reported in manufacturing, not only is the standard cost developed for the design compared to the actual, but the family process for the family of parts corrected and or updated to reflect any logical changes. In this manner an estimating function is continuously making use of up-to-date data.

2. *Item Master*. The item master may be considered the key database where accounting stores cost data. Many systems are quite elaborate, computing and saving one or all of the following costs:

Last one or up to six historical actual costs
Standard cost
Average cost
First in first out cost (FIFO)
Last in last out cost (LIFO)

For assemblies, only an incremental cost for labor is usually maintained.

3. *Product Structure*. When a total assembly cost is required, the computer is used to perform a cost roll-up. To perform this function, the product structure in the form of the bill of material for the assembly or product is exploded down to its lowest level, and individual costs are then added one to another from bottom to top and rolled up to the assembly level requested. During this roll-up labor and material costs are maintained separately. Often different overheads are also found for labor and material and maintained as separate cost entities.

4. *Engineering Change*. An engineering change can change the cost of a part, an assembly, and a product. If these changes are expensive, it is necessary to reflect them at the proper point in time for future order costs. Consequently the "effectivity" of change is important to the costed bill of material and is reflected in the cost roll-up.

5. *Production Plan*. The production plan is maintained continuously by labor reporting as well as inventory reporting. Labor reporting is conducted through the use of simple handwritten labor tickets as well as by sophisticated labor collection systems which convey data through bar code readers, data collection stations, and other such equipment. Labor reporting also influences inventory reporting. The "last operation" report implies part completion and a corresponding increase in part inventory.

6. *Inventory Control*. Inventory control costs are maintained indirectly as a result of labor reporting. All manufactured parts are costed by part number through the labor reporting system and used to evaluate the inventory.

7. *Group Technology Cell*. Parts manufactured in a group technology cell may require slightly different labor reporting. The economies of group technology require similar parts to be manufactured at the same time. Recording small amounts of labor for each individual part passing through the same operation defeats this goal. Thus estimated labor reporting rather than actual labor may be required. In fact, no labor reporting other than general attendance may be required for these work areas.

8. *Generic Bill of Material*. Since a system using a generic bill of material should be considered a special case of manufacturing, all methods for labor reporting are the same as those used within other more standard systems. However, many of the labor costs reported may be on part quantities of one only. Thus average, actual, and standard costs have no real meaning for them. The total costs reported are, in fact, total costs for the order.

9. *Raw Material*. Since raw material is a significant part of a product cost (50% if purchased parts are included), a labor overhead for material handling is often artificially imposed on them. This cost must be maintained so that a cost roll-up is accurate.

While no action is reflected in electronic mail, labor and inventory reporting is more efficient if it is performed on line. Should this be more effective using the electronic mail system, it should be considered.

Inventory Reporting

During the past 20 years inventory control systems and reporting of inventory have become more complex and demanding. Government tax on inventory was a significant impetus to this increase of complexity, but the need to decrease the lead time between an order and the finished product, the mass production of ever more complex products, and the storage of several levels of assembly of products also contributed. For a long period of time accounting determined "inventory." This accounting was normally in dollar value only, so physical quantities and their corresponding dollar values did not always agree. Today physical part quantities and their corresponding dollar value must track in parallel through a series of different inventory classes. These classes of inventory are generally considered to be as follows:

Raw material
Purchased and manufactured parts
Subassemblies
Work in process
Finished goods

(It should be noted that job cost inventory is usually recorded in two categories only, stock parts and work in process.)

In a manner similar to data collection methods for labor reporting, finished manufactured parts and assemblies are reported and recorded in the inventory control system. In some cases, usually when there are extremely large, bulky, or heavy assemblies and parts, completion is recorded, but the parts may not be physically moved into an inventory area.

Purchased parts may pass directly through the receiving department into an inventory bin, or, for large and heavy parts, a storage yard. In some cases these parts are recorded as received and "in inspection," unavailable for use until examined.

The following actions are performed as a result of either a labor or inventory transaction:

1. *Family Process*. Since the family process may be considered the basis for estimating a new design cost prior to its actual manufacture, economical order quantities, cost of parts in inventory, and any other inventory analysis will be dependent upon this data.

2. *Item Master.* Inventory data should be considered an extension of the item master. Before engineering documentation was considered to be of equal importance to other portions of the manufacturing database, the item master was considered the basic inventory record.

3. *Product Structure.* The product structure is a significant tool for inventory transactions. The product structure in the form of a picking list is used to withdraw parts from inventory to manufacture an assembly. The picking list may be used to prepost or postpost inventory as it is withdrawn from inventory. In the first method, the inventory is checked to see if the parts are available before the picking list is issued. If the parts are available, they are reserved. In the second method, the inventory of parts is not reduced until they are picked. It should be emphasized that the picking list is, in reality, equivalent to an engineering parts list and is actually extracted from the bill-of-material system.

4. *Engineering Change.* An engineering change may change the components required within an assembly and also obsolete the design no longer required. To avoid excessive scrap, an effective data of change from one design to another may be implemented. This data is dependent upon the master schedule and any changes to it, the usage of the design being obsoleted, and the parts in inventory or on order. In addition, the lead time of both the old and the new design must be considered since it is impossible to implement the new design if the parts in inventory will not cover the period of requirements until the new design can be obtained.

5. *Production Planning.* Production planning is directly influenced by inventory reporting since finsihed designs are subtracted from requirements generated by the master schedule. Finished designs, when completed, are recorded in the inventory control system.

6. *Inventory Control.* Each stage of inventory control is maintained by inventory reporting. It is now claimed that a successful MRP system must have inventory quantity reporting which is 95% accurate. This statistic, combined with others, designates a Class A MRP installation.

7. *Group Technology Cell.* Parts routed through a group technology cell may be recorded in inventory slightly differently. Since many diverse parts may be in the same manufacturing process, all of them may be completed before the inventory report is made. This may have some slight disadvantage to a current inventory system depending upon the elapsed time involved in the manufacture of the group of parts.

8. *Generic Bill of Material.* The generic bill of material should be considered a special case of normal manufacturing processes and have little or no effect on inventory reporting.

9. *Raw Materials*. Since raw materials are often considered a separate inventory class, a bulk inventory reporting technique may be required. Pounds, feet, meters, kegs, reels, and other such units of measure may be used, as opposed to "pieces" or "each" normally associated with part and assembly designs. The relationship between raw materials and their corresponding part designs requires careful consideration of units of measure in the reporting of inventory. Material such as steel may be purchased in pounds, sorted as plates, and issued in square inches.

3
RESIDENT INFORMATION

RESIDENT INFORMATION

In one sense resident information can be considered input data also. However, the difference is that such data is developed and maintained as a series of closely integrated but often separate sets of data which are stored in various databases and are either the result of a mass input, or the result of activities in the use of the system.

The nineteen resident information segments are as follows:

Design standards	Traceability structure
Drafting standards	Microfilm microfiche
Design analysis	Process standards
Finite element meshes	Family processes
Design catalog	Work center parameters
Family geometry	Process polycodes
Parametric drawings	Transit times
Simulation models	NC tool paths
Configuration management	Labor material costs
Producibility tips	

Each manufacturing industry, each manufacturing company has specialties and variations to the norm. Thus it is almost inevitable that other data may be required by a company which does not appear in these lists. An analysis of such data and its optimum location in a database segment is necessary. It is obvious that, no matter how well defined, a database must be customized for the user's special requirements.

Please notice in subsequent sections that a table showing each of these database segments has been prepared with "O" or an "X" tabulated beside each of them for the subjects being discussed. This table is an extraction from the chart and shows whether there is a relationship between the item being discussed and that segment of the database.

DESIGN STANDARDS

DESIGN RETRIEVAL-------------O
FAMILY PROCESSES-------------O
PART FAMILY DATA--------------X
PROCESS POLYCODES-----------O
ITEM MASTERS------------------O
PERISHABLE TOOLS-------------O
PRODUCT STRUCTURES---------O
JIGS AND FIXTURES------------O
ENGINEERING CHANGES---------O
PRODUCTION PLANNING-------- O
PART GEOMETRY-----------------X
INVENTORY CONTROL---------- O
ASSEMBLY GEOMETRY------------X
GROUP TECHNOLOGY CELLS---- O
GENERIC BILL OF MATERIAL--- O
RAW MATERIAL INVENTORY------X
ELECTRONIC MAIL-------------- O

Design standards are often difficult to define. Most engineers still consider design to be a creative process. Engineering departments have, perhaps, allowed the designer, engineer, and drafter too much license to be different under the guise of creativity.

Two forms of design standards exist. One type of such standards will be described under producibility tips and may be considered the mutually accepted design and/or manufacturing practices which will be reflected in new designs. These interrelationships are important to manufacturing for economical production.

The second type of design standard is directly connected to the design criteria. In both cases many of them are passed from one designer verbally, implicitly documented on old designs, or documented in dusty manuals. Newly hired engineers and designers may be unaware of their existence or skeptical of their value, yet they represent past experiences and years of painful empirical testing, and "cutting and fitting." We must plan to make this data active, dynamic, and available to all. In addition, much of this empirical design should be replaced with more concise and effective design analysis methods. Data such as this is often difficult to convert to a database. Can we achieve this goal? Design standards are often defined by "seat of the pants" experience. Yet this experience can be converted to factual data if an effort is made to do so and there is a place to store and retrieve the data.

For example, steel liners in a rolling mill are made from simple plates each of which has a bevel so the moving hot steel will not catch its edge. One or more angle of bevel is more effective, thus all such liners should have this bevel. How will this knowledge be maintained? The liners will probably be found under one part family. A family drawing representing these parts can have, as a part of it, notes on the optimum and/or standard bevel and the reasons for them. (It might be questioned why such standards are important. In one actual example, different bevels require different milling cutters costing $12,000 per set, with a back-up set of cutters required also.)

If a retrieval system for families of parts is not available, then a database of design standards must be prepared and tabulated by function, design type, and other means by which the engineer can retrieve them. In any case, it will be important that these standards are available on line to the designer preparing new designs on a computer-aided design/computer-aided manufacturing (CAD/CAM) system otherwise there will be proliferation of design alternates and past experience will be ignored. It seems apparent that design analysis programs incorporating such design standards will be a major factor in implementing these important requirements.

The following parts of the database are affected by design standard information:

1. *Part Family Data*. Design standards are stored by part family data. The standards may appear in the form of actual dimensions or notes. Still another set of design data can also be stored. Many designs have failed in the past. Perhaps the radius of a slot was too sharp. Storing what not to do by family may be just as important as storing how to. Design optimization and technical data retrieval can thus be recalled and errors not repeated. In many cases, design standards may be considered global and thus stored for retrieval in a manner such that the designer can interrogate them as a separate set of entities.

2. *Part Geometry*. The design standards will be represented in each part geometry. In addition, a newly released design assists the checker in determining whether they have been followed correctly.

3. *Assembly Geometry*. The design standards will be represented in each assembly geometry. In addition, a newly released design may also create a list of design standards which will assist the checker in determining whether they have been followed correctly.

4. *Raw Material*. In many cases the difference between an expensive and an economical design is dependent upon the raw material used to fabricate the part. Thus the raw material inventory should be readily available to the designer. While it is the responsibility of the process planner to determine the ultimate raw material to use in fabrication, the designer can make this decision if he has some knowledge of fabrication methods and can convert his finished design into a required raw material which is available.

DRAFTING STANDARDS

DESIGN RETRIEVAL--------------O
FAMILY PROCESSES--------------O
PART FAMILY DATA---------------X
PROCESS POLYCODES------------O
ITEM MASTERS-------------------O
PERISHABLE TOOLS--------------O
PRODUCT STRUCTURES----------O
JIGS AND FIXTURES-------------O
ENGINEERING CHANGES----------O
PRODUCTION PLANNING---------O
PART GEOMETRY------------------X
INVENTORY CONTROL-----------O
ASSEMBLY GEOMETRY-------------X
GROUP TECHNOLOGY CELLS-----O
GENERIC BILL OF MATERIAL----O
RAW MATERIAL INVENTORY-------X
ELECTRONIC MAIL---------------O

Drafting standards should be considered as rules for making drawings of designs which are clear and concise, with the resulting geometric dimensions nonredundant and clearly defined. Unfortunately the present drafting standards in manufacturing companies have become confusing and inadequate. Companies have been making drawings for a long time. Many variations occur, but all insist theirs are the best. Most companies have not made an evaluation of their drawings for many years, and so the drawings have deteriorated badly. For "economy," so much shorthand, stylistic, compacted lines and notes versus dimensions have crept into drawings (not to scale), that they are almost incomprehensible.

Multiple parts on the same drawing, detailing of the assembly and components on one drawing, multiple assemblies on one drawing, tabulated part drawings, and drawings requiring reference to additional drawings all make use of a CAD/CAM system that is ineffective and highly inefficient. To correct these errors and poor quality, several changes must be made, including the use of piece part drawings. In all cases, however, the drawing content must be defined as a general set of drafting standards. If it is true that the drawing should be an accurate representation of the design, shouldn't it be clear, simple, consistent, and accurate?

The dimensioning of a design is an art by itself. Using a CAD/CAM system and preparing geometry for use by others no longer allows sloppy dimensions which will be "interpreted" by others. Thus dimensioning standards become an important part of the drafting standards. A means must be found to define and then readily retrieve them.

In almost all cases of dimensioning practices a two-dimensional drawing is considered. Using feature-based geometry including solid geometry will require new rules to establish surfaces rather than lines, axial locations rather than circles which imply holes, and more. If isometric views are prepared for the final drawing documentation, the rules must establish clear isometric dimensioning practices.

The following parts of the data base are affected by drafting standards information:

1. *Part Family Data*. Since drafting standards may be considered both global and yet specific to the type of design, it seems reasonable to assume that order can be made from chaos by use of a retrieval system which stores part family data. If for each family of parts a typical family drawing is prepared defining as well as showing the proper drawing content and the standard dimensioning process for the family, the resulting drawings will be clear and consistent. If, in addition, a parametric formula for creating new designs is attached to the family of parts, the proper dimensioning practices and content will be automatic. If such a parametric drawing is used, more attention can be paid to true rather than abbreviated drafting practices. In some cases, such as parts to be machined on NC machine tools, care can be taken to define the part geometry in the machining sequence on the family drawing so the designer will no longer have to rediscover how the dimensions should be sequenced and positioned each time a drawing is made which falls into such a family.

2. *Part Geometry*. The results of good drafting standards will appear on the documentation of each new design and its corresponding part geometry. (It should be noted that such onerous requirements as placing arrowheads and scribe lines set to national or government standards can become a standard utility of the CAD/CAM system.)

3. *Assembly Geometry*. In a similar manner the results of good drafting standards will appear on the documentation of each new design and its corresponding assembly geometry. However, it should be emphasized that the assembly drawing is also represented by a parts list, which, in turn, represents a product structure. If the engineering assembly and corresponding parts list do not represent the way by which manufacturing assembles, the drawing becomes meaningless, or manufacturing must redraw and restructure the design. Thus good drafting practices must include a compatible structuring of the product to meet the needs of manufacturing.

4. *Raw Material Geometry*. Since raw material inventory plays a critical part in the design, care must be taken in the preparation of drafting standards (as well as design standards) to recognize tolerance variations, for example, which occur in their use. Tolerances which are extremely tight may require machining of the surface of a plate, the straightening of an I-beam, the milling of an angle bar. Thus the special characteristics and availability of raw materials is critical to design.

DESIGN ANALYSIS

DESIGN RETRIEVAL--------------- X
FAMILY PROCESSES--------------- X
PART FAMILY DATA---------------X
PROCESS POLYCODES------------O
ITEM MASTERS-------------------- X
PERISHABLE TOOLS--------------O
PRODUCT STRUCTURES----------- X
JIGS AND FIXTURES------------- O
ENGINEERING CHANGES----------- X
PRODUCTION PLANNING--------- O
PART GEOMETRY------------------X
INVENTORY CONTROL----------- O
ASSEMBLY GEOMETRY-------------X
GROUP TECHNOLOGY CELLS-------X
GENERIC BILL OF MATERIAL---- O
RAW MATERIAL INVENTORY----- O
ELECTRONIC MAIL--------------- O

Of all engineering segments of CIM, design analysis has been utilized the most and the longest. Unfortunately, design analysis programs have been developed separately, with relatively poor documentation, and with little thought to the integration of design result to any other part of the CIM system. However, integration of design analysis programs will make one of the more powerful tools available within CIM.

Design analysis is the forerunner of a complete model simulation. Until parts of a product can be designed and understood through mathematics, it is difficult to create a complete mathematical model for the entire process create a complete mathematical model for the entire process through which a product may act. In addition, the integration of a design analysis program with documentation of the design in the form of engineering drawings from a CAD/CAM system will dramatically improve accuracy and consistency of design.

As design analysis programs proliferate, it will be necessary to find means to retrieve, alter, and document them so that they can be used by more than a select few specialists.

The following parts of the database are affected by design analysis information:

1. *Design Retrievals*. Design analysis programs should be classified in the design retrieval system according to the part families for which it can be used as gears for gear analysis, solenoids, springs, and others. However, many general purpose analysis programs, such

as heat transfer and stress and vibration analysis, may be considered global and classified so these algorithms are related to many part families.

2. *Family Process.* Since the family process may be designed to automatically accept dimensional and other data to prepare a new process and routing automatically, this database should be tied explicitly to the specific design analysis program. It should be recognized also that many designs which are created by a design analysis program can be used to create the NC tool path automatically if this discipline is tied to the program.

3. *Part Family.* As a specific set of data for a part family, the design analysis set of programs should be tied to them for ease of retrieval. A family drawing which is also tied to a part family may be helpful, and could be the means by which design data is entered into the program. In addition, many designers may be unaware of the programs available and/or their restrictions. Thus, when a new design is being considered, knowledge of such a program will be vital.

4. *Item Master.* As part of the design analysis program, a new item master should be generated automatically, including the part number, when the design is completed.

5. *Product Structure.* Many assemblies are also conducive to design analysis. When such a program is created, not only should item masters for all new parts be created, including that for the assembly, but also a parts list should be prepared for the product structure database.

6. *Engineering Change.* Since an old design may be altered using a design analysis program, the engineering change procedure should be tied to the design analysis programs.

7. *Part Geometry.* A design analysis program should make use of geometry features and, once completed, the resulting part geometry stored.

8. *Assembly Geometry.* If a new assembly is designed, all the geometry transforms, the geometry features, and other data normally prepared by the designer for an assembly designed manually should be prepared. The resulting assembly geometry should be stored.

9. *Group Technology Cell.* It is possible that the part or assembly designed with an analysis program is one which will be manufactured in a group technology cell. Thus the new design should be checked for compatibility and the cell updated to accommodate the new design.

DESIGN CATALOG

DESIGN RETRIEVAL---------------X
FAMILY PROCESSES---------------X
PART FAMILY DATA---------------X
PROCESS POLYCODES------------O
ITEM MASTERS------------------O
PERISHABLE TOOLS--------------O
PRODUCT STRUCTURES----------O
JIGS AND FIXTURES-------------O
ENGINEERING CHANGES--------- O
PRODUCTION PLANNING--------- O
PART GEOMETRY---------------- O
INVENTORY CONTROL----------- O
ASSEMBLY GEOMETRY----------- O
GROUP TECHNOLOGY CELLS----- O
GENERIC BILL OF MATERIAL---- O
RAW MATERIAL INVENTORY----- O
ELECTRONIC MAIL--------------- O

The design catalog should be considered a tabulation of all design analysis programs in the database. Programming standards should be devised to make certain it can be altered by persons other than the creator. A brief abstract of its use, limitations, variables, interfaces, and requirements should be prepared for each. Security of the program should be such that the original parameters cannot be changes without authority.

The following parts of the database are affected by the design catalog:

1. *Design Retrieved*. As mentioned previously, each design analysis program should be classified in the design retrieval system according to the parts for which it should be used. In addition, and as part of the database, a complete documentation of the program should be made available on line, including a user-oriented input format for data input, "help" features, and an abstract of the program for quick review.

2. *Family Processes*. Since the family processes including the NC tool paths will be closely associated to the design analysis program, the design catalog should be closely tied to them.

3. *Part Family*. The design analysis program becomes a segment of the part family data for fast retrieval.

FAMILY GEOMETRY

DESIGN RETRIEVAL--------------- X
FAMILY PROCESSES--------------- X
PART FAMILY DATA---------------X
PROCESS POLYCODES------------- X
ITEM MASTERS-------------------O
PERISHABLE TOOLS--------------O
PRODUCT STRUCTURES----------O
JIGS AND FIXTURES------------- O
ENGINEERING CHANGES----------O
PRODUCTION PLANNING--------- O
PART GEOMETRY------------------X
INVENTORY CONTROL----------- O
ASSEMBLY GEOMETRY-------------X
GROUP TECHNOLOGY CELLS----- O
GENERIC BILL OF MATERIAL---- O
RAW MATERIAL INVENTORY----- O
ELECTRONIC MAIL--------------- O

Family geometry is a significant new database in CIM. Storage, maintenance, and retrieval are closely associated with the design retrieval system. As discussed previously, geometry features have been developed on a global basis and catalogued in a manner for fast retrieval also. Once these geometry features are available, a series of them can be combined for any part family designated.

Several ways to combine geometry features into family geometry can be considered. If, for each part family, a family drawing can be prepared, the family drawing could represent the actual associativity of the geometry features. In some instances the same geometry feature may be used in more than one part of the design as holes, different diameters, threads, etc. Combining them into one family drawing will save part of the work later in making a new design since the basic associativity will have been preplanned. In another section we will discuss parametric drawings which are the natural result of this logic.

Another means for combining geometry features into family geometry is to simply create the associativity of the features for the family and present them on the screen for manipulating as the designer chooses. While more complex and time consuming, there may be several part families which are not conducive to a literal family drawing. Under such conditions this may be the only reasonable way to utilize the geometry features.

Still another method to consider is storing, as a parts list, the geometry features and also the geometry transforms as text data.

When a new design is required for a specific part family, the part geometry, either associated into a family drawing or as separate entities, can be retrieved and drawn. Storage of the family geometry is thus eliminated but may be troublesome if repainting and recalculation are slow.

The following parts of the database are affected by the family geometry:

1. *Design Retrieval*. It is assumed that all geometry features have been classified in a design retrieval system. In addition, it is assumed that the family geometry, no matter how stored, is classified by a design retrieval system also. Through this means a concise, logical, and rapid method for generating new designs will be achieved.

2. *Family Process*. As an important part of the family geometry, the manufacturing geometry and the corresponding family process should also be considered. Its inclusion will assist the designer to understand the parameters and limitations within the type of design upon which he is working. As will be described later, a parametric drawing should be such that the manufacturing geometry is varied according to the new design requirement also.

3. *Part Family*. The family geometry is considered a segment of the part family data. This data can be stored in the form of a parts list and other text or separate segments of geometry tied to the family with a parts list, or grouped as a complete entity.

4. *Process Polycodes*. The process polycodes are a specific set of features for a part family. Thus their inclusion within the family geometry will also assist in developing the specific requirements for a new design. Some polycodes may be global as well as family oriented. (Present methods usually create these global codes.) The significance of part families is that polycodes may be customized by family rather than globally, eliminating the need to carry codes for all parts which are essentially unnecessary.

5. *Part Geometry*. The family geometry can be considered a "piece part," with its own part number, item master, and so on. Thus the data for it can be stored in the part geometry in a manner similar to an actual design.

6. *Assembly Geometry*. The family geometry may be for an assembly and considered to have its own part number, item master, parts list, and so on. Thus the data for it can be stored as assembly geometry in a manner similar to an actual design.

PARAMETRIC DRAWINGS

DESIGN RETRIEVAL--------------- X
FAMILY PROCESSES--------------- X
PART FAMILY DATA---------------X
PROCESS POLYCODES------------O
ITEM MASTERS-------------------O
PERISHABLE TOOLS--------------O
PRODUCT STRUCTURES----------O
JIGS AND FIXTURES------------- O
ENGINEERING CHANGES----------O
PRODUCTION PLANNING--------- O
PART GEOMETRY------------------X
INVENTORY CONTROL----------- O
ASSEMBLY GEOMETRY------------- X
GROUP TECHNOLOGY CELLS----- O
GENERIC BILL OF MATERIAL---- O
RAW MATERIAL INVENTORY----- O
ELECTRONIC MAIL--------------- O

A CAD/CAM graphics system is an expensive T-square. Skilled operators may make engineering drawings rapidly, even economically. In the final analysis, however, the purpose of a CAD/CAM system should be to draw ideas, not drawings. In addition, feature-based geometry for use in making new designs can be clumsy to utilize if it is not organized properly. Consequently, the use of parametric drawings becomes a significant tool to enhance design.

In its simplest form a graphics program can be written for a generalized wire frame (even two dimensional) part or assembly. The program essentially "draws" a design. When properly programmed, the computer program stops and calls for entry of dimensions for the design as A, B, and so on. Obviously such a program is not cost effective unless many similar parts can be created using the same program. Thus a means for finding and reusing such parametric drawings is required. If a classification and design retrieval system is used to segregate part and assembly design by similar form, shape, or features a family parametric drawing may be created. During this process each dimension of the part will be replaced by a "neutral" dimensioning schema, as A, B, C, and so on. These letters will be replaced with actual dimensions as the user requires. In addition, of course, additional notes, minor changes, or additions to the design may be necessary. The program should be prepared in a manner such that, once entry of dimensions has been completed, a new design, drawn to scale, is prepared automatically.

Probably more significant and as a more universal parametric drawing concept, the family drawing should be prepared using feature-based geometry for solids. Thus, instead of general primitives, the geometry features should be engineering oriented. Some, but not all of these features are as follows:

Single round cylinder	Cube
Multiple round cylinder	Thread
Round hole	Bent "plate"
Keyway	Keyway
Reentry cut	"More"
Rectangle	

Retrieval of these features has already been described and is accomplished through the use of a parts list, assuming each of these features has its own item master data and is classified in the retrieval system.

There is reason to believe that a new design made in this manner should be drawn completely with isometric views only. The final part design will be clearer and will follow the principles originally prepared for engineering drawings, which were then replaced with "simplified" drafting techniques. Thus the quality of design expression should increase.

It should be expected that this principle will not be practical for at least 10% to 20% of designed parts in a manufacturing company, the percentage dependent to a great amount upon the products being manufactured. Even here, however, special conditions may be used to find solutions. While assemblies are not nearly so conducive to this practice, when they are, substantial time savings will be achieved.

It should be noted that almost all good retrieval systems which place designs into families of parts are today augmented with a family sketch. These can be replaced with the generalized family drawing.

It should be emphasized that a good classification system will divide 80,000 designed parts into no more than 1200 families. Thus only 1200 parametric drawings are required.

The following parts of the database are affected by parametric drawings:

1. *Design Retrieval*. Each parametric drawing and its accompanying program must be classified in a design retrieval system for ease of retrieval. It should be noted that some of the parametric drawing programs will be intimately tied to design analysis programs.

2. *Part Family Data*. Either in the form of a parametric drawing or a sketch, each family will have, as a part of the part family data, a picture of the general design so that entry into the proper family can be confirmed by the designer. In addition, this family drawing will be coordinated with the catalog of parts for the family so that critical dimensions may be examined.

3. *Part Geometry*. The geometry of part families may be stored with their own individual part numbers within the part geometry database if the entire geometry for a family of parts is stored as an entity.

4. *Assembly Geometry*. The geometry of assembly families may be stored with their own individual part numbers within the assembly geometry database if the entire geometry for a family of assemblies is stored as an entity.

TRACEABILITY STRUCTURE

DESIGN RETRIEVAL------------- O
FAMILY PROCESSES--------------O
PART FAMILY DATA------------- O
PROCESS POLYCODES------------O
ITEM MASTERS-------------------- X
PERISHABLE TOOLS--------------O
PRODUCT STRUCTURES----------- X
JIGS AND FIXTURES------------- O
ENGINEERING CHANGES----------- X
PRODUCTION PLANNING-----------X
PART GEOMETRY---------------- O
INVENTORY CONTROL-------------X
ASSEMBLY GEOMETRY----------- O
GROUP TECHNOLOGY CELLS----- O
GENERIC BILL OF MATERIAL---- O
RAW MATERIAL INVENTORY-------X
ELECTRONIC MAIL--------------- O

The need for design and material traceability is normally dependent upon government requirements. In the design of nuclear reactors, sensitive electronic equipment, and space vehicles, it has been common to trace parts and raw materials from their source to the finished product. Tracing these sources through the various product levels is expensive, time consuming, and difficult. The need to trace materials also creates a paradox. For example, a batch of identical fuel pellets or diodes or some other design must be traced to the final assembly and product in which they are used. Each part requires (in some cases), an individual serial number. This means that designs with identical part numbers must be identified with different serial numbers. This procedure almost makes each part unique and reduces the ability to manufacture identical parts together easily.

Fortunately, however, most requirements call for lot traceability. Under this condition, traceability functions can be made to fit in a CIM system, with some slight alteration to the principles of part number control. The following description of a means to control traceability is not necessarily applied in manufacturing companies. Today many (often unsuccessful) systems exist. The principles described here will, hopefully, assist in the development of a rational and economical method which can be used by all.

The following parts of the database are affected by traceability:

1. *Item Master*. Each part or assembly which must be controlled by traceability rules should be so categorized using a code found in the item master. In this manner a complete list of controlled parts may be generated. In addition, all transactions which affect these parts will be highlighted by use of this data.

2. *Product Structure.* All designs, whether piece parts or assemblies, will become part of a product structure. In all cases where such designs appear, a code showing them to be in the traceable category must appear.

3. *Engineering Change.* Because of the sensitive nature of designs under traceability control, no engineering change can be created before the effect of the change has been thoroughly examined. When such a change must occur, the relationship of parts in inventory under lot control (either single or multiple lots) and designs in process or already in some product must be analyzed and disposition meticulously planned.

4. *Production Planning.* As will be described more completely under inventory control, the production planning process must be altered to accommodate traceability. Parts issued for production may not be "borrowed" or transferred in any manner from one assembly or product to another. During the manufacturing process, a lot of parts must be maintained as a unit, scrap or rework documented, and the entire batch sent to inventory or the next operation as a group.

5. *Inventory Control.* The inventory control function is the mainstay of traceability. If the design is purchased, a specific lot size, when generated, will also include an assigned lot number. The designs, when received, will be documented both by part number and lot number. Each lot will be stored separately from another of the identical part number, and inventory records will be maintained in the same manner. In effect, multiple inventories will be maintained for identical designs separated by different inventory locations or bin numbers. The same segregation will be maintained for manufactured parts. When these designs are required for assembly, the picking list on which they are included will not only show the bin number from which to pick, but the lot number to be attached to them. The picking of the parts will be verified at all times. The picking lists will reflect a lot number and the corresponding bin location until one lot is exhausted. The next picking list will show a new lot number and corresponding bin location. In this manner tracing the designs from receipt to their ultimate use may be maintained.

6. *Raw Materials.* The most difficult traceability problem is controlling raw materials. Since the same raw material can be used for a host of different designs, the only certain method to control it is segregation. Thus it is recommended that all raw material which must be controlled have individual part numbers even if it is identical to other raw materials in common usage. While this will tend to increase the number of raw materials in inventory, it will lessen the cost of controlling traceability. Control of the raw material during manufacturing will follow the same principle described for part and assembly design. If the item master for designed parts contains the raw material requirements for it, coordinated control can be maintained.

CONFIGURATION MANAGEMENT

DESIGN RETRIEVAL-------------- O
FAMILY PROCESSES---------------- X
PART FAMILY DATA--------------- X
PROCESS POLYCODES------------ O
ITEM MASTERS-------------------- X
PERISHABLE TOOLS-------------- O
PRODUCT STRUCTURES----------- X
JIGS AND FIXTURES------------- O
ENGINEERING CHANGES----------- X
PRODUCTION PLANNING----------- X
PART GEOMETRY------------------ X
INVENTORY CONTROL------------- X
ASSEMBLY GEOMETRY------------- X
GROUP TECHNOLOGY CELLS----- O
GENERIC BILL OF MATERIAL---- O
RAW MATERIAL INVENTORY------- X
ELECTRONIC MAIL----------------- X

The following definition of configuration management may be used to assist in understanding the rules required to control engineering documentation: "Configuration Management is a systematic approach to identifying, controlling, and accounting for the status of the parts and assemblies required in a design (product) from the point of its initial definition throughout its entire life."

Configuration management should be considered a set of policies, standards, and/or procedures governing engineering actions. Unfortunately manual systems in engineering departments have followed these rules poorly. Thus in mechanized systems control of alphanumerical data, geometric data, and database control can be jeopardized until the rules are reemphasized and/or established.

As described in other segments, part number control and change, revision level changes, interchangeability rules, effectivity of change, drawing content, part numbering systems, single part drawings, and assembly drawings with their detached parts lists are all parts of the documentation which reflect original design and design changes. In addition, changes to the geometry of the design must be recorded and controlled in parallel to the "paper" documentation.

It should be emphasized that it is impossible for an integrated CIM system to operate and maintain its integrity without good configuration management.

The following parts of the database are affected by configuration management:

1. *Family Process*. In a manner similar to that used to record changes to a design, changes to a family process must be recorded. Since changes to a family process will usually be considered interchangeable, a change to the revision level of the process will usually

be sufficient. It should be noted, however, that two revision levels are required for a process, one for the part or assembly design for which it has been developed, and the process itself. Care therefore must be taken to determine if one or both of them must reflect changes.

2. *Part Family Data.* Configuration management rules are also reflected in part family data. Not only does each part in the family catalog have revision control, but the family drawing or geometry must be controlled as well.

3. *Item Master.* The item master should be considered the central record for a design. As such, documentation changes reflected in other records should be secondary when redundancy is required. Several different levels of configuration may have to be maintained based on effectivity of design changes.

4. *Product Structure.* In a like manner, the product structure should be considered the central record for a design composed of parts and subassemblies. Effectivity of change is recorded through changes to the assembly records. However, each component of an assembly will have its own individual configuration control. It should be emphasized that an assembly parts list should not be printed as output with the revision levels of the components as part of the document. If this occurs, a change to the configuration of any component will require a new output, which is unnecessary unless a part number change is needed.

5. *Engineering Changes.* An engineering change is the normal means of communicating configuration management rules when design changes occur. In an integrated system, however, all data increments related to a design are not under rules of configuration management. Thus care must be taken to define those data increments which are. Normally the data increments involved in design changes only should be considered under configuration management control.

6. *Production Planning.* When, through the application of configuration management rules, design changes occur designs in process must be considered in the production planning system. Many design changes cause complex interrelationship changes of individual designs which affect work in process and must be coordinated with manufacturing.

7. *Part Geometry.* While some engineering changes occur which do not affect the part geometry, those which do must be reflected in the database. Dependent upon the way in which individual designs are related one to another, a simple geometry change on a piece part design may require changes to the design geometry of several assemblies. The means by which this is performed is critical to the integrity of the geometric data base.

8. *Inventory Control.* In a manner similar to that discussed under production planning, parts which have been completed and stored in inventory control must be considered for rework or scrap under configuration management rules. On the other hand, many design changes may not require instant replacement of one design

with another, and the effectivity of change may depend upon the exhaustion of the parts in inventory.

9. *Assembly Geometry.* When a design change to an assembly is made, two conditions may occur. In the first case the piece parts used in the assembly may change. In the second case the assembly configuration may change. In either case the configuration management rules which apply must be reflected in the assembly geometry when geometry changes are required. Thus close coordination between the item master, product structure, part geometries, and the assembly geometry is required.

10. *Raw Material.* The raw material used in a design is normally considered the choice of industrial engineering. However, the chemical specifications are selected by engineering. Thus the raw material purchased for a design must conform fully to such material specifications, and the specifications must be documented and altered under the configuration management rules. In many cases coatings, plating, and other material deposits are placed on designs. The material specifications for this material are also considered to be under the rules of configuration management.

11. *Electronic Mail.* Since configuration management rules are closely associated with engineering changes, it is to be expected that their implementation will be observed and conveyed through the electronic mail.

PRODUCIBILITY TIPS

DESIGN RETRIEVAL----------------X
FAMILY PROCESSES--------------- X
PART FAMILY DATA---------------X
PROCESS POLYCODES------------O
ITEM MASTERS-------------------O
PERISHABLE TOOLS--------------- X
PRODUCT STRUCTURES----------O
JIGS AND FIXTURES------------- O
ENGINEERING CHANGES----------O
PRODUCTION PLANNING--------- O
PART GEOMETRY------------------X
INVENTORY CONTROL----------- O
ASSEMBLY GEOMETRY-------------X
GROUP TECHNOLOGY CELLS-----O
GENERIC BILL OF MATERIAL---- O
RAW MATERIAL INVENTORY-------X
ELECTRONIC MAIL---------------O

For many years little or no significant communications have existed between design and industrial engineers. More than occasionally designs are created in the vacuum of an engineering ivory tower with little regard to the economics of manufacturing. At one time many design engineers spent a considerable part of their apprentice days working in a manufacturing environment until they were familiar with the machine tools, fabrication restrictions, and tooling requirements and limitations. Today this knowledge is not made a part of design engineers' and drafters' curriculum. Consequently some means to replace this lack of knowledge is mandatory for the increases of productivity needed.

The creation of a CIM database is ideal for this opportunity. By first segregating designs into families of similar items, and then coordinating industrial and manufacturing engineering economies of production with such families, the manufacturing methods thus documented can become an almost automatic part of the design engineer's know-how and finished design.

Two types of producibility tips will be required, global and family. Family tips will largely be based upon the peculiarities of such a family and possibly related families of parts. Global tips will express general economies, such as hole size standardization, bevels, tolerances, and more.

The following data base parts are affected by producibility tips:

1. *Design Retrieval*. The design retrieval system should be designed to accommodate producibility tips. It is no doubt more rapid to see rather than read producibility tip data. Consequently as many as possible should be expressed visually. No such data should limit creative design. Instead the tips should be expressed in comparative or economic terms, allowing the designer to conclude for himself the optimum utilization of alternates.

2. *Family Processes*. Family processes will be affected by producibility tips, perhaps reflected in alternate or optimum routings and machining or fabricating practices.

3. *Part Family Data*. Part family data should become the vehicle to store and retrieve the producibility tips. When they are global, refrence to them should be maintained in family data.

4. *Perishable Tools*. Perishable tools and their inventory will be directly affected by producibility tips. Studies have shown that the highest variations in design occur in hole and tap sizes. Thus the standardization of processes for holes can create a significant change to the tool packages required on NC machine tools as well as normal machining. This should serve only as an example of the economies available through the use of producibility tips.

5. *Part Geometry*. Part geometry will directly reflect producibility tips. Bevels on ends of round parts rather than radii, for example, will be directly reflected in geometry. Bend diameters, hole sizes, finishes, and more become part of the geometric database and can also be reflected in the feature-based geometry stored for use by part families.

6. *Assembly Geometry*. Assembly geometry will reflect use of producibility tips in a way similar to that described for part geometry.

7. *Raw Material Inventory*. Raw material inventory will be affected by producibility tips also. Today many fabricated parts are designed without any reference to the availability of standard raw materials. For example, a piece part which could be produced from a 4 in. bar may have a $4\frac{1}{4}$ in. dimension arbitrarily selected. This minor deviation will require the part to be burned from a larger plate or sheared on both the length and width. These "minor" additional fabricating operations add to large costs over a period of time. Thus coordination of good fabricating and machining practices with either standardization of raw materials or selection of economical raw materials will have a significant effect on manufacturing economies.

FINITE ELEMENT MESHES

DESIGN RETRIEVAL--------------- X
FAMILY PROCESSES--------------O
PART FAMILY DATA---------------X
PROCESS POLYCODES------------O
ITEM MASTERS-------------------O
PERISHABLE TOOLS--------------O
PRODUCT STRUCTURES----------O
JIGS AND FIXTURES -------------O
ENGINEERING CHANGES----------O
PRODUCTION PLANNING--------- O
PART GEOMETRY------------------X
INVENTORY CONTROL----------- O
ASSEMBLY GEOMETRY-------------X
GROUP TECHNOLOGY CELLS----- O
GENERIC BILL OF MATERIAL---- O
RAW MATERIAL INVENTORY----- O
ELECTRONIC MAIL ---------------O

As a special case of design analysis, finite element meshes are used to analyze stress, torsion, compression, and other reactions to loads on a design. In other design examples, such as heat transfer and fluid dynamics, the design analysis is normally prepared through the use of formulas. In the case of stress analysis, designs must be converted or simulated by a series of nodes and elements represented by lines and intersections of lines. In addition, these nodes and elements must be numbered so that analysis results can be tabulated by location. Since many thousands of nodes and elements may be required, the preparation of a model can be tedious and full of potential errors even with automatic numbering systems. In addition, only a limited number of designers are capable today of developing a mesh which accurately reflects the actual design.

While several technical experts feel predesigned meshes are potentially dangerous as too general, and while others may feel analysis of piece parts is yet too expensive, the elimination of excess material, confirmation of sufficient material or design configuration, and optimizing design generally have a significant economic benefit. If a finite element mesh is designed for a family of parts, greater care to develop integrity of design can be achieved. It is even possible to suggest that each geometric feature for fabricated parts have an individual mesh developed for it. (This is already being done by some companies.)

The following parts of the database are affected by finite element meshes:

1. *Design Retrieval*. The design retrieval system should be designed to accommodate meshes developed by families of parts or assemblies or by geometric feature.

2. *Part Family Data*. Finite element meshes should be stored and retrievable through part family data.

3. *Part Geometry*. Consideration should be given to maintain the results of finite element analysis for each piece part design, where it is utilized, with the part geometry.

4. *Assembly Geometry*. Consideration should be given to maintain the results of finite element analysis for each assembly design, where it is utilized, with the assembly geometry.

SIMULATION MODELS

DESIGN RETRIEVAL--------------- X
FAMILY PROCESSES--------------O
PART FAMILY DATA---------------X
PROCESS POLYCODES------------O
ITEM MASTERS-------------------O
PERISHABLE TOOLS--------------O
PRODUCT STRUCTURES----------O
JIGS AND FIXTURES------------- O
ENGINEERING CHANGES----------O
PRODUCTION PLANNING--------- O
PART GEOMETRY------------------X
INVENTORY CONTROL----------- O
ASSEMBLY GEOMETRY-------------X
GROUP TECHNOLOGY CELLS----- O
GENERIC BILL OF MATERIAL---- O
RAW MATERIAL INVENTORY----- O
ELECTRONIC MAIL--------------- O

The meaning of simulation models will vary, depending on the type of manufacturing and the products involved. Two general types of simulation may be considered, ones which simulate the design in terms of adequacy and performance, others which might be used by marketing to illustrate the product in actual applications.

A typical and well know example of simulation is in the testing of airframes. It is no longer required that all simulations take place in a wind tunnel. Many are performed better and more accurately using simulation techniques. However, many products requiring lesser accuracy of design do not have similar simulation models. Some products which would be greatly improved in operating characteristics and economy include rolling mills, cranes, hoists, tractors, and lift trucks.

Design analysis programs generally consider the design of a small segment of a product. Through combinations of these with additional integrated analysis programs, simulation models should grow in a company once their importance is seen.

For marketing, simulations will assist in sales. By combining modular design and the bills of materials for these modules with the various design alternates and their operating characteristics and limitations, it should not be too difficult to demonstrate the operation of a product on the screen in front of the potential customer, prepare a specific bill of material, compute the cost and delivery date, and get an order!

The following parts of the database are affected by simulation models.

1. *Design Retrieval*. The design retrieval system should be planned to include pointers and/or other data for simulations of both parts and assemblies, and finished products and their combinations.

2. *Part Family Data*. Part family data should include the simulation model variables, rules, and restrictions. In this case part family data means finished product families in particular.

3. *Part Geometry*. The part geometry should be so designed that, when applicable, it can be used in the simulations required.

4. *Assembly Geometry*. The assembly geometry should also be configured so that it can be used properly in simulations. Since assembly geometry means also the finished product and modules, special attention should be given these particular assemblies.

MICROFILM MICROFICHE

DESIGN RETRIEVAL--------------- X
FAMILY PROCESSES--------------- X
PART FAMILY DATA---------------X
PROCESS POLYCODES------------O
ITEM MASTERS-------------------- X
PERISHABLE TOOLS--------------O
PRODUCT STRUCTURES----------- X
JIGS AND FIXTURES------------- O
ENGINEERING CHANGES----------O
PRODUCTION PLANNING--------- O
PART GEOMETRY------------------X
INVENTORY CONTROL----------- O
ASSEMBLY GEOMETRY-------------X
GROUP TECHNOLOGY CELLS----- O
GENERIC BILL OF MATERIAL---- O
RAW MATERIAL INVENTORY----- O
ELECTRONIC MAIL--------------- O

While it would be nice to believe that "paper" will not be required once an integrated CIM system has been implemented, the facts seem to be otherwise. Many old designs will continue to require reference to old drawings. Many new designs will be prepared without the benefit of a CAD/CAM system. Perhaps a logical and economical breakthrough will be achieved where all old designs can be digitized and enter into a database. Such a conversion will also require a large and economical database storage. Until that time, microfilm and microfiche systems seem a rapid, cost effective, and logical system to merge the old with the new. Indeed, with some of the newer computer output on microfilm (COM) systems, which merge graphics and text directly from the computer, the need for plotted drawings may become less economical. Thus the need for such systems is apparent.

The microfilm system requires that a film chip be mounted on a punched card and then data keypunched into it. (For engineering, microfilm is normally filed by part number. Each revision to a design is stored after the original sequentially.)

Microfiche are prepared in several ways. A flexible and economical method is to use a master with pockets so additional chips can be added and the entire unit photographed.

The following parts of the database are affected by microfilm and microfiche:

1. *Design Retrieval.* Since a microfiche is the most effective way to document the designs of a family of parts, the design retrieval system should be used as the means for cataloging designs on them.

As a secondary file, or for master files used in other departments, a copy of the master microfilm file sorted by family is a reasonable alternate.

2. *Part Family Data.* Adding part family data as well as the design should be considered a possibility for storage on microfiche.

3. *Family Process.* While today manufacturing processes are maintained as a database on the computer, a family process should be considered for placement on microfiche.

4. *Part Geometry.* A significant merging of old and new designs will occur when all new designs produced on a CAD/CAM system are merged with old designs and placed both on microfilm by part number and on a microfiche by family number. The part geometry of both old and new designs are all available equally when a retrieval system is interrogated.

5. *Assembly Geometry.* In a manner similar to that for piece part designs, assembly geometry should be merged with old assembly designs on both microfilm and microfiche. It should be noted that newer versions of microfilm and microfiche preparation may take designs prepared on a CAD/CAM system and directly convert them. This will make plotters candidates for obsolescence.

PROCESS STANDARDS

DESIGN RETRIEVAL------------ O
FAMILY PROCESSES--------------- X
PART FAMILY DATA---------------X
PROCESS POLYCODES------------- X
ITEM MASTERS--------------------O
PERISHABLE TOOLS--------------- X
PRODUCT STRUCTURES----------O
JIGS AND FIXTURES---------------X
ENGINEERING CHANGES----------O
PRODUCTION PLANNING--------- O
PART GEOMETRY---------------- O
INVENTORY CONTROL----------- O
ASSEMBLY GEOMETRY----------- O
GROUP TECHNOLOGY CELLS-------X
GENERIC BILL OF MATERIAL---- O
RAW MATERIAL INVENTORY----- O
ELECTRONIC MAIL--------------O

Process standards are an important and necessary first step for improving productivity in manufacturing as well as for considering flexible manufacturing or group technology and machining centers or "cells."

The process is for manufacturing what a bill of material is for engineering. Since the processes for parts and assemblies are initiated by manufacturing engineering, it can be considered the source database of information for HOW a part or assembly is manufactured. (The difference between a routing and a process may be found under manufacturing definitions.)

To standardize processes in a manufacturing concern which has had a manufacturing facility for years is not simple. Many similar parts have both minor and major variations to their manufacturing process. Each current process has often been prepared in isolation by a variety of industrial engineers over many years. Rather than inventing a standard process, it is often more practical to first examine processes which have been prepared and make an amalgamation of their best methods. This standardization of processes is enhanced with part families generated through a classification system. Process standards must consider the following variables:

Work centers
Perishable tools
Engineering design
Manufacturing practices
Standard time data
Production volume
Jigs and fixtures
Numerically controlled machine tools

Process standards can be developed individually through examination of the present manufacturing methods. However, the more logical and rapid method is to examine similar parts to determine those processes which can be and are identical. Through comparative analysis, this will assist in determining the best. If process standards are to be prepared using part families, the following steps are required for each of them:

1. Simplify operation sequence, eliminating and/or combining them on the "best" work center and for the "best" method.
2. Develop standard descriptions of the process step for global use in other families.
3. Develop a retrieval system which allows rapid retrieval of standard text and operation descriptions. (It should be noted that the retrieval system for standard text can be conceived as a portion of work center data which is described in a separate section.)
4. Simplify tool references, operation codes, and standard times. As noted in other sections, tooling and fixture part numbers are required.
5. Standardize set-up times and descriptions.
6. Determine, by family, whether the routing and process will include all potential machining requirements, or whether

the composite routing will show only main manufacturing features.
7. Extract all standards developed for a family of parts for use both globally and by other families as a set of process standards.
8. Analyze producibility tips by part families both for global process standards and for family process standards.
9. Examine the feasibility of generative versus variant process planning.

The result of this exercise will be standard processes. In general the standard processes may fall under the following categories, depending upon the manufacturing requirements:

- Machining
  - Flats
  - Contours
  - Drilling
  - Rounds
  - Grinding
- Joining
- Forming
- Cutting
- Assembly, test, and inspection
- Miscellaneous
- Unplanned operations
- Overhead operations

An example of standard text for forming is as follows:

- Bend
- Roll
- Press
- Forge
- Cast
- Mold

The following parts of the database are affected by process standards:

1. *Family Processes*. The family processes will be created through the development of process standards. As discussed above, it is feasible to consider developing family processes and process standards in parallel for a manufacturing company with a long history of manufacturing. Thus which comes first, the process or the standards, may be a moot question.

2. *Part Family*. Just as family geometry, parametric drawings, and other engineering data associated with similar parts reside in

the part family database, so should family routing and the corresponding process standards. These process standards may be stored in the form of a family process, but will also be stored as a separate set of data for analysis, reference, and global use.

3. *Process Polycodes*. Process polycodes are often the result of determining a set of process standards. Some, but not all, of them will be utilized to determine optimum sets of parts for group technology and machining cells for flexible manufacturing.

4. *Perishable Tools*. When standard processes are combined with producibility tips and standard manufacturing techniques, the results will be reflected both in quantity and variations of the perishable tool requirements. A set of perishable tools in a tool package will be critical for eliminating excessive changes of them on a NC machine tool, eliminating set-up time. The process standards will directly relate also to potential changes in engineering design to facilitate standardization of perishable tools.

5. *Jigs and Fixtures*. Jigs and fixtures will be standardized not only because families of parts will tend to use similar tools, but standard operations will reduce the variety of tools required because operations have been performed previously on different machine tools or using different methods.

6. *Group Technology Cells*. Process standards are mandatory to allow the creation of group technology cells of machine tools. Either through the use of part families or polycodes defining a specific set of operations, the practical use of group technology is defined, together with appropriate machine cells.

WORK CENTER PARAMETERS

DESIGN RETREIVAL-------------O
FAMILY PROCESSES---------------X
PART FAMILY DATA------------ O
PROCESS POLYCODES------------ X
ITEM MASTERS------------------O
PERISHABLE TOOLS-------------- X
PRODUCT STRUCTURES----------O
JIGS AND FIXTURES--------------X
ENGINEERING CHANGES----------O
PRODUCTION PLANNING-----------X
PART GEOMETRY--------------- O
INVENTORY CONTROL---------- O
ASSEMBLY GEOMETRY---------- O
GROUP TECHNOLOGY CELLS-------X
GENERIC BILL OF MATERIALS--- O
RAW MATERIAL INVENTORY----- O
ELECTRONIC MAIL--------------O

Each machine tool or other type of work center has limitations, capabilities, and capacities. Work center criteria must be defined in such a manner as to expedite the implementation of a process for a new design (as well as old) economically.

Maximum size and/or weight of the design, operations which can be performed, length and width of bed travel and tool travel, minimum and maximum turning rpm, and others are just some of the criteria which require definition. Often two categories of tools are defined. Some tools always reside at the work center and may be specialized for that location. Other tooling is global and can be used at several work centers. Jigs and fixtures may be designed specifically for one or several designs and be prepared for a specific work center or may be used globally as well.

An example of a set of work center parameters follows:

CNC MACHINE CENTER

QUANTITY----1
TYPE-------TWIN SPINDLE HORIZONTAL
MINIMUM HORSE POWER--16HP/SPDL
NUMBER OF TOOL STATIONS---90/SPDL
SPEED-------0-04500 RPM
CLOSET TOLERANCE---0.001" DIA.
CONTROL---FANUC/GENERAL NUMERIC
TAPE STORAGE---1000FT (250" MIN)
PALLET CHANGER WITH INDEX OF 1 DEGREE
LARGEST TOOL---6" DIAMETER
LARGEST PART---9" × 9" × 22" (85 POUNDS)

Work center data should be considered important reference material as new processes are prepared as well as for use in the development of family processes.

The following parts of the database are affected by work center parameters:

1. *Family Processes*. The family processes are directly dependent upon work center parameters since standardizing them is dependent upon optimizing their use.

2. *Process Polycodes*. Work center parameters are directly related to process polycodes. By correlating work center parameters of the operations which can be performed to the processes required for part families or individual designs, an optimum and standard process can be prepared.

3. *Perishable Tools*. Perishable tool requirements are determined by the work center parameters, which include determination of tool packages resident at the work center site.

4. *Jigs and Fixtures*. Jigs and fixtures may be related to work centers, particularly when they are designed for a specific one.

5. *Group Technology Cells.* Group technology cells are developed by determining those similar parts which are most conducive to being manufactured as a unit. Since such parts may require a group of machine tools to perform all operations, the work center parameters must be used to determine which ones are suitable for groups of parts.

TRANSIT TIMES

DESIGN RETRIEVAL-------------- O
FAMILY PROCESSES--------------- X
PART FAMILY DATA------------- O
PROCESS POLYCODES------------- X
ITEM MASTERS------------------- O
PERISHABLE TOOLS-------------- O
PRODUCT STRUCTURES---------- O
JIGS AND FIXTURES------------- O
ENGINEERING CHANGES---------- O
PRODUCTION PLANNING----------- X
PART GEOMETRY---------------- O
INVENTORY CONTROL------------- X
ASSEMBLY GEOMETRY----------- O
GROUP TECHNOLOGY CELLS------- X
GENERIC BILL OF MATERIAL---- O
RAW MATERIAL INVENTORY----- O
ELECTRONIC MAIL--------------- O

There are many data increments within a CIM system which could be discussed as thoroughly as transit times. However, transit times have often been overlooked and not considered as influential on manufacturing as they really are. To accentuate the value of valid transit times, they have been included as a specific category.

Transit times are time intervals between operations in the manufacture of a part or assembly. Generally transit times are divided into three segments, wait time, travel time, and queue time. Wait time occurs when, during an operation on the design, the total group of parts upon which the operation is being performed "wait" until the entire batch is completed. Travel time is that which occurs as the parts are moved from one machine or assembly area to another for the next operation. Queue time is that which occurs after the part has been moved to the next operation but must wait its turn before in line.

The importance of accuracy in developing these time intervals is becoming more apparent in the use of "just in time" (JIT) manufacturing. Today transit times are often defined quite crudely as one day between operations within the same department, and two days if the operations occur in different departments. This tends to bloat work in process, overstate work load, and extend lead times of a product unnecessarily.

The following parts of the database are affected by transit times:

1. *Family Processes*. The most important place to consider transit times is in developing family processes. Since similar parts should have similar transit times, a set of parts can be considered together. It must be remembered that wait time is directly dependent upon batch quantities to be manufactured and will therefore vary based on requirements. It should also be remembered that set-up time becomes a part of the queue time.

2. *Process Polycodes*. The value and size of transit times are a strong consideration in the development of group technology and the corresponding requirements for process polycodes. Such operations as milling, drilling, and boring are useful as polycodes only if there is some advantage to combining them in a family of parts for group technology and manufacturing. Transit time of parts compared with these polycodes will allow analysis of this criterion. On the other hand, if plant capacity and work center capacity are to be evaluated, these polycodes, tied to processes and transit times, will enable the development of gross amounts of machining or assembly type operations to be computed, which can then be compared to work center capacities.

3. *Production Planning*. Production planning is significantly influenced by transit times. If transit times are defined incorrectly, it is even possible to believe a large expansion of a manufacturing facility is necessary to allow additional capacity. Consequently the definition of transit times must start with the objective to create a concise, practical, and logical production planning system.

4. *Inventory Control*. Inventory control is also directly based on lead time. Parts and assemblies are manufactured to meet the schedule of materials generated by material requirements planning. If the lead times for parts and assemblies are incorrectly defined, these parts and assemblies will be manufactured and stored in inventory prematurely. Since, for every dollar of inventory, it requires $.40 per year to maintain and store it, the value of transit times becomes quite evident.

5. *Group Technology Cells*. Group technology cells are planned and thought economical for several reasons. However, one of the more important reasons is to eliminate or reduce transit times. An NC machine tool which combines operations formerly performed on several machine tools is an example of this elimination. If, however, a set of machine tools is to be used as a cell to perform various operations of a series of parts, the evaluation of this set-up depends on reduction of transit times. If transit times are not defined

accurately prior to such an evaluation, improper machine cells may be designated.

FAMILY PROCESSES AND PROCESS POLYCODES

DESIGN RETRIEVAL--------------- X
FAMILY PROCESSES--------------- X
PART FAMILY DATA---------------X
PROCESS POLYCODES------------- X
ITEM MASTERS-------------------O
PERISHABLE TOOLS--------------- X
PRODUCT STRUCTURES----------O
JIGS AND FIXTURES---------------X
ENGINEERING CHANGES----------O
PRODUCTION PLANNING--------- O
PART GEOMETRY---------------- O
INVENTORY CONTROL----------- O
ASSEMBLY GEOMETRY----------- O
GROUP TECHNOLOGY CELLS-------X
GENERIC BILL OF MATERIAL---- O
RAW MATERIAL INVENTORY----- O
ELECTRONIC MAIL ---------------O

The development of family processes presupposes a classification system which allows the compilation and segregation of parts into those with similar form, shape, or function. The part family must be conducive to the development of a process which encompasses all of the parts or assemblies with the same manufacturing characteristics.

Once a family of parts has been defined, it is necessary to examine their present processes to set standards as described previously. Once these have been established, they may be used to develop a family process. It should be emphasized that consideration should be given to integration of the process with the part or assembly geometry and with the design analysis program which defines the part. If this can be achieved, it is possible to create a generative process as opposed to a variant process. It is almost certain that some families of parts will lend themselves to generative process planning.

If we consider two types of classification systems, a monocode for design retrieval, and a polycode for manufacturing attributes, we must be careful to determine whether both are necessary, and define carefully the objectives of each. If we examine the contents of a process and the extracted routing, it can be seen that a work center designation and an operation number are always required. If these two data increments are "coded" (particularly the operation number), a code of sorts has been created. Thus if "shear" is always 34, "burn" is always 39, and "drill" is always 62, for example, manufacturing data can be extracted using these operation numbers. This approach is possibly more economical and practical since operation numbers are always used in preparing a process, standard text can be developed for each of them, and it is not necessary to develop a separate polycode which may not encompass requirements found necessary later.

In addition, a polycode does not necessarily bring together parts with several similar operations, only one or a few specific ones. Thus a polycode system for group technology may collect parts with widely varying shapes and manufacturing processes for the benefit of combining them for one specific operation only.

It is important, at this time, to consider alternate routings. The ideal combination of operations and work centers should be prepared but alternates should be designated where felt necessary. This will avoid the old problem of random solutions when changes must be made in the future.

The following parts of the database are affected by family processes and process polycodes:

1. *Design Retrieval.* The design retrieval system will have been designed not only based upon classification of designs for engineering retrieval, but with the process of part families in mind. It could be emphasized that both requirements are important for integration of the CIM system.

2. *Family Process.* The family process will be designed as described. Since preparation of a family process is time consuming, priorities may be established for which families are created first, based on number of parts, part usage, and/or dollar value.

3. *Part Family Data.* The part family data will include the family process, an index for perishable tools and jigs and fixtures, and critical dimensional and raw material data which might effect the work centers upon which the part family can be routed.

4. *Process Polycodes.* The process polycodes will be developed as the classification system is defined and implemented. However, refinements to the polycodes may be necessary as family processes are prepared. The polycodes may, in fact, be included in the operation numbering system and then combined with the work center numbering system without developing a separate polycode for attributes.

5. *Perishable Tools.* Since the family process cannot designate specific perishable tools, a combined table of perishable tools used for each specific part in the family should be prepared. In addition, it should be relatively simple to extract any one design from the family and itemize all tools used on it.

6. *Jigs and Fixtures.* The availability of jigs and fixtures should be designed in a manner similar to that described for perishable tools.

7. *Group Technology Cells*. Group technology cells will be best defined after a family process is determined. The description of the database for group technology cells best described some of the rules necessary for implementing such a cell.

NC CUTTER CENTERLINES

DESIGN RETRIEVAL--------------O
FAMILY PROCESS------------------X
PART FAMILY DATA---------------X
PROCESS POLYCODES------------- X
ITEM MASTERS-------------------O
PERISHABLE TOOLS--------------- X
PRODUCT STRUCTURES----------O
JIGS AND FIXTURES---------------X
ENGINEERING CHANGES----------- X
PRODUCTION PLANNING--------- O
PART GEOMETRY------------------X
INVENTORY CONTROL-------------X
ASSEMBLY GEOMETRY-------------X
GROUP TECHNOLOGY CELLS------ X
GENERIC BILL OF MATERIAL---- O
RAW MATERIAL INVENTORY------O
ELECTRONIC MAIL--------------- O

The use of NC machine tools should normally be considered another portion of the process planning activity. However, NC tools have often been isolated from normal manufacturing procedures, used as a machine for mass producing only a few parts, and generally treated as a poor relative to mainstream manufacturing. Implementation of a CAD/CAM system as a portion of CIM can, and will, change this radically. Thus the description of an NC tape cutter centerline database is not meant to isolate but, rather, to reinforce the need for integration of it into the mainstream of CIM.

In the past 20 years thousands of paper tapes, thousands of "macro" machine tool instructions, have been prepared. Often each new tape is prepared with relatively antiquated equipment and developed in relative isolation one from another. Many industrial engineering groups have given up trying to keep up with the vast new design releases which should be machined on NC tools because of the difficulty in preparing and checking new tapes rapidly. Computer Numerical Control or Direct Numerical Control is not utilized by many companies. Consequently each new design to be machined has its own paper tape which must be manually mounted on the controller of the machine tool.

Through the use of part families and design analysis programs, combined with producibility tips, a great majority of machining operations can be prepared as the new design is created. If the geometry of the family has been rigorously defined, a family cutter centerline can be established. The new design dimensions can then be entered into the family trace, and a new tape for a machine tool prepared automatically.

If part families are not available, developing new NC tapes on a CAD/CAM system is relatively faster than older methods where all the geometric data had to be written down meticulously by a parts programmer and a test piece made to certify the program. The tool path is generated on the screen of a CAD/CAM system in as many views as the programmer desires, including an isometric view. This versatility goes a long way to speeding up the accuracy and completion of a new cutter centerline.

Combined with the ability to store cutter centerlines on the computer and the advent of G (general) postprocessors use of direct numerical control or computer numerical control will be implemented more easily. This will tend to increase the use of numerically controlled machine tools for smaller lots of parts and also decrease set-up times between parts machined on such tools.

The following parts of the database are affected by NC cutter center line data:

1. *Family Processes.* As family processes are developed, care must be taken to develop optimum routing for NC tools and for developing standard sequencing of operations for them so that a family cutter centerline can be developed. It must be emphasized that a family process will require the cooperation of design engineering to establish standard dimensioning formats and standard operations, such as countersinks, bevels, hole sizes, and tolerances.

2. *Part Family Data.* When examining the database of part family data, cutter centerline information should be readily available and, in fact, integrated where possible so that development of a new design in a family will transfer design and machining data automatically into a new cutter centerline.

3. *Process Polycodes.* It may be important to develop special combined process polycodes for parts which will be machined on NC machine tools. This will tend to group parts from more that one family into logical groups which can be processed economically on an NC machine tool.

4. *Perishable Tools.* It is important to establish a tool package for each NC machine tool to minimize set-ups between parts. Thus the perishable tools used for a family of parts should be equated to one or more NC tools for standardization.

5. *Jigs and Fixtures.* Jigs and fixtures should be designed too for use on as many parts as possible in a family or families of parts.

While simple hold-down clamps may be sufficient for many parts, complex machining can be enhanced by such a standardization.

6. *Engineering Changes.* Engineering changes must be examined to determine any machining changes caused by them. If all cutter centerlines are stored in the computer rather than on paper tapes, changes can be expedited, the new tool path simulated on the screen, and the changed designs machined more quickly. It should be noted that those parts which have been designed in an integrated system where the routing and NC tape is prepared as a byproduct of the design will automatically create a new cutter centerline.

7. *Part Geometry.* The part geometry is directly related since it must be reflected in the cutter centerline. If the part geometry is in two or three-dimensional wire frame, requiring interpretation by the part programmer, care must be taken that the proper interpretation is made. If the part geometry is a solids model and used automatically to generate the cutter centerline, care must be taken that all geometry has been defined geometrically and not with notes or other implied machining.

8. *Assembly Geometry.* In a manner similar to that described for part geometry, assembly geometry must be either examined for interpretation or described carefully for those portions which will be machined using a cutter centerline.

9. *Group Technology.* An NC machine tool can be considered a group technology cell. In any event, the use of an integrated system to prepare both family and specific cutter centerlines will enhance the use of NC tools in a production mode other than mass production.

LABOR MATERIAL COSTS

DESIGN RETRIEVAL--------------O
FAMILY PROCESSES---------------X
PART FAMILY DATA---------------X
PROCESS POLYCODES------------O
ITEM MASTERS--------------------X
PERISHABLE TOOLS--------------O
PRODUCT STRUCTURES-----------X
JIGS AND FIXTURES-------------O
ENGINEERING CHANGES-----------X
PRODUCTION PLANNING-----------X
PART GEOMETRY---------------- O
INVENTORY CONTROL-------------X
ASSEMBLY GEOMETRY----------- O
GROUP TECHNOLOGY CELLS------ X
GENERIC BILL OF MATERIAL------X
RAW MATERIAL INVENTORY-------X
ELECTRONIC MAIL---------------O

Labor and material costs are the result of both engineering and manufacturing activities. Obviously they will be reflected in the financial statements of the company and be used to determine profit and loss. Our interest here is the basic requirements for cost accounting as they directly affect manufacturing.

Inventory and labor transactions are used to obtain material and labor costs. However, maintenance of several costs may be required. Some companies maintain a standard cost, an average cost, and several historical actual costs. This data is transferred into many accounting functions, but are also maintained to generate costed bills of material, inventory evaluation, and more.

The following parts of the database are affected by labor and material costs:

1. *Family Process*. In the development of a family process, care should be taken to examine costs associated with parts which are in the family to aid in eliminating excessive machining or assembly operations. On the other hand, care must be taken that the cost figures are developed rationally so that an expensive new machine is not abandoned for an old due to a lesser overhead cost. Cost accounting must become a partner in these evaluations setting up equitable standards for use by manufacturing and industrial engineering.

2. *Part Family Data*. If an "average" cost for manufacturing a part or assembly can be developed, storing such data by family could be highly important to marketing or others who estimate the cost of a future product. When a new product, a new design, or a new order requires a cost estimate, exploding a "pseudo" bill of material and using part family data will create a reasonably accurate estimate rapidly before the design is completed.

3. *Item Master*. The item master is the normal place to store and retrieve labor and material costs, usually by part number. In the past, these costs were considered part of the inventory file. If we now consider the item master to be a complete record not only of engineering data, but also manufacturing, the two former separate files will have been merged into one.

4. *Product Structure*. The product structure portion of the bill-of-material system is used to perform a cost roll-up to generate product or assembly costs. Normally total costs for an assembly or a finished product are not stored with the part numbers for these items. Rather, labor costs only for assemblies are stored, and the cost roll-up is used to add the material and labor of lower levels of the bill of material to the labor cost of the assembly and obtain total labor and material costs at any level of the bill of material.

5. *Engineering Changes*. Most engineering changes affect costs. In fact, it is important to estimate the cost of major engineering changes before they are implemented to be certain of their

affect on product costs. Thus the engineering change is an important factor in the cost of the product and the cost accounting procedures.

6. *Production Planning*. Production planning directly affects labor costs in that the quantity of parts produced absorbs the set-up costs. In this manner, also, group technology becomes a significant factor. In a like manner, the quantity discounts for purchased parts are balanced against surplus inventory carrying costs. Thus the production planning establishes basic cost factors as well as efficient plant loading and absorption of factory overhead.

7. *Inventory Control*. Inventory control is significantly influenced by material and labor costs, since the cost of carrying inventory is computed by them. Work in process, finished stores, and finished goods are determined by the contrast between quantity discounts for purchased parts and labor and material costs for fabricated parts and assemblies to the cost of carrying these items in inventory. A typical and important evaluation of these inventories is examined through the use of an economic order quantity formula contrasting carrying costs with return on investment. In addition, an ABC inventory analysis examines the value of each part with all others to establish those with the highest yearly usage in dollars. This evaluation determines inventory policies. A items are considered those which must be obtained only as needed on a day by day basis. B items are considered those which are obtained using the economic order quantity, while C items are normally considered floor stock and are obtained on a six-month or yearly basis. Lately group technology or JIT principles have been studied and implemented to further reduce inventory costs.

8. *Group Technology*. As mentioned previously, group technology is being utilized to minimize inventory costs. However, this principle will often change accounting practices. Normally, accounting desires a detailed cost for each operation on each manufactured cost. If several different parts are manufactured together in batches using the same operation, it is difficult, if not impractical, to separate operation time for each part. In fact, this separation would tend to increase the cost of each part, which defeats the purpose of the method. Thus groups of parts manufactured using identical operations will require more accurate estimating and a different cost accounting method to spread costs by individual parts.

9. *Generic Bill of Material*. If a generic bill of material is used, family costs as well as actual costs become quite important. The generic bill of material may have been used to establish cost and price for an order. Progress cost and payments may be initiated as the generic bill of material is converted to the actual design. The family cost could be considered a standard cost used to develop the estimate.

10. *Raw Material.* Raw material is often as much as 50% of the total product cost. Thus efficient costing methods and inventory control methods are important. Since raw material may be purchased and stocked in large sizes and used in small segments, the units of measure both of procurement and disbursement are important to maintain accuracy.

4
DATABASE SEGMENTS

The databases designated on the chart have not been segmented in a classical manner but are shown to emphasize both new types of data which must be recognized for a CIM system and to emphasize segments of the database which have been considered of lesser importance in the past. The segregation shown should in no way be considered rigorous in that, through analysis, segments of these databases may be combined, added to, or made subordinate one to another. Thus the databases shown here are more for purposes of explanation of data and its relationship to CIM rather than a description of the optimum means for storage and manipulation of data.

The seventeen database segments are as follows:

- Design retrieval
- Family processes
- Part family data
- Process polycodes
- Item masters
- Perishable tools
- Product structures
- Jigs and fixtures
- Engineering changes
- Production planning
- Part geometry
- Inventory control
- Assembly geometry
- Group technology cells
- Generic bill of material
- Raw material inventory
- Electronic mail

(Some readers may be disappointed that there is not a more detailed analysis of the production planning database segment. Probably thousands of pages of information and numerous articles have analyzed this area at considerable detail. All descriptions of it and references to it have been made to place it in proper context, without emphasizing it.)

Please notice in chapter two, chapter three, chapter six, and chapter seven, that a table showing each of these data base segments has been prepared with a "O" or an "X" discussed. This table is an extraction from the chart and shows whether there is a relationship between the item being discussed and that segment of the database.

DESIGN RETRIEVAL

DESIGN RETRIEVAL	:X:X:X:X:X:X:X:X:X:X:X:X:X:	
	: : : : : : : : : : : : : :	FAMILY PROCESSES
PART FAMILY DATA	: : : : : : : : : : : : : :	
	: : : : : : : : : : : : : :	PROCESS POLYCODES
ITEM MASTERS	: : : : : : : : : : : : : :	
	: : : : : : : : : : : : : :	PERISHABLE TOOLS
PRODUCT STRUCTURES	: : : : : : : : : : : : : :	
	: : : : : : : : : : : : : :	JIGS AND FIXTURES
ENGINEERING CHANGES	: : : : : : : : : : : : : :	
	: : : : : : : : : : : : : :	PRODUCTION PLANNING
PART GEOMETRY	: : : : : : : : : : : : : :	
	: : : : : : : : : : : : : :	INVENTORY CONTROL
ASSEMBLY GEOMETRY	: : : : : : : : : : : : : :	
	: : : : : : : : : : : : : :	GROUP TECHNOLOGY CELLS
GENERIC BILL OF MATERIAL	: : : : : : : : : : : : : :	
	: : : : : : : : : : : : : :	RAW MATERIAL INVENTORY
ELECTRONIC MAIL	: : : : : : : : : : : : : :	

Although design retrieval system is a significant segment of CIM, many manufacturing companies are not even aware that formal systems can be designed and implemented. Once acquainted with the idea of a design retrieval system tied to CIM, some persons often express the belief that a "canned" general purpose system should be installed. Unfortunately no successful universal classification system exists. The U.S. government has attempted to design one with dramatically unsuccessful results. In fact, the government has tried several times. One result generated a 40-digit "common" part number.

All attempts at universality of classification, including amateur systems, generally ignore the one most significant feature—ease of retrieval. It makes little or no difference how a classification system is devised if the speed, accuracy, and objectives of retrieval are ignored.

Another significant feature overlooked is in the use of a coding system in the retrieval system. Often a long series of descriptors or categories are defined in the system and each is provided a separate set of numbers or alpha characters. When these are added together into a total "code," it becomes difficult or impossible to initiate and maintain because of its length. Recently it was claimed that the computer can now take care of numbers, thus the length of a code is insignificant. Those who preach this gospel have never classified 80,000 drawings and have not been responsible for the continued process of classifying as many as 7000 new designs yearly. Neither have they had the opportunity to take such a set of numbers and attempt an orderly analysis of all parts to derive important analyses from them.

Segregation of designs in a classification and retrieval system requires an ingenious set of associated nouns and adjectives which succinctly and correctly describes the various features of a design in an orderly sequence. Not only should the sequence be devised which will logically segment designs into ever smaller segments, but these segments must be logical, exclusive, and small enough to require only a small amount of analysis to either find an old design which meets new requirements or a family of designs closely similar to the new requirement.

During the past 20 to 30 years two types of professional classification systems have grown popular. The principles involved can be defined under the terminology of polycode and monocode. Their definitions follow:

> A monocode is an integrated progressive coding system in which each digit qualifies the information contained in the previous digit. Each digit may contain more than one "descriptor" or characteristic.
>
> A polycode system is a set of independent sequential codes, each digit or set of digits describing individual characteristic variations. Polycodes are often described as "feature" codes.

Many of the features of both these types of systems have been described in other places. However, a separate section has been prepared for polycodes as they are reflected in the use of family processes. The use of a monocode seems ideal for the retrieval of designs. In fact, a considerable amount of discussion has been developed to show how such a system might be tied to feature-based geometry. With these explanations in mind, let us look at the design retrieval database to see how it may be integrated with the CIM concept.

During the past few years, many monocode classification systems have been installed and implemented using a computer as part of the system. While none are yet integrated with geometry, the following set of conditions seems to be the path which further integration will follow:

1. The classification system will be on line, with user-oriented prompts that lead the user a step at a time to the family of parts which he or she is using.
2. Once the family of parts has been found, a sketch or a parametric drawing will appear on the screen to confirm the user has found the proper family.
3. A catalog of parts within the family will appear on a split screen format. Since the parametric drawing will have dimensions A, B and so on, defined on the family drawing which are found in equivalent tabulations in the catalog, the user can quickly compare old designs with his new design requirements.
4. If a design is seen to be close to that required, the drawing for it will be shown on the screen for confirmation. (It is possible that all of the drawings for a family will appear as a matrix on the screen with the ability to zoom in on any and all of them for comparative analysis of the part family.)
5. If no old design is found adequate, the parametric drawing will be used to create another. Automatic prompts will allow the entry of dimensional data to replace all alphabetical characters. If additional dimensions or additional geometry are required to customize the part, provision for adding additional geometry to the part family will be provided.
6. The new part will be "released" to the system, meaning the item master will be updated by utilizing a menu design to enter the required data. For example, the part number and the part description will be entered automatically on the drawing and into a temporary working database as the design is prepared. The additional data required for the item master will be entered through an appropriate menu.
7. A classification system for assemblies may or may not exist.

Where it does, an identical procedure for retrieving and utilizing a parametric drawing or family sketch will be utilized.

8. Once a new design has been released, a key element of such a release is auditing the new design to make certain it has been classified properly in the right family. Once audited, the new design, with appropriate data, is entered into the classification catalog so others may find and use the design if it meets requirements.

PART FAMILY DATA

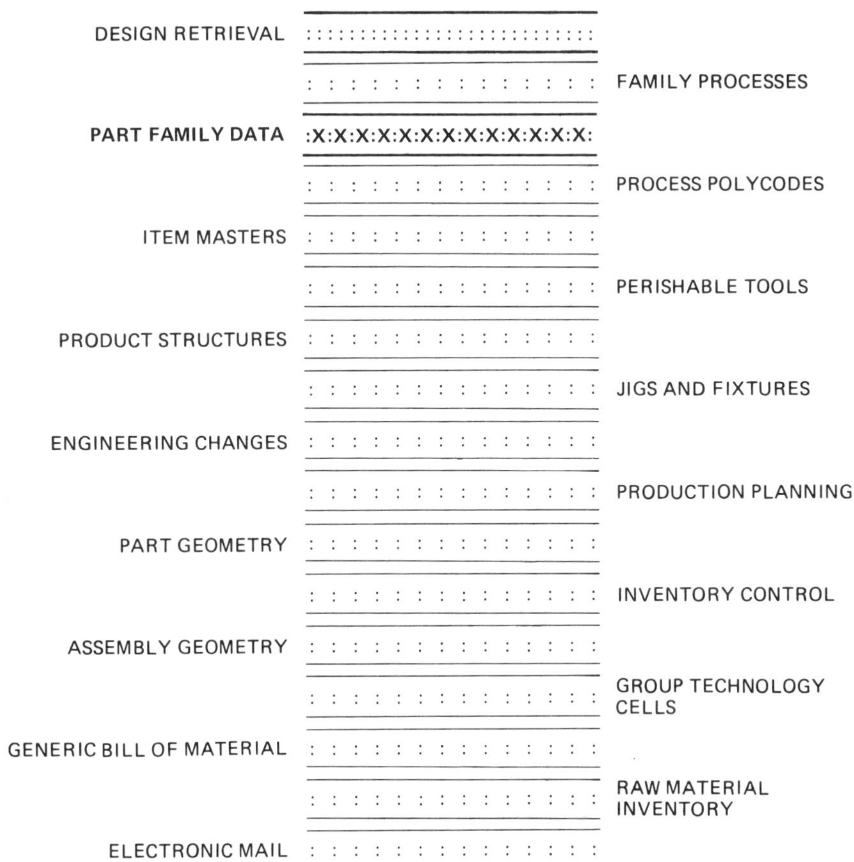

A part family database is the logical extention of a classification and design retrieval system. Collecting data of various types about similar parts into one database creates an extremely powerful information set.

Further, this segregation of data eliminates the need to look at thousands of individual parts to define engineering, design, drafting, and other standards. It also provides a small number of part families for use in the analysis of designs.

Once a design retrieval system with the corresponding family of parts has been established, a significant array of part family data may be stored to expedite both standardization and engineering design benefits. The following are some: design standards, drafting standards, producibility tips, design analysis parameters, parts catalog, tooling catalog, NC cutter centerline, parametric drawings, and geometry.

Design standards by family seems more logical and customized than creating generalized and bulky sets of such standards which have only general application. It is possible to equate these standards with a parametric drawing in the form of standard notes, dimensions or dimensioning, or simply by referencing a text data source as the family is examined. Additional geometry requirements, such as surface finish, flatness, and parallelism, are possibilities. Design analysis programs also are a way to define design standards since the algorithms within such a program are parameters, limitations, and definitions of design requirements.

Drafting standards are often different between designers and drafters. Consequently the tendency is to duplicate designs, one by designers in layout form, the other in a "different" drawing format. If this difference is not resolved, the two departments will insist on developing different geometry format. By mutually preparing drafting standards for each family of parts, both groups should be willing to abide by them. In addition, these standards are conducive to clarity in industrial engineering and for shop practices.

Producibility tips are methods of design which affect the manufacturing process. Through mutual agreement, these tips are itemized for use by engineering as a new design is prepared. Often producibility tips are represented by or in design and drafting standards. However, additional features of a design may be expressed as global requirements such as standard hole sizes, threads, countersinks, raw material sizes, chemistry specifications, and radii. Each family of parts, however, may have their own customized producibility tips, such as the exact dimensioning protocol, economical finish, bolt centers, and bending radii. In any case the use of producibility tips achieves a more consistent, economical, repeatable, and logical design.

Design analysis parameters are important for families of parts which are created using some design analysis computer program. These families include gears, solenoids, and bent tubes. Not only should the design formulas be maintained for each family of parts, but the maximum and/or minimum acceptable loads, sizes, and such documented. This data will allow engineers who did not generate the analysis program to achieve a higher degree of comfort when using it on their own design. It is also a more mature means for documenting a design analysis program and its parameters and limitations.

As previously indicated, a parts catalog of data is associated with each part family. The catalog contains a cross-index between the family code number and the part numbers of the designs meeting the family criteria. In addition, a set of critical dimension or other features of the family are catloged, such as chemistry, special notes, thread sizes, and special treatments like plating. The tabulated dimensions should be equivalent to those found on the parametric drawing. Certainly the dimensions thus tabulated should be sufficient to a designer so he can either accept or reject all the old designs in the family as potential candidates for use in his new design. This tabulation of data is an additional method to assist persons achieve rapid retrieval of designs to avoid unnecessary proliferation.

Reuse of tools required to manufacture designs can be greatly enhanced if they are tied to part families, including perishable tools and jigs and fixtures. If the tools required to make each part are tabulated, and since that part has been classified into a part family, all tools used to manufacture all parts in a family can be located and analyzed as a group. This indirect method of tool classification has some benefits in that it would appear obvious that similar parts should be manufactured using similar tools. Thus newly designed parts can be analyzed to see if old tooling could be altered slightly to manufacture the new design.

NC tape preparation for manufacturing designs has been active for many years. Considerable study has been devoted to methods for more rapid and accurate means to do so. It would appear feasible, economical, and practical to consider a family of parts as having a generalized cutter centerline. If this cutter data was left blank as a "swiss cheese," differences in dimensions, material, and the like could be entered to customize the NC data for a new design rather than recreating the entire set of data. Less errors, rapid turnaround, and more consistency would result.

Storing parametric drawings and part geometry is probably the most exciting potential for a part family database. If geometric features or entities are associated with a family of parts, once again consistency of design and manufacturing will occur. Using geometry features to develop the parametric drawing will result not only in a realistic description of the family, but a realistic set of finished geometry for a new design when the parameters for it are entered. It might be said that the geometric features combined together for a specific family represent the best of design standards, manufacturing standards, drafting standards, and tooling requirements. In essence, a text database has been converted into a geometric representation. Of course no family of parts is defined so meticulously that other new geometric features are not sometimes added for a specific part in the family. However, each such addition can be considered as "special" and not made a part of the parametric drawing, or added for a more clear representation of the family. It is even probable that several sets of geometry and corresponding parametric drawing will be required for one part family due to the variety of geometry variations. In a like manner, the same geometry and parametric drawing may be adequate for several part families which have been separated for ease of design retrieval but are governed by identical geometry.

In any case part family data makes analysis, standardization, and new design release and costing relatively more rapid and consistent.

ITEM MASTER

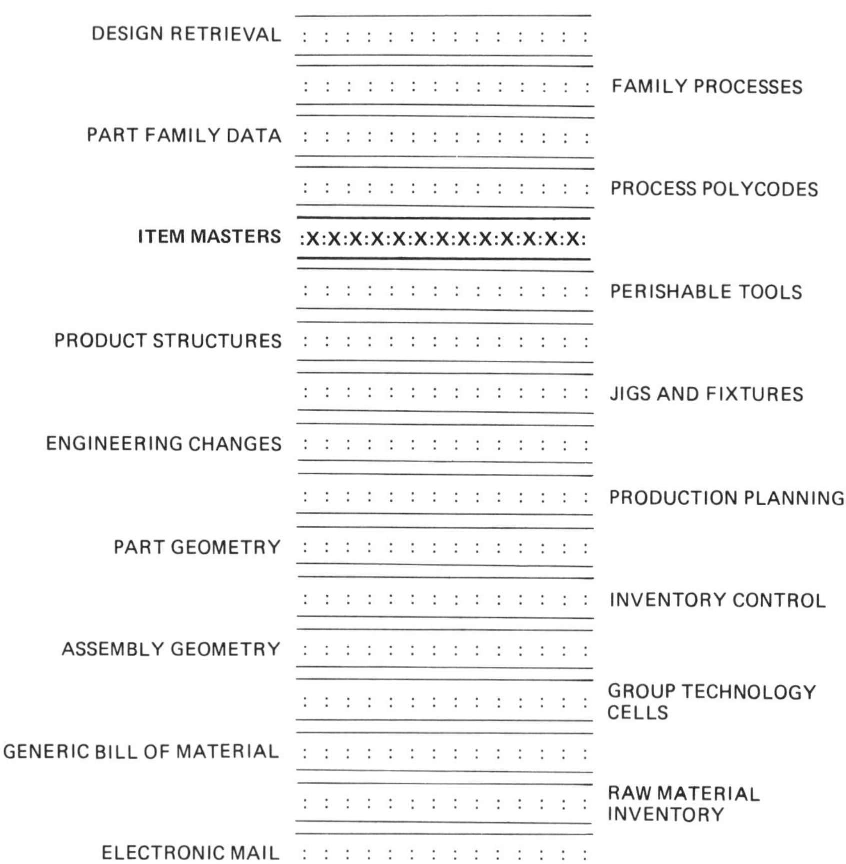

The item master segment of the database is considered a major requirement for an engineering manufacturing business system. Although data about each design resides here, it has long been considered more inventory control and manufacturing oriented. As the emphasis to mechanize engineering data has increased, the item master has been expanded to include such data. Thus engineering documentation is now established as a significant portion of this segment.

Data maintained in the item master is usually considered under two categories, static data, or the "pedigree" of the design, and dynamic data, such as inventory quantities and quantities on order. Initiation of a new item master is closely associated with the creation of a new design whether a part or assembly. In many integrated systems, the design release procedure is initiated by the preparation of a new item master. The part number log system is tied to both the release procedure and the item master.

Some of the more generally recognized data increments found in the item master initiated and required by engineering are as follows:

Part number
Revision level
Unit of measure
Description
Part type
Drawing size
Weight
Release date
Material specification
Design status
Replaces part number
Classification code
Drawing available code
Spare part code
Critical dimensions

The critical dimension data reflects a merger of data from the catalog of the part family database with the item master. This portion of the part family database could be considered a direct part of the item master in the future.

It should be remembered that over 200 data increments may be found in the item master segment, including data from manufacturing engineering, inventory control, tooling, and accounting. With CIM, the item master may be considered the major CAD/CAM and business system data set and index.

As feature-based geometry is utilized, an item master will be expanded to include data about geometric features, and an index to their geometry. The classical use of the item master data, however, is in conjunction with the product structure data segment to create bills of material and other related information. With such an integration, data about a design maintained in the item master need not be repeated each time it is used and data redundancy is avoided.

PRODUCT STRUCTURES

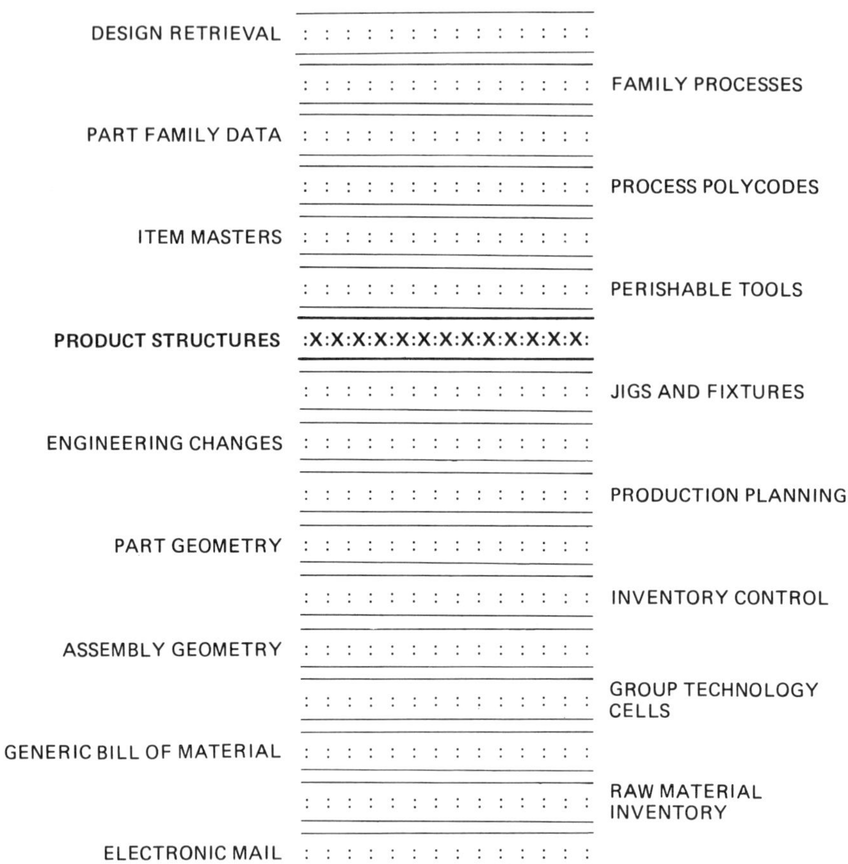

The product structure data segment represents one of the first relational data sets that was required in a business system. The parent-component relationship of an assembly and its components is represented in a parts list, an indented bill of material, and in a product cost roll-up.

The product structure data represents a significant interaction between all parts of a manufacturing company. As a designer prepares a new design, it is documented as a series of piece parts and assemblies. The assembly component relationship should represent the way in which manufacturing wishes to assemble. The assembly thus defined is also then documented by an assembly drawing and a detached parts list which may also be called a single level bill of material. The following data increments are normally found in the product structure database and are those which pertain only to the parent-component relationship:

Assembly part number
Release date
Assembly revision level
Component part numbers
Component "find" numbers
Quantity per component
Unit of measure
Effectivity in/out
Effectivity date
Issue sequence
Assembly operation number
Scrap percentage

An effective bill-of-material system is based on the product structure database. Many significant changes are required in engineering documentation to use a mechanized bill of material. The parent-component relationship should be first documented by creating a general product structure that establishes a mutually agreed-upon but generalized way in which a product will be assembled.

When a customer order is received, or the new product requires documentation, specific product structures are prepared and recorded as assembly drawings and detached parts lists. These detached parts lists replace the old material lists which were often found on the body of the assembly drawing. When these material lists are made into a separate document, it is necessary to coordinate them. This is performed through the use of find numbers, or bubble numbers, which appear on the face of the drawing and are tabulated together with the part numbers of the components on the parts list.

The product structure is a level-by-level or single level bill of material. Many computer programs are designed to utilize this data, including preparation of summarized, indented, and costed bills of material.

It should be emphasized that the product structure data can be inverted and used to create a "where-used" or used on file of data or report. This information is useful to the designer making an engineering change replacing one part for another, for example. It is then possible for him or her to discover all assemblies and products affected by such a change.

The bill of material is also a source document for material requirements planning (MRP).

Control of the product structure database is related to a configuration management system. Interchangeability rules, revision level control, and the implementation of an engineering change system are required. Configuration management rules also dictate data changes to the item master data system.

ENGINEERING CHANGES

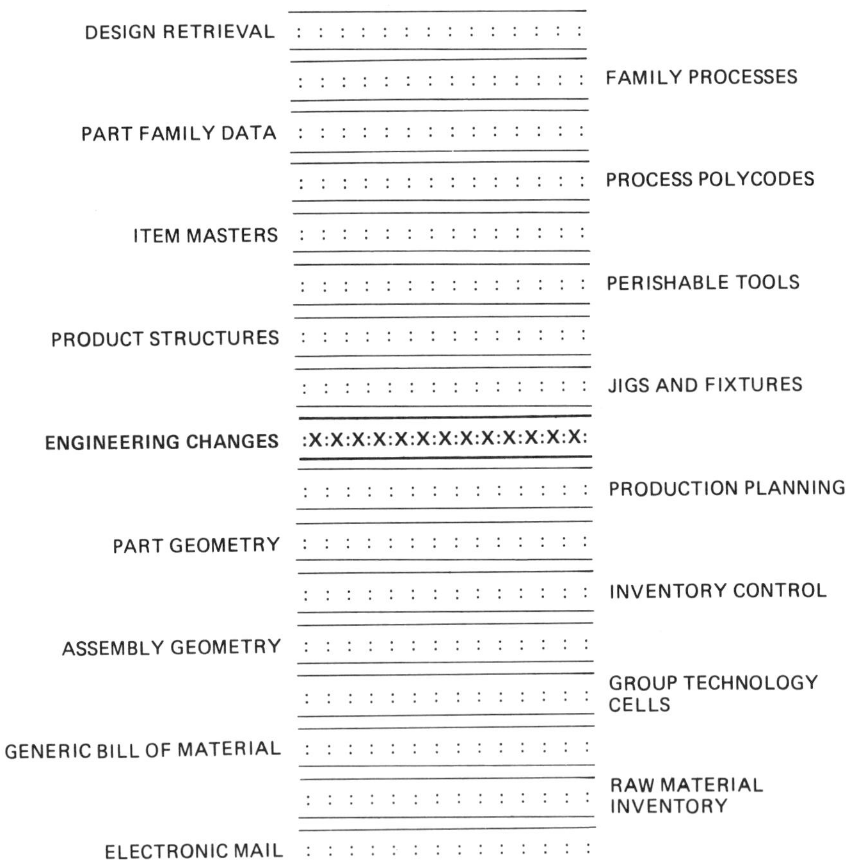

Most persons probably consider engineering changes as a series of activities reflecting design changes rather than a database. In fact, this database segment represents not only changes in progress, but also the historical configuration of the product and the results of configuration management. The data is invaluable to reflect quality, design costs, and effectivity of changes.

The engineering change database segment represents all changes to the design configurations of products. Since this data does reflect engineering changes, the results of such changes can affect all other database segments.

Prior to the consideration of integrated engineering and manufacturing systems, each engineering change could be used to inform persons within various disciplines of the occurrence and appropriate adjustments made. Today an engineering change not only notifies, it also signifies that many of the changes previously prepared serially and manually have been completed. For example, the interchangeable but dimensional change to a piece part design now also means the geometry of the part has been automatically updated. This may mean the tool path to machine the part and the tools and fixtures required have been changed.

In actual practice, of course, these peripheral changes are not allowed until approval has been given by the appropriate personnel. However, the means to so operate is now available, making the integration of changes and their recording ever more important.

PART GEOMETRY

DESIGN RETRIEVAL
FAMILY PROCESSES
PART FAMILY DATA
PROCESS POLYCODES
ITEM MASTERS
PERISHABLE TOOLS
PRODUCT STRUCTURES
JIGS AND FIXTURES
ENGINEERING CHANGES
PRODUCTION PLANNING
PART GEOMETRY :X:X:X:X:X:X:X:X:X:X:X:X:X:X:
INVENTORY CONTROL
ASSEMBLY GEOMETRY
GROUP TECHNOLOGY CELLS
GENERIC BILL OF MATERIAL
RAW MATERIAL INVENTORY
ELECTRONIC MAIL

The part geometry database represents one of the newer concepts of a database segment. The part geometry may be considered a set of general geometric features or attributes combined and defined with specific dimensional data to make one specific new design.

A family of parts database which uses general geometric features to define it may have been used in conjunction with a design retrieval system and a parametric drawing program to generate the specific design.

Feature-based design is an exciting concept. It implies that the designer will make use of familiar shapes, defined mathematically as a solid, but named with engineering titles such as hole, thread, and slot. By also relating the scribe lines, tolerances, finishes, and other design attributes to the geometry, a complete definition of the design will be available.

It should be noted that feature-based design presumes that the designer starts with a basic shape, such as a solid block or a cylinder, and subtracts the features from the original shape. This closely follows the manufacturing process and is thus quite compatible to the process an industrial engineer might consider. For example, a boss on a rectangular shape might look as though it was added to the rectangle. It real life, unless the boss is welded as a separate entity to the rectangular block, it is, in fact, machined from a still larger block by removing material. Thus the principles of feature-based design follow traditional and also modern manufacturing practices.

ASSEMBLY GEOMETRY

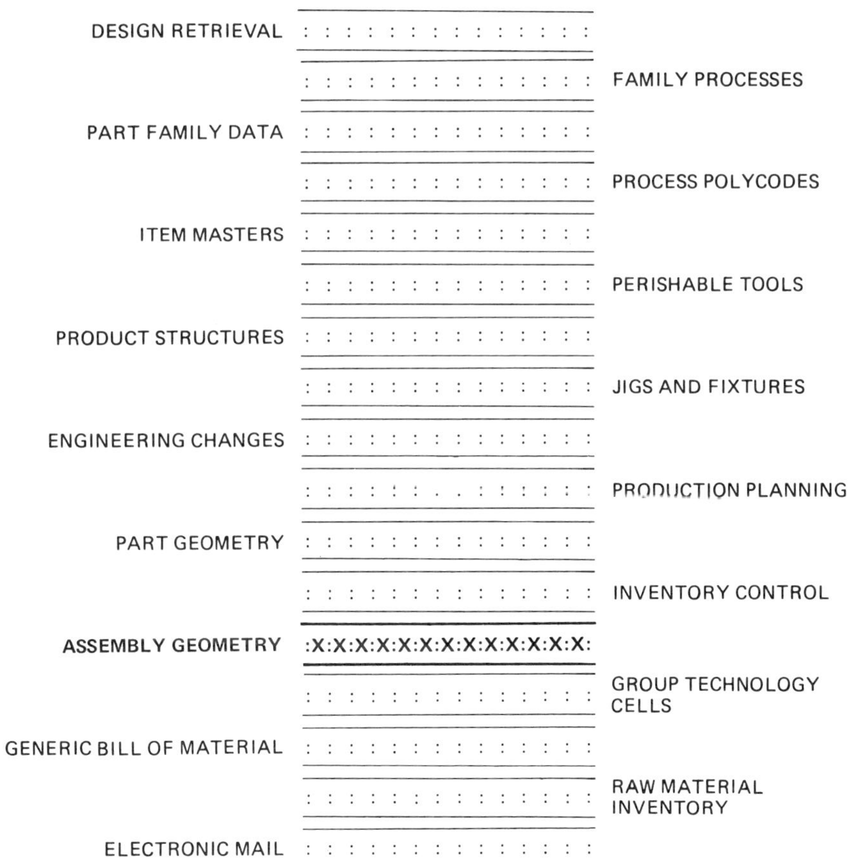

With the implementation of feature-based geometry, assembly geometry will become even more significant. It is likely that the parts which make up an assembly design will be stored separately, and "assembled" only when needed. This is almost the same principal by which a piece part is created by "assembling" features. Thus an assembly which goes into an assembly must be stored only once.

The parts which compose the assembly can be generated through either a parametric program or by collecting features subtracted from a basic shape. In turn, these parts can be oriented in the assembly database, so that, when they are required, the parts may be pictured in the proper relationship one to another. As mentioned previously, these geometry transforms can depict the parts oriented like an exploded isometric view, as they should appear in a complete assembly, or even disoriented as though in trays or bins waiting for assembly. The potential power of this principle is almost endless. If an assembly must in turn be pictured in an assembly, the same methods can be used as just described. The association of the higher level assembly to lower assemblies and/or additional parts may be stored as geometry transforms and the higher level assembly. Thus redundant data need not be stored and the most complex set of assemblies may be maintained as a single set of entities not an ever more complex set of redundancies.

If machining is to be performed once the assembly is completed, the industrial engineer is faced with a mathematical solid model of the parts in combination. In addition, all sorts of simulations for stress, temperature, and more may be made on the assembly since the entire geometry set and tolerances are available.

GENERIC BILL OF MATERIAL

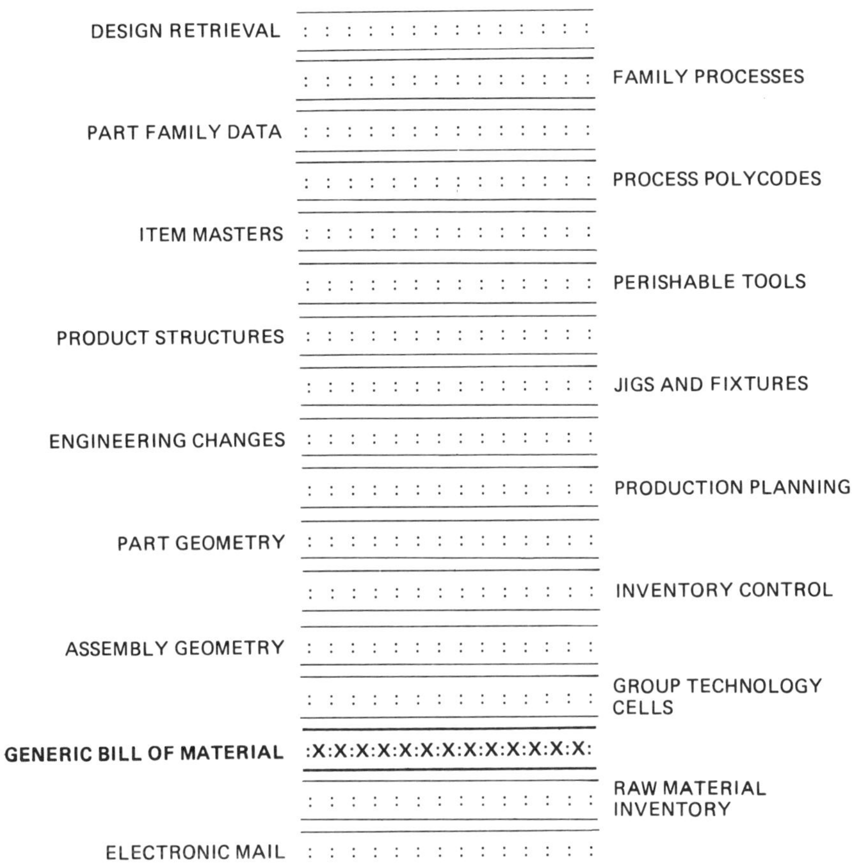

The generic bill of material is another exciting new concept, which makes use of proper bill-of-material structuring and modular design. Once a product has been defined in general, and each part and assembly classified, each of them may be given a family lead time, cost, design, and drafting time. With this data, a new order can be entered quickly and a reservation or simulation of new capacity requirements generated. Estimated cost, total lead time, and an engineering schedule are also available.

The generic bill of material may represent a product with all the "bells and whistles" possible. Thus, when an order is obtained or an estimate required, the generic bill of material may need to be tailored to fit the new requirement. In addition, there will be variance in cost, lead time, and other data, based on the similarity of parts or assemblies within the family. Thus a statistical analysis of the factors may be made to show the degree of variance. Even though this procedure is inexact, it is a planning tool of some power.

It should be noted that the generic bill of material may be converted a step at a time as the new order or product is designed. As this conversion occurs, inevitably many parts, and perhaps some assemblies, will be reuseable from prior usage. Once they are entered into the database, they may be scheduled.

It is clear that this database of generic bills of material will be used extensively by master scheduling, production control, and engineering design scheduling.

ELECTRONIC MAIL

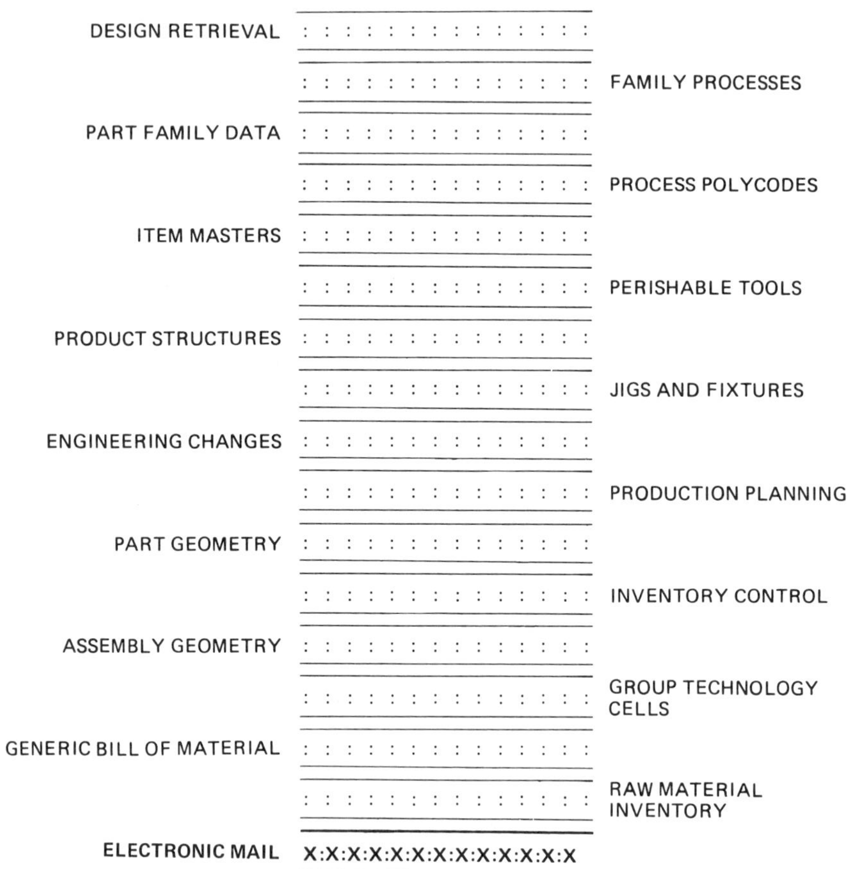

Electronic mail is growing in importance to enhance a CIM system. More and more data is being input to the system on line. While the data thus entered is available more readily than ever before, the proper persons must be notified of its existence. This requirement is ideal for electronic mail. Two important examples are the present use of an engineering release notice and the engineering change notification procedure. Both these systems are enhanced by quick reaction to the initiation of either a new design, or a change to an old. This database holds such data until it is recognized and acted by the users.

It is not unreasonable to assume that pictorial data as well as text data will be transmitted from various sources to multiple terminals. This ability will further enhance a CIM system. It may be that the electronic mail system will become the major repository of data in flux and data in process. Thus care must be taken to make this segment of the database compatible to all others.

FAMILY PROCESSES

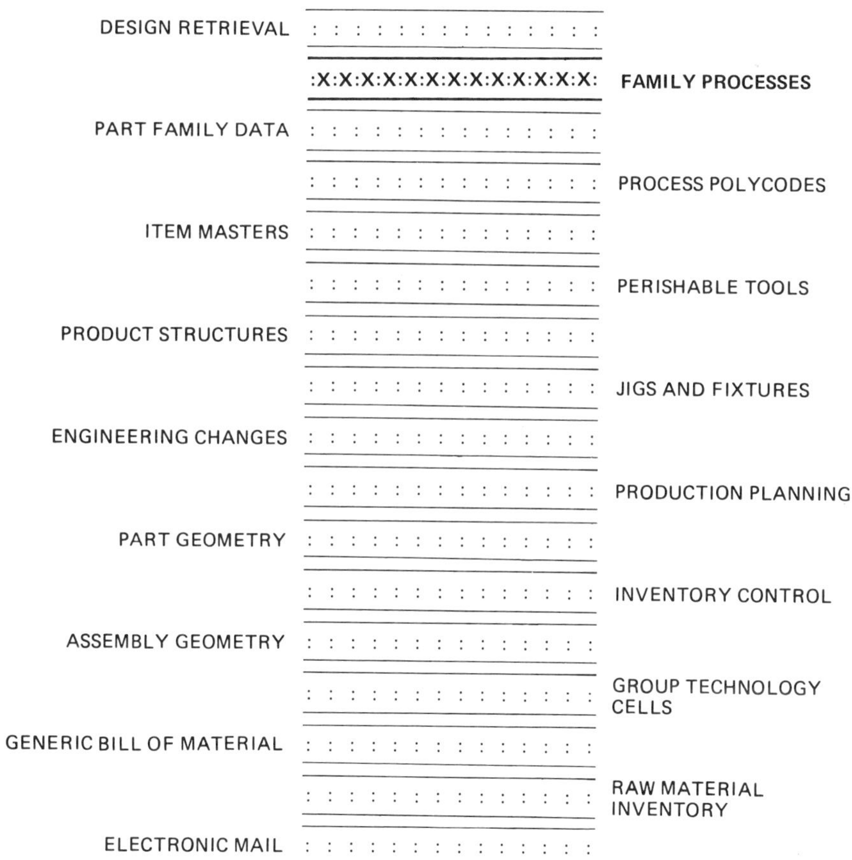

As described previously, a family process can be generated for a part family utilizing a combination of design or manufacturing attribute retrieval systems. Retrieving the appropriate one when a new design is released will save considerable time and avoid multiple and nonoptimized outings. The family process database may be considered as significant to manufacturing as the item master and product structure and generic bill of material are to engineering.

Family processes are similar to generic bills of material because they may be used as a preanalyzed optimum. Many decisions on best machine tool, perishable tooling, and jigs and fixtures can be made ahead of time.

In addition, many of the producibility tips generated to assist design engineers will result from developing this database. Family processes are truly the culmination of a coordinated evaluation of designs and facilities to make just-in-time (JIT) manufacturing, flexible work cells, or group technology economically feasible and practical.

PROCESS POLYCODES

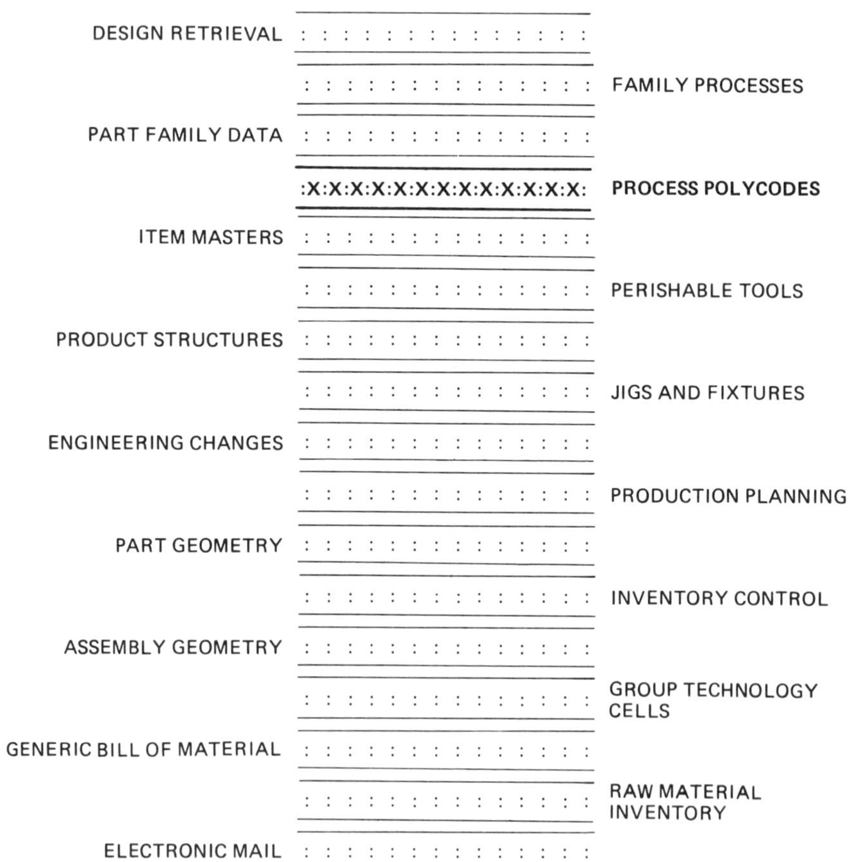

The process polycodes are developed when a manufacturing attribute classification system is developed. This set of codes fine tunes the design retrieval system so that a family of parts may be segmented into practical manufacturing families or a series of design retrieval families may be combined into one manufacturing family.

It is recommended that universal polycodes be replaced with sets of polycodes applicable to one of, or a series of, families, not the entire universe of all parts. This may be a bit more complex to establish. However, if the design retrieval system is developed first, an analysis of these families may make such a segregation more logical and easier.

A polycode cannot be developed economically for each design or manufacturing attribute even though they may be considered permanent characteristics. Thus it is possible that, from time to time, additional polycodes are added or deleted from a family set of characteristics. The design retrieval system should remain relatively permanent, however. Thus this database segment may require periodic changes.

PERISHABLE TOOLS

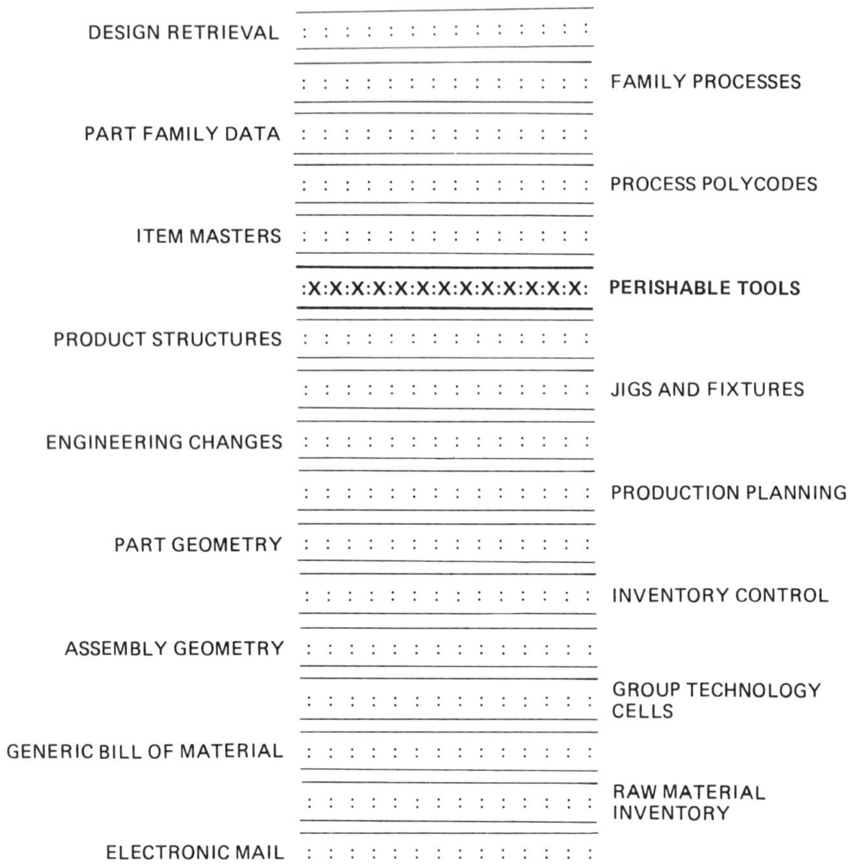

Considering perishable tools part of the CIM database may be a new concept. However, lack of control of perishable tools has hidden both their high cost and unbelievable variety. It is not only necessary to establish the varieties to begin eliminating those which are unnecessary, but it is also necessary to create tool packages for NC tooling, and to develop. It has been discovered that the machine tool operator may retrieve a new tool before the previous one has become unusable, thus proliferating expensive inventories both at a work center and the tool crib.

One company made a study of hole size variations including countersinks and counterbores, and found that a significant amount of inventory investment for perishable tools could be eliminated through standardization. The perishable tool database will assist in eliminating production bottlenecks as well as reduce variety and thus inventory costs while increasing tool availability.

JIGS AND FIXTURES

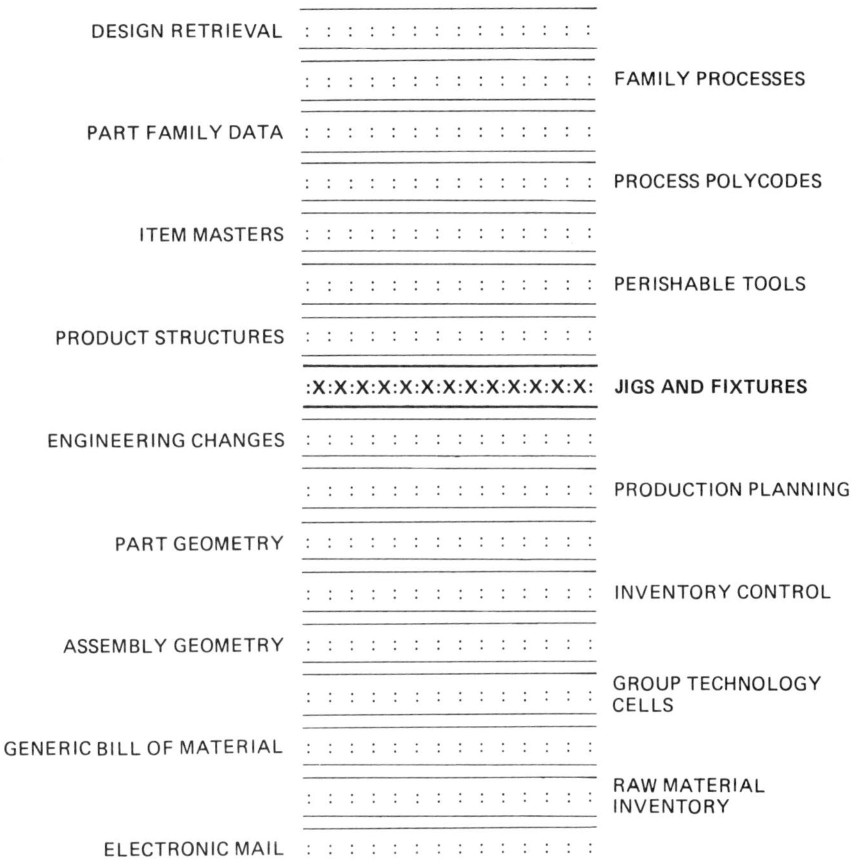

In a manner similar to that used for perishable tools, a jigs and fixtures database should be maintained. Utilizing the principles of a bill of material, a "whereused" for jigs and fixtures may be generated, thus allowing the retrieval of these tools for similar parts. Using these expensive tools with minor alteration for similar parts can cut lead time for the initiation of a new design by as much as six months, and, often, eliminate expensive tooling check-out.

It is recommended that tools be classified also, including the parts required to make them. This will decrease the lead time to manufacture them and also decrease their cost.

It should be considered important to design these tools in the same manner as designs are prepared on a CAD/CAM system. The jigs and fixtures database will then have pointers to their geometry. Thus simulation of tool-part combinations can be made. This combination in conjunction with the generation of NC tool paths will increase manufacturing planning productivity.

PRODUCTION PLANNING

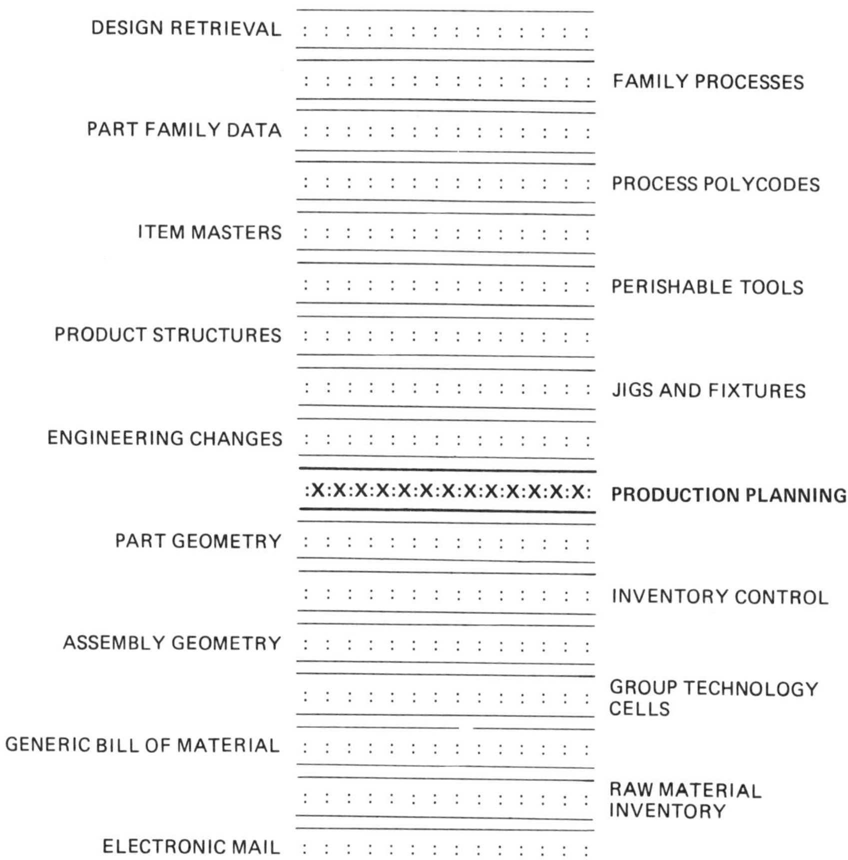

The production planning segment of the database is not defined here in a classical manner. Within this segment are found several familiar sets of data or standard routings, gross requirement data for material requirements planning, work center loads, capacity requirements plans, cost accounting data, and other information normally considered part of manufacturing planning.

It is, of course, possible to think of each of these as additional database segments. However, they are so closely associated one to another, they have been combined here to emphasize this relationship. Inventory control and purchasing data are also closely related, but can be, and often are, maintained as a stand-alone system.

There are some who may disagree with this segregation of data. It will probably be because of the many charts developed over the years showing boxes marked "material requirements planning," "detailed capacity requirements plan," "product and job costing," "shop floor control," and more. These boxes show applications of the database we are describing here.

If we will stand back and look again at database concepts, we will see the need to, once again, define the data needed to fulfill the applications for which we have requirements. Thus the title of this database segment.

INVENTORY CONTROL

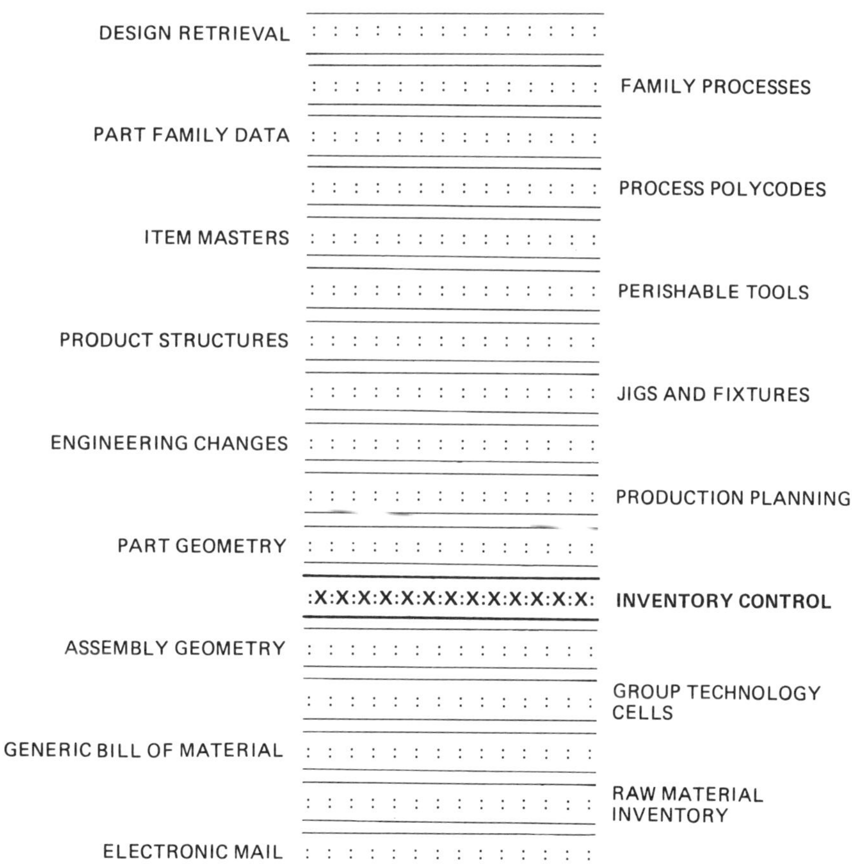

Inventory control was one of the first parts of manufacturing to be controlled by a computer. Consequently it is a familiar database. However, the implementation of material requirements planning (MRP) did alter the concept. Inventory was maintained by past requirement statistics and called "order point control." Only stock inventory was maintained. A traveling requisition informed purchasing of new requirements for purchased parts, the majority of items in an inventory system. While the accuracy of inventory was poor, the vast majority of designs were controlled manually.

Today, all parts must be controlled using an inventory control system if MRP is to operate correctly, including work in process, items in inspection, rework, and on order but not yet delivered.

Accuracy for a class A MRP system is considered insufficient until 95% or better has been reached. This requirement has put a severe strain on the past methods of conducting business, requiring closed inventory areas, more accurate receiving, cycle counting, and a host of other disciplines which did not exist previously.

This segment of the database is combined with the production planning database to activate MRPII.

The purchasing department has been forced to make many adjustments also. Part requirements are now designated monthly, weekly, daily, even hourly. Since these requirements can change just as rapidly, the vendors of a manufacturing company can be jerked back and forth uncontrollably without some form of damping. Thus purchasing departments may create yearly orders with minimum lots for delivery, or even suggest a vendor build a plant or warehouse nearby.

GROUP TECHNOLOGY CELLS

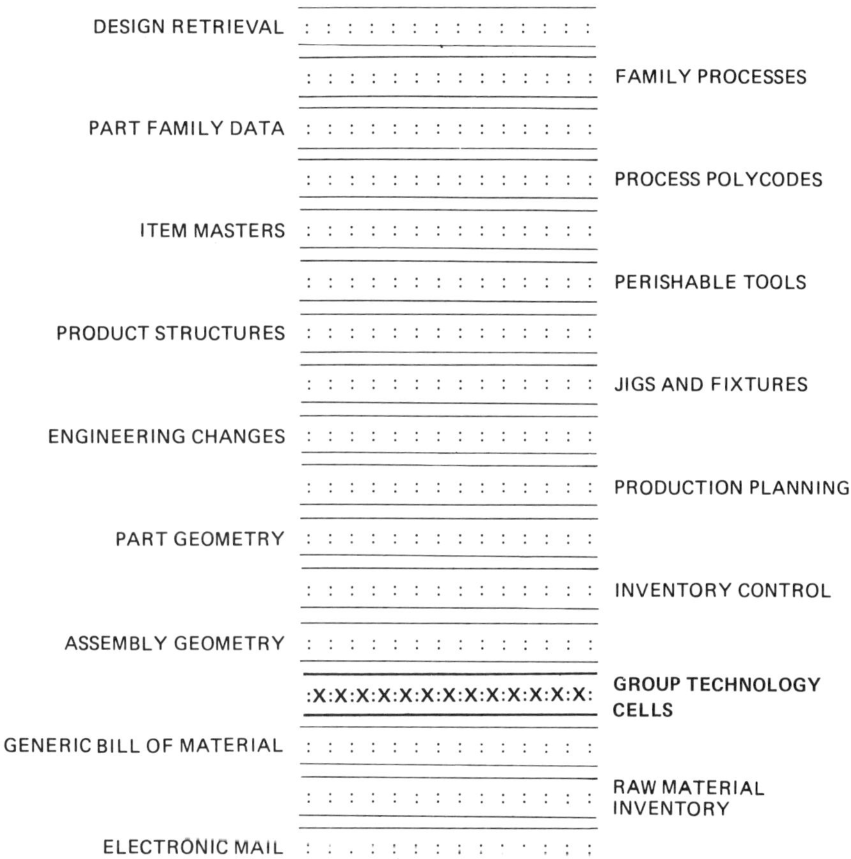

As a special database for production planning, group technology cells and the data required for them must be maintained. Each cell will have peculiarities of machine tools, fixtures, special routings and processes, and NC machine tool data. Data for these cells may be provided from process polycode data, machine tool capability data, and family of parts material requirements planning.

The use of cell data in conjunction with process planning, manufacturing resource planning, and production planning will change the concept of manufacturing logistics. Each cell may be developed as a separate entity, being derived from several studies that combine families of parts or families of identical processes.

The derivation of this database may also result from an ABC inventory analysis by family which determines the most frequently used set of parts, their similarities, and their compatibility to the group technology process.

If we consider the segments of data which may be called the engineering database, and combine them with the manufacturing dynamic data, the result will be an engineering manufacturing planning database of part families.

RAW MATERIAL INVENTORY

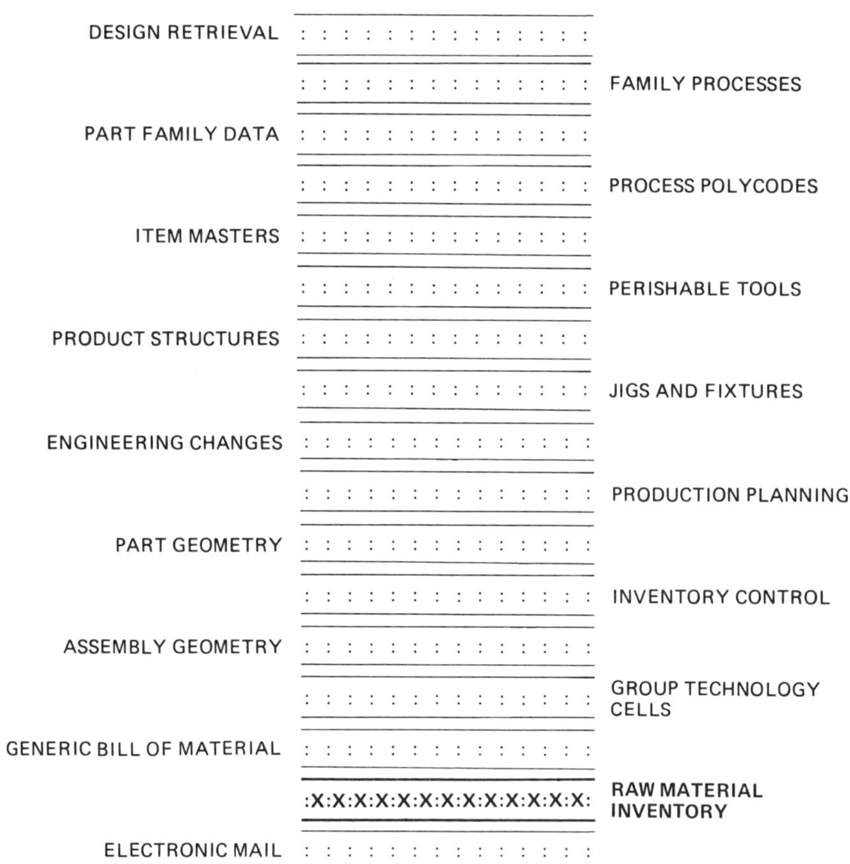

The raw material inventory database represents a segment which has been normally considered a portion of the inventory control database. It has been shown as a separate unit here to emphasize the differences between it and other inventories.

Often the units of measure, the product structure, and the responsibility for buying and using raw materials are different from those of other inventories and determination.

Engineering design should document the "what" of requirements, while industrial engineering should document the "how." In addition, the purchasing requirements and units of measure may be different from the inventory storage units of measure as well as the disbursement units of measure.

For example, the design of a part 2 in. in diameter, 6 in. long made from S.A.E. 1020 may be purchased in four different ways, as a bar with a specific length, such as 20 ft, a random length, in multiples, or in random shorts. Each of these other than the specific length has different costs, and different purchasing unit of measures. They may all be stored in feet, but the specific lengths could be stored as "each."

Purchase of these different types may well depend upon the parts to be manufactured from them, including the length, the quantity, and the material cost. Thus the industrial engineer plays an important part in this determination.

The industrial engineer prepares a process and routing to make the part. Additional material may be needed for each part to saw, cut, or otherwise extract the finished part from the original material purchased. It may also be more economical to manufacture the finished parts in groups of four, say, sawing a length of bar for them with additional material for holding, sawing, or clamping. It may also be necessary to utilize raw material which is larger in diameter than the finished part to machine it to tolerances required. Some parts of irregular shape may be "nested" on bar or plate, and require a "first piece extra." Thus the combination of engineering design, part requirements, means of manufacturing, and economics of purchase significantly affect raw material purchase and inventory. The raw material inventory is thus a reflection of design requirements converted into economical manufacture.

5
ORGANIZING DATA FOR CIM APPLICATIONS: FEATURE-BASED DESIGN

ROSS S. CULVERHOUSE

INTRODUCTION

In the early 1800s, Eli Whitney introduced interchangeable assembly by producing firearms with similar piece parts. Each part was made by precision, the method of producing interchangeable parts by duplication using matched or common fixtures and tools to hold close tolerances. As designs became more complex and more parts were being produced and assembled at many different locations, accuracy became the primary method of producing interchangeable parts. Accuracy is the degree of conformity of a measure to a standard or true value. Accuracy requires complete, exact, and unambiguous communication of part information between designers and manufacturers to ensure the design integrity of a part during manufacturing, the proper function and performance of a part with respect to its mating parts.

In the industrial world, the engineering drawing has been used as the primary communication device between design and manufacturing. The engineering drawing is a graphical representation of a part specifying its size and shape. Geometric dimensioning and tolerancing is an engineering design method used to specify the functional relationships of part features on engineering drawings. It has been developed since World War II as a communication medium of universal interpretation to provide clear and complete information on drawings.

Today, the drawing is produced on the computer and is still used extensively throughout manufacturing industry and within individual companies. The main CIM data chart illustrates how extensively part and assembly designs are used in a computer integrated manufacturing (CIM) environment. The information on the drawing is used by many different people who perform design, manufacturing, and inspection functions. One of the main goals of CIM is to automate these engineering functions.

Drawings produced on current computer-aided design/computer-aided manufacturing (CAD/CAM) systems inherently do not facilitate automation of many of the engineering functions. A drawing is basically a human communication device. The part information of a CAD drawing is typically stored in the database in geometric form; points, lines, circles, cylinders, and so on. This is a geometry-based system. Therefore, it requires that people read the drawing to extract the part information, the dimensions and tolerances of features. So it is difficult at best to automate any engineering function that requires part information from a drawing. Someone would have to read it and reenter the part information into another system.

The classification of parts by similar features is an example of a manual process that could be automated. Currently, one is prompted to answer questions about a part, so we read the drawing and enter the answers something like:

> Computer—Does the part have any holes?
> Me—(Look at drawing. Yes, there are some holes) ENTER (Y OR N) > Y
> Computer—How many?
> Me—(Look at drawing again and count up holes 1, 2, . . . 25) ENTER > 25
> Computer—What is the maximum diameter?
> Me—("Oops, there are 23 holes. I hope it doesn't matter. Let's see, maximum diameter?")

Imagine doing this for 10,000 drawings. The explicit information of the part is not captured in the computer database. If it were, then the above dialog could have taken place between a classification program and the part database (hopefully the program would count features better than myself).

A CIM environment requires that part information be in the database in a form that facilitates automating design and manufacturing engineering functions. Engineers already know how to design using dimensions and tolerances to define part features, and how to use this information for many process planning, production, and inspection functions. If the part information is stored in a well-designed engineering database, then automation of many engineering functions would be greatly simplified. This will result in better designed parts in less time and at less cost.

This chapter describes a concept called feature-based design that uses familiar engineering entities related with solid modeling for the purpose of designing parts in a CAD/CAM system. The intent is to capture and store in the CIM database explicit part information in order to facilitate implementing future CAD/CAM applications. It is not enough to just draw and transfer drawings electronically as this represents only a small change from the drafting board. Without better methods of capturing engineering information in the database, the CAD/CAM computer systems are nothing more than expensive drafting boards.

DESIGNING WITH FEATURES

Feature-based design is a system for designing parts using features. It is a system that raises the level of parts modeling to the level of the design and manufacturing engineer; from a geometry level to an engineering level. It is a system that allows the engineer to define a feature using familiar engineering terminology and then to use the feature in many different part designs.

A feature, like beauty, is in the eye of the beholder. In feature-based design, the questions that come to mind provide the definition of a feature. If a designer wants to design a hole in a part, with a certain depth, diameter, surface roughness, and then specify a positional tolerance, then the system allows him or her to define the feature and then design parts using the feature in precisely those terms. The information of the feature is retained so that it can be used for other design, manufacturing, and inspection functions.

Features are specific physical portions of a part, such as hole, slot, surface, face, countersink, knurl, or screw threads. Parts are normally thought of as having features. However, in today's CAD databases, part information is normally stored as geometry. For example, Figure 1 illustrates a simple part with a through hole. In the typical CAD database the information is not stored as a hole with dimensions and tolerances to size the hole and position it relative to the part. Indeed, there is no notion of a hole stored in CAD systems. Rather, a collection of lines, circles, and text is captured. When engineers think of the feature hole, a flood of design and processing information immediately comes to mind. A feature inherently carries more engineering information than does the circle. The feature information should be captured as well as the geometric representation.

Features may be individual or interrelated. Geometric dimensioning and tolerancing provides a method that is widely used to specify the functional relationships of part features. We can explicitly define a feature in as great detail as we require. Features can be described with combinations of dimensions and tolerances. The sizes of feature shapes can be specified with size dimensions and tolerances. Features can be related to each other with positional

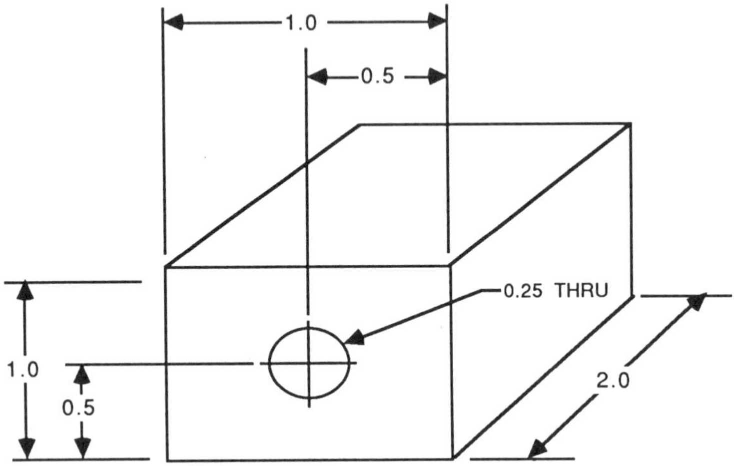

FIGURE 1 Part with a thru hole. (Or is it a box with a cylinder?)

dimensions and geometric tolerances. Dimensions and geometric tolerances are used to relate mating parts in an assembly.

Although there are many variations on the dimensioning and tolerancing standards, designers and manufacturers can usually agree upon how a certain feature should be dimensioned and toleranced. With current computer database, software, and hardware technology it is realistic to contemplate the implementation of the engineering design methods on the computer and use it as the basis for specifying, storing, and retrieving part information.

TOLERANCE BUILD-UPS

Dimensions and tolerances form the basis for defining features and designing parts with features. Dimensions are numerical values expressed in appropriate units of measure. Tolerance is the total amount by which a specific dimension may vary. They are used to define the size and geometric characteristics of a part feature and to relate features on a part. For example, a feature of size may be a cylindrical surface, or a set of two plane parallel surfaces, each of which is associated with a size dimension. The size dimension for the cylinder would be its diameter and for the two plane surfaces the size dimension is a linear size dimension. Each of these size dimensions has a tolerance specified.

A good test of a dimensioning system used in a CAD/CAM database is if it can compute dimension and tolerance stack-ups. This is a vital tool that a designer or production engineer must have in order to control tolerance build-ups. Performing a stack-up manually across three or 10 parts (pulling out the drawings, finding the dimension circuit path from one extension line to another, converting dimensions and tolerances to equal bilateral form, adding long columns of the plus and minus direction dimensions, adding up all the tolerances, do it one more time just to be sure) is not the most creative job around, but one that probably should be done more often than it is.

Fortunately, a stack-up is a straightforward procedural engineering function and therefore could be automated. It would only require that the database representation of dimensions and tolerances be designed to facilitate implementing the functions that the dimensions and tolerances should perform. Dimensions and tolerances are used by design and manufacturing to specify and communicate design intent and to design production processes.

The stack-up problem is basically finding the dimension path between any two parallel planes. A dimension circuit path is the series of dimensions that must be traversed from one plane to another. A tolerance build-up is the sum of the dimensional tolerances that lie on the dimension path. Once the dimension path is known, the tolerances and dimensions are to be converted into equal bilateral form and added to produce the dimension stack-up and tolerance build-up.

Linear dimensions are unambiguous; there is one and only one interpretation that can be made of any valid dimensioning scheme. Valid dimensioning schemes are conceptually easy to create and maintain by following one simple rule; that there be one and only one dimension path between any two parallel planes. Linear dimension "points" to two different parallel planes, as illustrated in Figure 2A. This would essentially be a linked list in the computer. By knowing which dimensions point to which planes, it would be possible to determine the dimension path between any two planes. Then take the sum the tolerances on the path, Figure 2B.

Creating a valid linear dimension scheme is straightforward. If every new extension line is created a relative distance away from an existing extension line and a linear dimension is created between these two extension lines, then the dimension is valid because there is one and only one dimension path between the new extension line and the one it was made from. The dimension scheme remains balanced, not over dimensioned or under dimensioned. "Overdimensioned" means that there is more than one path between two extension lines. "Underdimensioned" means that an extension line has no dimension relating it to another extension line. Dimension integrity is maintained by the system at all times.

Finding a dimension path between two extension lines can be used for functions other than dimensional and tolerance stack-ups.

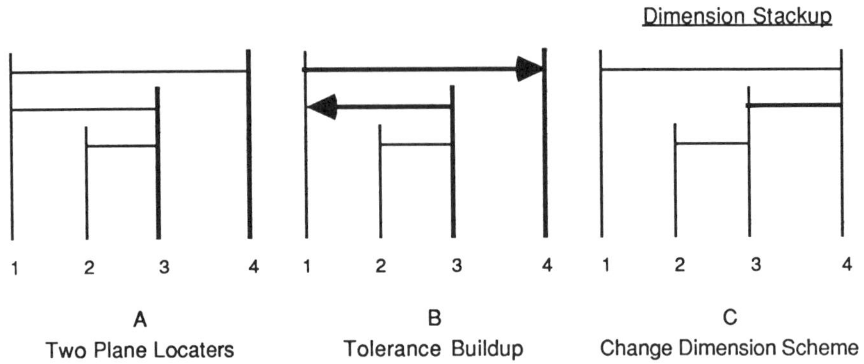

FIGURE 2 Figures showing (A) a dimensioning scheme with plane locaters 3 and 4 selected, (B) dimension path between planes 3 and 4, (C) dimension between 1 and 3 eliminated and reassigned to between 3 and 4.

Suppose that I change my mind and decided that I wanted to change the dimensioning scheme in Figure 2B and have the system maintain the dimension integrity. If I were to select the two extension lines between which I want my new dimension (3–4), the dimension path could be displayed. If a new dimension were added, it would close the dimension path loop, causing the part to be over dimensioned. One of the existing dimensions would have to first be eliminated, then it would be reassigned to its new extension lines (Figure 2C). This maintains the dimensioning integrity, and the rule has not been broken. The linear dimensions capture design intent, functional relationships between features. They are required for process planning, NC programming, tool and fixture design, gage design, inspection, tolerance charting, and many other CAD/CAM functions.

The dimension path can also be used to determine which extension lines should be moved and which ones should not when a dimension value is changed. If the dimension (3–4) in Figure 2C were changed, extension lines 2 and 3 would move. The origin (line 1) does not move. If for each extension line the changed dimension was on the dimension path between the line and the origin, then the location of the extension line is translated by the change in the dimension value.

Changing a dimension must always cause all the right things to happen, and many times a lot of things will happen. Designing with dimensions means that a single dimensional change could result in many extension lines, and geometry, being relocated. Complex

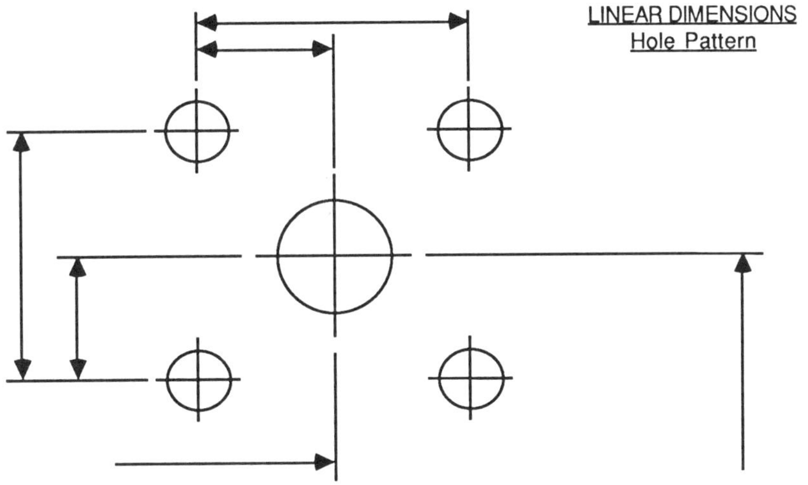

FIGURE 3 Four hole pattern.

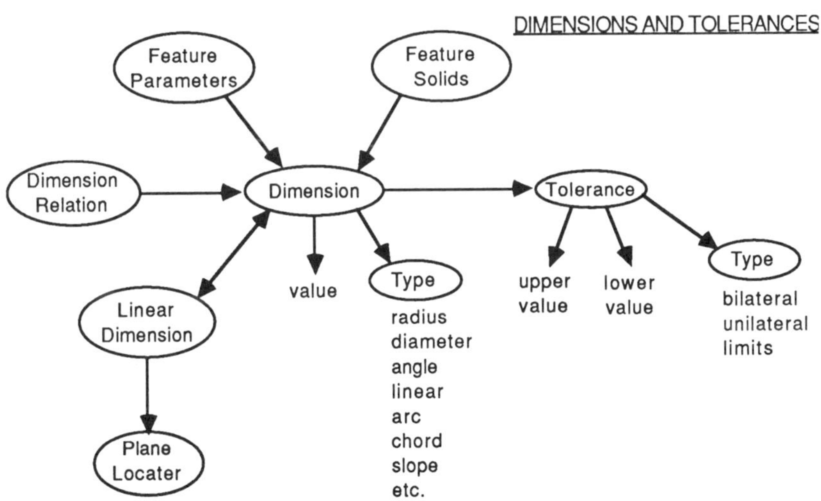

FIGURE 4 A graphical layout of dimension relationships for a data-base record.

features, such as the hole pattern in Figure 3, can be repositioned by changing one of the two dimensions that position the pattern.

Figure 4 illustrates some of the dimension relationships that can be defined in the database. The dimension has a value, a type, and a tolerance. The tolerance has an upper and lower value which specifies the variation of the dimension value, and the type of tolerance the values represents. The linear dimension can be associated with a dimension and with two planes. Feature parameters and solids can extract the appropriate dimensional information they require from the database. It is important that the dimension data realtionships are designed to facilitate implemention of the types functions that dimensions and tolerances are required to do—for example, tolerance stack-ups, tolerance analysis, tolerance charting, and machine and tool section.

DATUM REFERENCE FRAME

A datum reference frame is a set of three mutually perpendicular planes. The datum reference frame exists in theory only and not on the part. When an actual feature of a part is used to establish real datum, it is called a datum feature. Datum features are usually planes and axes. The perpendicular relationship provides direction as well as the origin for dimensions and measurements.

The extension line that I referred to so much represents an infinite plane at a distance away from one of the three planes. Figure 5 illustrates a linear dimension that is the distance between two parallel extension lines normally seen on drawings. The extension lines on a CAD drawing usually dimension the distance between two parallel lines. However, the reality is usually that the lines represent the edges of two parallel planes seen on edge. This is typical of most CAD systems; the geometry is stored but what is not stored is what the geometry represents. Figure 5 shows a linear dimension that is referring to two parallel planes. The locations of the planes can be controlled and specified by dimensions.

Figure 6A illustrates two mutually perpendicular planes that form the axis of a cylindrical feature. In this case two position dimensions relate the position of the axis to the sides of the rectangular box. Figure 6B shows the normal two-dimensional projection we normally view on a drawing.

Figure 7 illustrates how a point can be defined using three mutually perpendicular planes. Since each plane is referred to by a dimension, it is a dimensioned point. These can be used as datum target points.

Figure 8 illustrates a linear size as the distance between two parallel planes. Point, axis, and linear size can be defined in terms

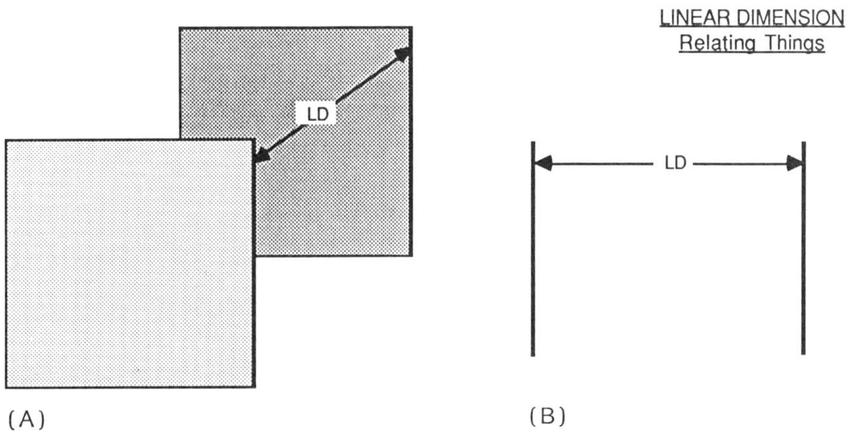

FIGURE 5 Linear dimension (A) the distance between two infinite parallel planes, (B) as normally displayed on drawings with extension lines representing the parallel planes.

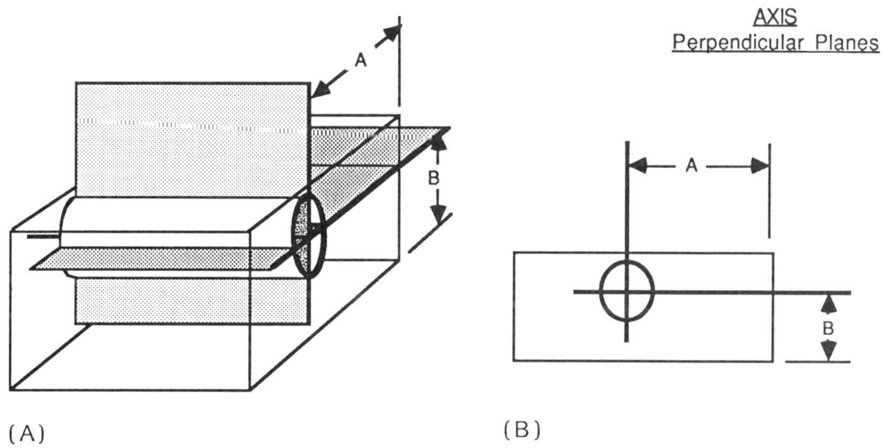

FIGURE 6 Two mutually perpendicular planes that form an axis. (A) Linear dimensions position planes that form axis, (B) 2D projection of axis normally viewed on a drawing.

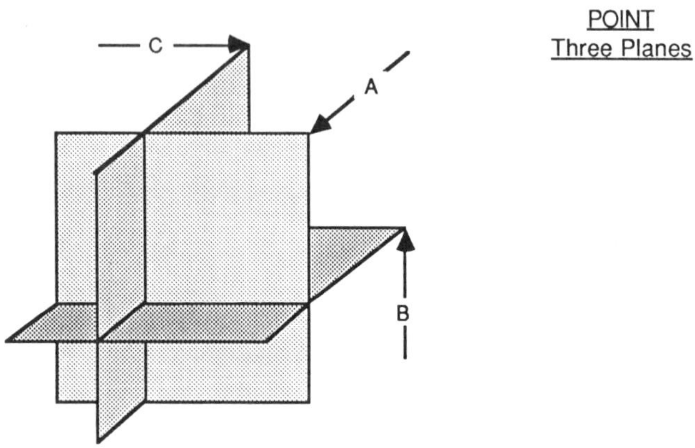

FIGURE 7 Point defined by three mutually perpendicular planes. Three linear dimensions position point.

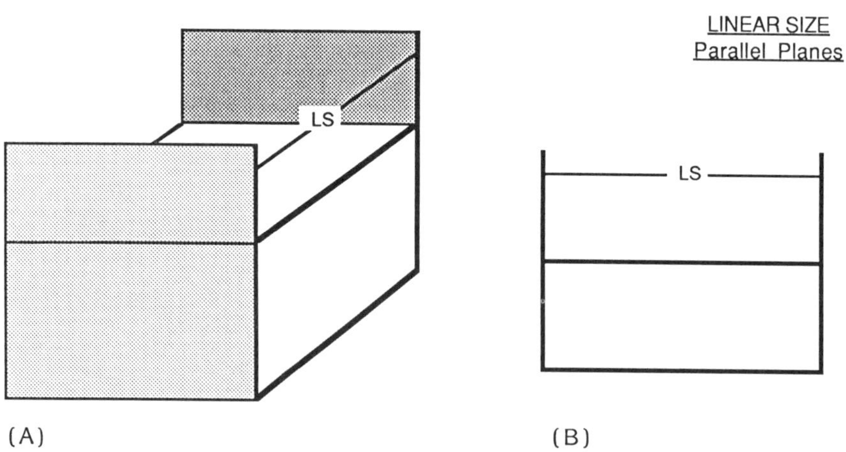

FIGURE 8 Linear size (A) distance between two parallel planes, (B) 2D projection of linear size and planes with extension lines. A linear size may result from a stackup of linear dimensions, or there may be a linear dimension defining the size explicitly.

of combinations of planes in a datum reference frame. Each of these can become a datum feature or controlled by a geometric tolerance.

DIMENSIONS AND SOLID MODELS

Solid modelers provide a very convenient resource for geometrically representing parts and part features. Many features can be geometrically modeled using current solid modeler primitives. Figure 9 illustrates six commonly used solid primitives; there are many others. Figure 10 illustrates the types of dimensions that each primitive can be defined by. For example, the cylindrical feature can be defined by an axis, a linear size for its length, and a radius or a diameter dimension. The dimensions that define the features can be associated and translated into a form that a solid modeler defines its primitives. The solid modeler might have the cylinder defined by a base point, a three-component vector for the axis, and a radius. The translation from feature-defined dimensioned shapes to solid modeler points and vectors is straightforward. The inverse may not be possible because much of the feature information would be lost in the translation to geometry; the geometry has no provisions to store dimensions and tolerances to translate back again.

This one-way translation means that a feature-based system could act as a front end to geometry-based systems. A feature-based system could automatically generate solid models of parts defined with dimensioned feature shapes. The diameter of a cylindrical feature would be changed by changing its diameter dimension and then updating the solid modeler cylinder. It also means that any change

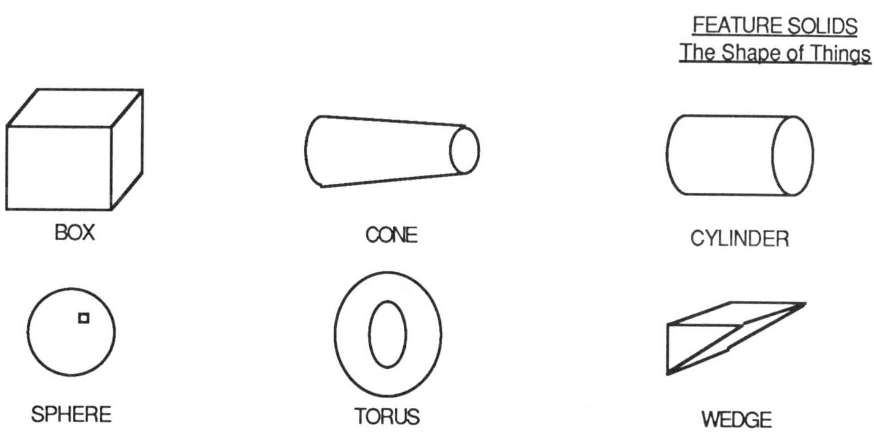

FIGURE 9 Six common geometric solid shapes.

made by methods other than with feature dimensions to the solid model would not be known to the part defined with features.

If the cylinder is in any way moved or its size changed by methods other than the feature dimensions, the feature definition would not know about it. There is currently no practical method to pass that change back to the feature definition. Well, maybe if all we were doing were cylinders, but when you consider the wide variety of solid primitives and the large amounts of geometry to keep track of; then there is no current practical method.

Designing with features means that design changes are made by changing dimensional values and letting the system automatically keep the solid model and the rest of the feature database current. To do this means that the solid model would have to have dimensions displayed, such as in Figure 11. Then by selecting a dimension and changing its value, appropriate dimensional and geometric changes can be made.

The dimensions in Figure 11 are graphically represented in a manner similar to a dimensioned drawing. The dimensions are used to define the geometry; dimensions control the size, position, and orientation of the geometry. The American National Standards Institute (ANSI Y14.5M-1982) suggests that for incorporating dimensional changes for CAD applications, dimensional changes of any magnitude must be made to the database which requires graphic reiteration. This must be done to ensure mathematical accuracy of changed values and to maintain the integrity of the CAD/CAM database.

Consider Figure 11 again. If the 3.000 dimension were changed, the size of the block would change, and the slot would move to stay

	SOLID PARAMETERS					
FEATURE SOLIDS	FEATURE GEOMETRY			FEATURE DIMENSIONS		
SOLID NAMES	AXIS	POINT	LINEAR SIZE	DIAMETER	RADIUS	ANGLE
BOX			√ √ √			
CONE	√		√	√ √	√ √	√
CYLINDER	√		√	√	√	
SPHERE		√		√	√	
TORUS	√			√ √	√ √	
WEDGE			√ √ √			√
			FEATURE PARAMETERS			

FIGURE 10 Table illustrates the combinations of dimensions that can be used to define solid primitives. For example, a BOX requires three linear sizes (three pairs of parallel planes).

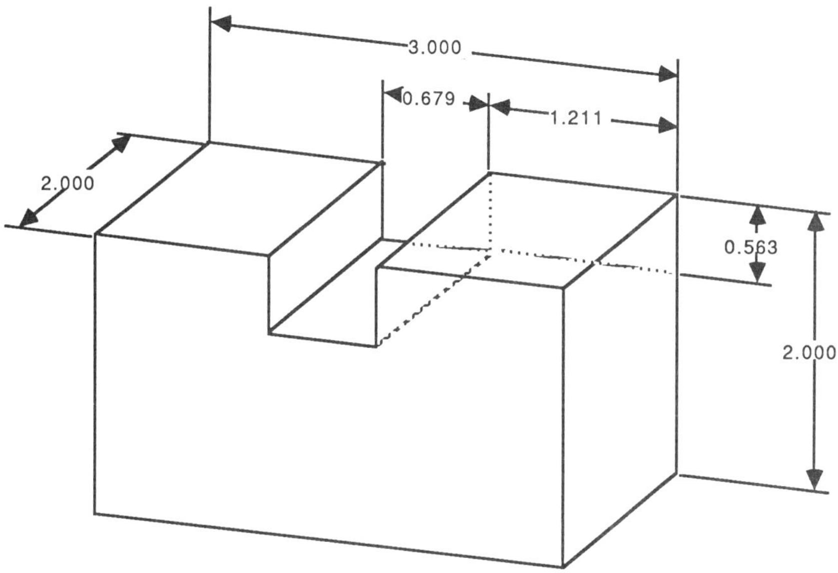

FIGURE 11 Isometric dimensioning of a solid model.

1.211 in. away from the right side of the block. This represents a one-to-many relationship between a dimension and the geometry that is affected when the dimension value is changed. By having the dimension control the geometry, the geometry always reflects the dimensional state. Also, the designer can specify the exact change he had in mind rather easily. If geometry is not associated with the dimensions, then each piece of geometry would have to be modified by the designer, as well as the dimension text. The process of having to change geometry is not only time consuming for a simple dimension change but also very error prone for complex parts.

Another CAD/CAM application that could be realized by having dimensions control geometry is that of the CAD system using the dimension and tolerance information to construct material condition models of the part. There are at least two types of material condition models that could be constructed, least material condition (LMC) and maximum material condition (MMC) models. For example, if an MMC model of the part in Figure 11 were to be constructed, the upper limits of those dimensions that size the block and the lower dimension limits of the slot dimensions would be used to construct the solid model. This requires that each dimension must know whether it is sizing positive or negative features. The MMC solid model could be used for worse case interference checking of mating part features.

SURFACE TEXTURE

Surface texture is the repetitive or random deviations from a nominal surface and is a result of the manufacture processing method. Figure 12 illustrates the surface texture symbol displayed on a drawing. It is used to designate control of surface irregularities to meet design requirements for the functional performance or appearance of the part. Determination of design requirements involves consideration of such things as contact area, load, speed, direction of motion, lubrication, temperature, material, bearing surfaces, gear surfaces, dimensional tolerance control, coating adherence, marked surface legibility, appearance, gas and fluid flow, and heat and light reflectivity.

Most surfaces do not require any control of surface texture beyond what is obtained by meeting necessary dimensional specifications. For example, to make a dimensional measurement of a surface, it is necessary to produce the surface with a suitably small surface roughness. Otherwise, the dimensional tolerance will be taken up by roughness height. Generally, the surface roughness should not exceed one-eighth of the dimensional tolerance. Where there are no other surface requirements, surface roughness can be specified by dimensional tolerance requirements.

A possible surface texture database representation, illustrated in Figure 13, has four constituents, roughness, waviness, lay designation, and material removal designation. This represents the same information as displayed in the surface texture symbol in Figure 12. The symbol is associated directly or indirectly with a surface on a drawing. In the database, surface texture can be associated with a surface of a feature or with a diameter dimension, which might define the cylindrical surface of a hole.

Roughness is the finer irregularities of surface texture. As was previously mentioned, there is a relationship between roughness and

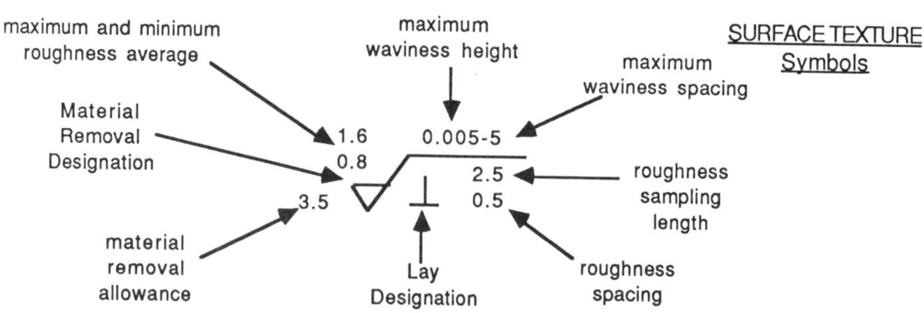

FIGURE 12 Surface texture symbol displayed on a drawing.

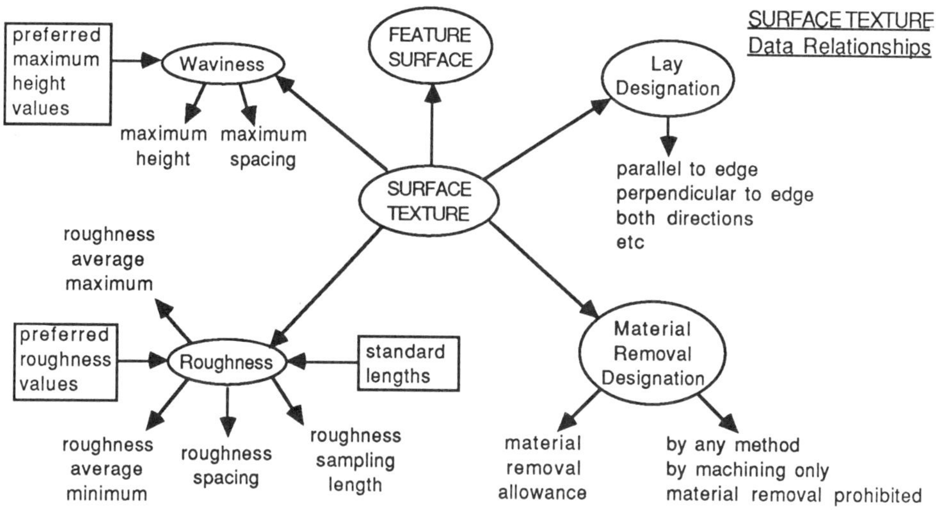

FIGURE 13 Possible surface texture database representation. Square boxes represent standard and preferred values.

dimensional tolerance. A minimum roughness for each surface can be determined based on the dimensional tolerances associated with each surface. Standard lengths and preferred roughness values can be selected from the database.

Waviness is a more widely spaced component of surface texture than roughness. Roughness is considered superimposed on a wavy surface. The preferred series of maximum waviness height values can be selected from the database. Waviness usually results from factors such as machine deflections, vibration, chatter, or heat treatment. As a result, when control of waviness is required it has an impact on the process selection and the tool and fixture design.

The lay designation specifies the direction of the predominant surface pattern. Lay is produced by the production method. When a lay is specified, it puts a constraint on process, tooling, and feed directions. When it is not specified, then any lay produced is assumed to be acceptable as long as other part specifications are met. The lay is designated on the surface texture symbol with a lay symbol. The material removal designation modifies the surface texture symbol when it is necessary to require or prohibit removal of material.

We mean to store the surface texture information explicitly in the database so that CAD/CAM functions can be developed to aid the designer in specifying surface texture requirements. Surface texture requirements are usually specified by the designer based on past experience or experimental results. It would be highly desirable if this experience were captured in the database in such a way that all designers could have this information at their fingertips to use as a guide in specifying the proper surface texture for a particular function the surface must perform.

This information is also used by CAM applications to design the process by which the part is to be manufactured. The selection of the proper combination of feed, speed, tooling, machine tools, and the like will result in the proper surface texture called out for the part. Automation of the process design will require explicit surface texture data representation in the database. Indeed, explicit surface texture information about the part is important data that must be stored in the database so that it will be available for CAD/CAM applications.

GEOMETRIC TOLERANCE

Geometric tolerance is a general term applied to the category of tolerances used to control form, profile, orientation, location, and runout of part features. It is used as a means of specifying dimensions and tolerances of a part with respect to the function of the feature on the part and can aid in determining the most economical manufacturing method. This method also provides that there is one, and only one, interpretation of the part.

The tolerances are generally determined by the designer to assure interchangeability of mating parts of an assembly. Maximum production tolerances can be attained, and the design integrity of the part can be maintained during manufacture. The information is used by many CAD/CAM applications such as process planning, tool and fixture design, and for determining inspection procedures. Capturing this data in the database is a prerequisite to implementing CAD/CAM design and manufacturing functions for engineers.

Figure 14 illustrates a feature control frame specifying a position tolerance normally seen on a drawing. A position tolerance defines a zone within which the axis or center plane of a feature is permitted to vary from true position. It is considered to be the heart of the geometric tolerancing system. Figure 15 illustrates a possible database representation of the feature control frame in Figure 14. The position tolerance relates two features; one is positioned relative to the other. For instance, the datum feature could be a plane that the axis of another feature is positioned in.

The position dimensions would be basic dimensions, and the position tolerance would be assigned to the feature being controlled.

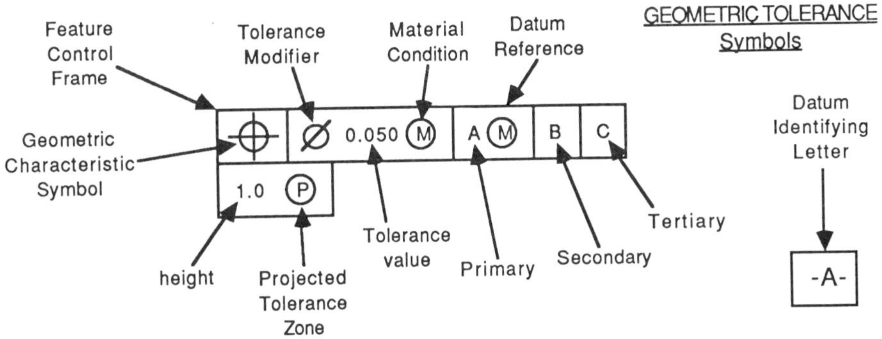

FIGURE 14 Feature control frame, graphical representation of a geometric tolerance seen on a drawing.

The location of one feature has no effect on any other feature. As a result, there is no tolerance build-up; chain and baseline dimensioning would give the same results. This would have to be taken into consideration during a tolerance stack-up. Tolerance charting, however, converts positional tolerances into equal bilateral tolerances to determine machining dimensions and tolerances.

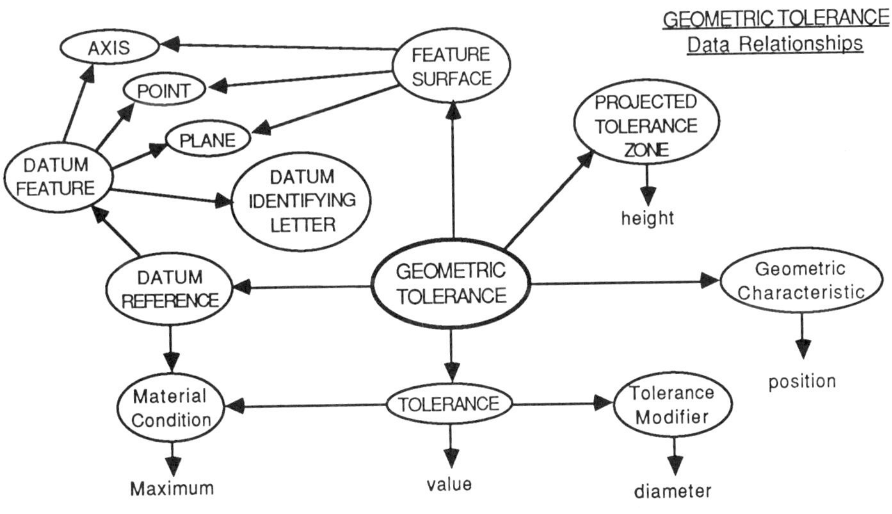

FIGURE 15 Database representation of a geometric tolerance.

CONCLUSION

There is a big job making CIM work. Parts and designs are becoming more complex; we need to build them right the first time; and, as always, they were needed yesterday. For a long time, there has been that wall between design and manufacturing. The wall will crumble as we improve communication between design and manufacturing. This chapter discussed some concepts and ideas in the hope that they may aid in improving some of the communication problems that exist in CAD/CAM.

REFERENCES

1. American National Standards Institute, "Dimensioning and Tolerancing," ANSI Y14.5M–1982. New York: American Society of Mechanical Engineers, February 15, 1983.
2. American National Standards Institute, "Surface Texture Symbols," ANSI Y14.36–1978. New York: The American Society of Mechanical Engineers, 1978.
3. Society of Manufacturing Engineers, *Tool and Manufacturing Engineers Handbook. Volume 1: Machining*. Dearborn, Michigan, 1983.

6
OUTPUT INFORMATION

Output information should be considered information which may be examined, analyzed, or interrogated on an informal basis or used as a master output file of data for back-up. The data may, however, never appear in the form of hard copy but be available on line at a terminal as required. In the tables shown in this chapter, each "X" represents which data segment may contribute to the output being described. The Item Master report, for example, represents a complete record of all parts and associated data of the Item Master database. Combined with data from other database segments, (marked with an "X"), a variety of reports can be generated. Not only could these be considered back-up documents, they may also be considered information which is on line for interrogation. The 20 sets of output information are as follows:

Item master
Parts catalog
Old drawings
Indented bill of material
Level-by-level bill of material
Summarized bill of material
Where-used bill of material
Parametric drawings
Job shop bill of material
Layouts design models simulations
Group technology planning
ABC inventory analysis
Purchase order status
Material requirements
Production schedule
Shop floor control data
Inventory control data

Purchasing data
Costed bill of material
Quality assurance data

Nothing is more variable than the form such outputs take. As they are described here, please remember that personal or company preferences dictate the contents, the formats, even the method by which such data is made available.

ITEM MASTER

DESIGN RETRIEVAL-------------- X
FAMILY PROCESSES-------------O
PART FAMILY DATA---------------X
PROCESS POLYCODES------------ X
ITEM MASTER-------------------- X
PERISHABLE TOOLS-------------O
PRODUCT STRUCTURES---------O
JIGS AND FIXTURES------------ O
ENGINEERING CHANGES----------X
PRODUCTION PLANNING-------- O
PART GEOMETRY--------------- O
INVENTORY CONTROL---------- O
ASSEMBLY GEOMETRY---------- O
GROUP TECHNOLOGY CELLS----O
GENERIC BILL OF MATERIAL--- O
RAW MATERIAL INVENTORY---- O
ELECTRONIC MAIL--------------- X

The item master is an important database which is often not fully utilized. As a history of active designs, the report can be sorted in a variety of ways to analyze designs, retrieve old ones, categorize designs by types, and generally reflect the history of products. In this discussion, only constant data is considered. However, the same reports mentioned combined with active data, such as inventory, production activity, and other manufacturing data, can be the focal point of an engineering manufacturing planning database analysis.

The following interrelationships and information may be required with this report:

1. *Design Retrieval System.* The design retrieval system is used to extract the family number of each part in the item master. If the item master is then sorted by this number and interrogated, similar parts can be detected easily. If critical dimensions and other data found in the design retrieval system are also merged

with the basic item master, a cogent and reliable set of data is available for analysis and corresponding action. It is a method to extract a small amount of meaningful data from a large database of designs.

2. *Part Family Data.* Part family data is implied in the paragraph above. However, new technology will eventually allow the merging of geometry, design data, tool path data, and other family data, including cost, lead times, and more. This will make an even more powerful set of data to analyze.

3. *Process Polycodes.* Process polycodes are another set of classifications for the parts in the item master. Once again, their merger with the data from the item master can create a significant set of data for analysis.

4. *Item Master.* The item master can be considered a storehouse of design information. Think of this data as a means to analyze the varieties of designs. Of course, such data is usually considered stored and examined in part number sequence. However, segregation of this database by the following criteria can be important also:

Part type code	Make or buy
Lead time	Manufacturing polycodes
Classification code	Cost
Part description	Use analysis
Product used on	Drawing size or availability

5. *Engineering Changes.* Since the revision level of a design controls its configuration, all engineering changes must be reflected in the design database. Each document for each design should have the same revision level attached to it, and the item master database should be the master repository of such information. As an integrated system grows more complex, an index of documents may be required for each design to maintain the revision level and/or configuration integrity. As more and more engineering (and other) documents become mechanized, this record becomes ever more important. Some of the possible documents which may be tied to or reflect a portion of the design documentation are as follows:

Engineering drawing
Schematic
Parts list
Outline drawing
Family tree
Specifications
Artwork
Numerical control (NC) tapes
Manufacturing process
Manufacturing routing

Tooling
Jigs and fixtures
Product specifications
Customer specifications
Generic bills of material
Instruction manuals
Modules and attachments
Marketing decision trees
Quality control documents
Inspection procedures

The coordination of changes must be reflected through these documents so all of them may have their integrity maintained.

6. *Electronic Mail.* Electronic mail will become an important means to release new designs as well as notify all engineering and manufacturing personnel of design changes. It is also a means to request specific data from the item master for on-line analysis or just to obtain information about a design quickly. In the not-too-distant future, such a request will also include the actual geometry of all segments of the design, including the NC tool path and the jigs, fixtures, and other tools related to it. The item master may be considered a central clearing house for all engineering data both as a record, and a means to maintain the integrity of the data through a coordinated set of changes.

PARTS CATALOG

DESIGN RETRIEVAL-------------- X
FAMILY PROCESSES-------------O
PART FAMILY DATA--------------X
PROCESS POLYCODES-----------O
ITEM MASTER-------------------- X
PERISHABLE TOOLS-------------O
PRODUCT STRUCTURES---------O
JIGS AND FIXTURES------------O
ENGINEERING CHANGES----------X
PRODUCTION PLANNING-------- O
PART GEOMETRY--------------- O
INVENTORY CONTROL---------- O
ASSEMBLY GEOMETRY---------- O
GROUP TECHNOLOGY CELLS---- O
GENERIC BILL OF MATERIAL--- O
RAW MATERIAL INVENTORY---- O
ELECTRONIC MAIL--------------O

A spare parts catalog might be considered a special case of the item master. However, spare parts activity and sales functions are usually considered an offshoot of design. Many parts used for service may have become obsolete in current designs. Thus the maintenance of spare part documentation may require different rules.

If a separate customer bill of material is maintained, the usage of service parts by customer may be maintained. In addition, economics in the manufacture of these parts may rely heavily on the use of part family data. An accurate list of spare parts by customer will enhance the sale of these parts when they are required.

The following interrelationships and/or information may be required with this report:

1. *Design Retrieval*. The design retrieval database can be used as a source for segregating spare parts into families. It is possible to imagine finding a similar part that could be altered slightly to make a spare part. The design retrieval system will combine similar parts to allow quicker response to a customer inquiry by determining estimated cost and lead time faster. When spare part sales are forecast and then combined by the retrieval system, it is possible to analyze requirements as a group of parts, not individually.

2. *Part Family Data*. Through analysis, part family data might show it economical to manufacture spare parts by family in a cellular manufacturing arrangement. Thus the spare part catalog can reflect part families for quick reference. In addition, obsolete tools, fixtures, and processes, routings, and NC tapes can be maintained to assist in the rapid response to a customer order.

3. *Item Master Data*. The item master data is essential to record the history of a design and to maintain current records which can be coordinated with the parts catalogs a customer might have. It is also possible a spare part may be replaced with a later design. Thus there needs to be a cross-reference between the old part design and the new one. Many companies have changed part number systems. Older spare parts catalogs in the hands of customers do not reflect these changes. Thus it may be important to replace a customer parts catalog with one which shows both the old and new.

4. *Engineering Changes*. Engineering changes must be analyzed to determine how they might affect a customer parts catalog, and all the items mentioned above must be analyzed.

OLD DRAWINGS

```
DESIGN RETRIEVAL--------------- X
FAMILY PROCESSES--------------- X
PART FAMILY DATA---------------X
PROCESS POLYCODES------------ O
ITEM MASTER---------------------X
PERISHABLE TOOLS--------------O
PRODUCT STRUCTURES----------- X
JIGS AND FIXTURES------------- O
ENGINEERING CHANGES----------- X
PRODUCTION PLANNING--------- O
PART GEOMETRY------------------X
INVENTORY CONTROL----------- O
ASSEMBLY GEOMETRY-------------X
GROUP TECHNOLOGY CELLS----- O
GENERIC BILL OF MATERIAL---- O
RAW MATERIAL INVENTORY----- O
ELECTRONIC MAIL--------------- O
```

The term *old drawings* does not imply they are also obsolete. In an integrated system, designed for implementation for today and the future must take into account designs and their corresponding data which has not been prepared on a computer system. Many thousands of old designs have been documented with standard drafting and design practices. Conversion of them to a part and assembly geometry data base may be economically impractical. However, it would be senseless to "start over" by ignoring these designs or by making them obsolete. Thus they must be considered usable in the future.

Later we will discuss how the microfilm system plays a part in combining designs made on a CAD/CAM system with older drawings prepared conventionally.

1. *Design Retrieval*. The design retrieval database will be used to retrieve old designs as well as new designs prepared on a CAD/CAM system.

2. *Family Processes*. Family processes are as logical for use with old designs as they are for new.

3. *Part Family Data*. Part family data is compatible to old designs as well as new.

4. *Item Master*. The item master data is as important for old designs as for new. However, the integration of geometry is missing.

5. *Product Structure*. The product structure data will be represented in the form of an assembly detached parts list but will not be used to combine part geometry into an assembly.

6. *Engineering Changes*. Engineering changes will follow the same pattern and be documented in the same method for old drawings as for new.

7. *Part Geometry*. Many old drawings for part geometry may be represented as multiple parts under one drawing number, or an assembly may have been generated which has the detailed parts appearing on the same document. This practice is not conducive to an integrated system. When such parts are still current and will be used continually, it may be necessary to redraw them as individual parts. This may be performed economically using a CAD/CAM system and preparing parametric drawings to expedite the conversion.

8. *Assembly Geometry*. In a like manner, assembly geometry may require a format change. These assembly drawings may be economically converted using a CAD/CAM system, but using the part crawings converted previously to expedite the conversion.

INDENTED BILL OF MATERIAL

DESIGN RETRIEVAL--------------O
FAMILY PROCESSES--------------O
PART FAMILY DATA------------- O
PROCESS POLYCODES------------O
ITEM MASTER--------------------- X
PERISHABLE TOOLS--------------O
PRODUCT STRUCTURES-----------X
JIGS AND FIXTURES------------- O
ENGINEERING CHANGES-----------X
PRODUCTION PLANNING-----------X
PART GEOMETRY---------------- O
INVENTORY CONTROL----------- O
ASSEMBLY GEOMETRY----------- O
GROUP TECHNOLOGY CELLS----- O
GENERIC BILL OF MATERIAL------X
RAW MATERIAL INVENTORY-------X
ELECTRONIC MAIL--------------- O

The indented bill of material is an important tool for both engineering and manufacturing. The report is a reflection of a graphical product structure which, hopefully, has been prepared for each product through a cooperative effort between design and manufacturing planning. The format of this report is, usually, a list of all parts in a product but with the "level" upon which each part is used designated. On the left of the report a series of columns are available to label these levels. For example, the first part number is the final product, and a 1 appears in the first column. The next part number is normally a part or assembly, which is a level 2, and the number 2 appears indented in the next column. If this part number is that of an assembly, all parts and subassemblies will next appear in the listing, and the number 3 will appear indented in the proper column. This listing and the corresponding level of each design will continue until the entire product has been documented. Of course, if the same design is used in more than one assembly, it will appear several times in the proper relationship.

With experience, the user can quickly see how a product is assembled, whether too many parts have been placed in one large assembly, and how many levels of assembly are required.

1. *Item Master.* The item master is the source of all data which appears on the face of the indented bill-of-material report. It should be emphasized that the indented bill of material may take several forms. One listing may show engineering data, another manufacturing data—each taken from the same item master database.

2. *Product Structure.* The product structure database is used to prepare the part assembly relationships as pictured in the indented bill-of-material format. The item master data and the product structure data are merged together by the computer as the report is prepared. Thus there is no redundant data generated by the user, since the same data from the item master may be used over and over wherever the same design is used. If a manufacturing firm utilizes modular design techniques, no top level part number for a complete product may be available. To print a selective bill of material for a series of modules and attachments might require preparing a one-time part number which collects this data together and then use this collection to prepare an indented bill of material.

3. *Engineering Changes.* Engineering changes can be reflected in an indented bill of material. As will be noted later, effectivity of changes can be documented by date so that a current bill of material or one reflecting the structure in the future or in the past is documented. In some cases, for audits or for current information, the revision level of each design may appear on this report. This data is proper on the indented bill of material but would be dangerous and costly on a single level bill of material called a parts list. This will be explained later.

4. *Production Planning.* When production planning data is added to the bill of material, including cost and lead time data, a costed bill of material can be prepared using a cost roll-up procedure. This helpful not only to the accounting department, but also to sales and marketing to price the product and estimate mark-up and profit. Excessive costs in some area of the product might make it possible to analyze the manufacturing process or vendors of purchased parts to find more efficient methods or purchasing practices. A lead time chart can be prepared with lead time data tied to the

indented bill of material. Thus, in a job shop, for example, the delivery date of an order can be used to graphically portray the starting and completion data of each part and assembly in the order backed off by lead time. This report might be called the "poor man's" production schedule.

5. *Generic Bill of Material.* A generic bill of material is a special indented bill of material. As described previously, each product can be defined in general by a product structure which can be documented using a generic bill of material which consists of "phoney" part numbers. This documentation can be particularly useful to a job shop which may obtain an order that is only completely documented when it is shipped. If the phoney part numbers are based on the design retrieval system, lead times, costs, routings, and other data normally prepared for a "real" bill of material can be generated. In many data processing systems a "rough cut" process planning system is available. The generic bill of material is a more complete means to achieve the same result. The generic bill of material is also useful to sales and marketing since it can be used to show expected delivery date and costs of the finished unit if the phoney part numbers can be tied to families of parts which have estimated costs and lead times. Even if each family of parts from which the part numbers are taken has variable costs and lead times, the high and low can be documented to give a spread of costs and time for the estimate. If an estimated design time and an estimated drafting time are also added to lead time of each part number, the lead time chart mentioned previously could also show a schedule for design and drafting. When a generic bill of material is first used, costs, lead times, and design and drafting time will cover all parts in the order. Once portions of this bill of material are replaced with parts and assemblies that can be used for the specific order, these schedules will show corresponding changes.

6. *Raw Material Inventory.* The raw material inventory and corresponding part numbers are a natural extension of the indented bill-of-material report. Many manufacturing companies require that engineering designate the type and amount of raw material required for each designed part. This creates a major redundancy, since manufacturing planning must review these decisions and, normally, alter them to meet the manufacturing practices and raw material available. It is far more practical to divide this responsibility between engineering and manufacturing by allowing the engineer to make the design and determine the material chemistry only, such as SAE 1020, for example. The manufacturing or industrial engineer should convert the design requirement into the proper amount of raw material needed to create one part, and also determine the type and size of raw material which meets this requirement. The decision should then be documented as a part of the bill-of-material system in the form of a one-line

parts list. The indented bill of material will then show these raw material requirements one level below each of the piece parts.

LEVEL-BY-LEVEL BILL OF MATERIAL

DESIGN RETRIEVAL--------------O
FAMILY PROCESSES--------------O
PART FAMILY DATA------------- O
PROCESS POLYCODES------------O
ITEM MASTER--------------------- X
PERISHABLE TOOLS--------------O
PRODUCT STRUCTURES----------- X
JIGS AND FIXTURES------------- O
ENGINEERING CHANGES----------- X
PRODUCTION PLANNING-----------X
PART GEOMETRY---------------- O
INVENTORY CONTROL----------- O
ASSEMBLY GEOMETRY----------- O
GROUP TECHNOLOGY CELLS----- O
GENERIC BILL OF MATERIAL------X
RAW MATERIAL INVENTORY-------X
ELECTRONIC MAIL--------------- O

The level-by-level bill of material is one of the more important documents of the system. As will be discussed later, this will be used to generate and to create a parts list and a picking list.

Terminology is often used carelessly. A bill of material is always more than one level of a product documentation. A level-by-level bill of material always represents a set of one-level bills of material, and the more accurate description for them is assembly parts list, parts list, or detached parts list.

In the form of a level-by-level bill of material, the implication of such data is that it will be used to analyze a product or a segment of the product and will not be used in some formal manner. The following interrelationships and information may be required with this report:

1. *Item Master*. As in the case of all bill-of-material formats, the basic data will be obtained from the item master database.

2. *Product Structure*. In a like manner the level-by-level bills of material will be created by using the part assembly relationships from the product structure database.

3. *Engineering Changes*. The engineering changes, past, present, or future, can be incorporated in the report to determine assembly changes, and to establish effectivity of such changes.

4. *Production Planning.* Data from the production planning database can be used to compare requirements of parts and assemblies to inventory on hand or order, and each assembly can be costed using accounting data created in the production planning systems.

5. *Auditability.* If a generic bill of material is created in the form of a level-by-level bill of material, it can be audited for accuracy, cost, and lead time on an assembly-by-assembly basis. Thus this report may become an important means to keep the database accurate.

6. *Raw Material Inventory.* As mentioned previously, the raw material inventory requirements for designed parts can become a part of the bill-of-material system. However, a level-by-level bill of material means each product structure for each part—raw material relationship would be created. This might be considered excessive data. A means to forestall this amount of documentation is to enter all raw material data, including part number, unit of measure, and quantity for each fabricated part, into the item master of the part itself. While this does not create a product structure as known classically, programming logic can be created to simulate a product structure when needed. This can eliminate all the one-line parts lists which are required if a true product structure for raw material is created.

SUMMARIZED BILL OF MATERIAL

DESIGN RETRIEVAL--------------O
FAMILY PROCESSES--------------O
PART FAMILY DATA-------------O
PROCESS POLYCODES------------O
ITEM MASTER---------------------X
PERISHABLE TOOLS--------------O
PRODUCT STRUCTURES-----------X
JIGS AND FIXTURES-------------O
ENGINEERING CHANGES-----------X
PRODUCTION PLANNING-----------X
PART GEOMETRY----------------O
INVENTORY CONTROL-----------O
ASSEMBLY GEOMETRY-----------O
GROUP TECHNOLOGY CELLS-----O
GENERIC BILL OF MATERIAL------X
RAW MATERIAL INVENTORY-------X
ELECTRONIC MAIL---------------O

The summarized bill of material is another important form of data derived from the bill-of-material system. In this report, all structure relationships are ignored, and the total requirements for each part, assembly, and raw material are listed, usually by part type, and in part number sequence. Before more sophisticated means for developing material requirements were created, this summarization was used to order parts and materials—particularly in a job shop environment. Even today, this practice can be used for small manufacturing companies, and thus becomes a viable means to determine not only requirements but even total material cost for an order or product.

As each of these alternate means for displaying bill-of-material data is described, it should be emphasized that all of them come from the same database, but the data is manipulated by the computer into the various formats described. In the past, many of these reports were tediously and meticulously prepared individually, creating numerous errors and costly preparation time.

1. *Item Master.* The item master data is, once again, used for this report.

2. *Product Structure.* The product structure data is used but combined by part number to prepare this report. All actual structures are avoided as the bill of material is utilized.

3. *Engineering Changes.* Engineering changes are incorporated into the report as described previously.

4. *Production Planning.* Production planning data as cost and lead time may be incorporated into this report.

5. *Generic Bill of Material.* A generic bill of material can be studied if this report is created. As a new design is made which replaces a generic bill of material, a cross-check can be made to be certain that a new design has been incorporated correctly into all places where the part is used. This cross-reference can be enhanced by creating an indented bill of materil and giving each line of such a report a line number. By sorting the data by part number it becomes, essentially, a summarized bill of material which can be used to make certain all new designs are incorporated.

6. *Raw Material Inventory.* Raw material inventory data may be incorporated into the summarized bill of material. However, this data will serve to determine total raw material requirements and the number of raw material variations. If these variations are extensive, an analysis to consolidate them may be a logical extension for analysis.

WHERE-USED BILL OF MATERIAL

DESIGN RETRIEVAL--------------O
FAMILY PROCESSES--------------O
PART FAMILY DATA------------- O
PROCESS POLYCODES------------O
ITEM MASTER--------------------- X
PERISHABLE TOOLS--------------O
PRODUCT STRUCTURES----------- X
JIGS AND FIXTURES------------- O
ENGINEERING CHANGES----------- X
PRODUCTION PLANNING-----------X
PART GEOMETRY---------------- O
INVENTORY CONTROL----------- O
ASSEMBLY GEOMETRY----------- O
GROUP TECHNOLOGY CELLS----- O
GENERIC BILL OF MATERIAL------X
RAW MATERIAL INVENTORY-------X
ELECTRONIC MAIL--------------- O

For some unknown reason, the where-used aspect of a bill of material does not seem to be emphasized in the engineering documentation systems. However, it is an important and significant tool for engineering and manufacturing. Several formats exist.

The most common where-used report shows a single level "used-on." For each part number, the next (higher) level assembly where the design in used is listed. (It should be noted that where-used data is described by at least two other titles, "used-on," and "gozinta."

Another report, not used often because of time-consuming computer time, is a two-level used-on, showing not only the next level usage, but also the top level, or product part number usage.

Of course, it is possible to show all levels through which a part is related as required, but this can become so complicated and time consuming to prepare as to be almost useless.

The importance of this data cannot be emphasized too much. If, for example, a part design is dangerous, and the part must be replaced wherever it is used, the used-on report presents all these usages. On the other hand, this report creates a problem which engineers did not have formerly. If an old design is replaced by a new one, the engineer who performs this activity usually has only his designs in mind. Picture his plight when he, through the mechanized system, discovers "his" design is used in 20 or 30 other designed assemblies. This discovery will cause him to require changing designs not under his jurisdiction, a job he will not necessarily like. Consequently, we must remember this kind of data is the sign of a good, integrated system.

1. *Item Master.* The item master data is, once again, used for this report.

2. *Product Structure.* The product structure is the key element in preparing the where-used report. In some large systems, there is a separate where-used database, at least for part number relationships. This avoids taking product structures and inverting them each time a where-used report is prepared. An on-line system could be enhanced with this procedure for rapid response.

3. *Engineering Changes.* Engineering changes are incorporated into the report as described previously.

4. *Production Planning.* Production planning data, such as cost and lead time, may be incorporated into this report.

5. *Generic Bill of Material.* A generic bill of material converted to a used-on is another method for making sure each new design that converts a generic bill of material to a real one has been accommodated.

6. *Raw Material Inventory.* A raw material inventory report showing what parts use the same raw material will be helpful in the analysis of proliferation of parts both in sizes and by chemistry.

PARAMETRIC DRAWINGS

DESIGN RETRIEVAL	X
FAMILY PROCESSES	X
PART FAMILY DATA	X
PROCESS POLYCODES	O
ITEM MASTER	O
PERISHABLE TOOLS	O
PRODUCT STRUCTURES	O
JIGS AND FIXTURES	O
ENGINEERING CHANGES	X
PRODUCTION PLANNING	O
PART GEOMETRY	X
INVENTORY CONTROL	O
ASSEMBLY GEOMETRY	X
GROUP TECHNOLOGY CELLS	O
GENERIC BILL OF MATERIAL	O
RAW MATERIAL INVENTORY	O
ELECTRONIC MAIL	O

Parametric drawings are created for each part family defined in a classification system. Their use decreases drafting and design time. However, the title should not be misconstrued to mean a typical engineering drawing only. The parametric drawing should represent a complete set of geometry, design features, and other attributes of a part family.

This database is a significant new adjunct to CIM and should be planned carefully. While parametric drawings are normally considered to be for piece parts, it is also possible that common assemblies could lend themselves to parametric programs. For them, individual part should still be recognized but made into an entity for convenience.

Not all designs can be defined using parametric drawings. However, it is felt that at least 80% of all designs lend themselves to this procedure.

1. *Design Retrieval.* The design retrieval system is the source of families which can then be used to determine the logical parametric drawings needed. In some cases, more than one parametric drawing may be needed for a family of parts due to some diversification of design features in the family. On the other hand, one parametric drawing may be appropriate for several part families.

2. *Family Processes.* If family processes are also tied to the parametric program, entry of design data for a specific design can be transferred into the routing and process data too. For example, a milling operation on the length of a part can be prepared so that the length dimension of the design can be entered into the appropriate place in the process description.

3. *Part Family Data.* The part family data can be used to enhance the parametric drawing by designating design features which apply to the family. In addition, drafting and design standards can be built into the parametric drawing to create a standard design. Where producibility tips have been determined, such as bevels, these dimensions can become fixed in a manner such that the designer cannot change them when a new design is created without specific approval.

4. *Engineering Changes.* Engineering changes may not directly affect parametric drawings. However, they should be examined to determine if the engineering change is due to changes in producibility tips or other parameters which affect all designs, including new machine tools, and other process data.

5. *Part Geometry.* The part geometry is the basic data which must be used to make a generalized parametric drawing. This is used to make a generalized parametric drawing. This is probably the most important reason for creating a design retrieval system, since a parametric drawing should cover many, not one, design.

6. *Assembly Geometry.* Assembly geometry is also important when parametric drawings are created. However, the frequency of practical use for assemblies will probably be lower. If design analysis is used, is may be possible that a parametric drawing tied to such analysis will make recording of a finished assembly practical.

JOB SHOP BILL OF MATERIAL

DESIGN RETRIEVAL--------------O
FAMILY PROCESSES--------------O
PART FAMILY DATA------------- O
PROCESS POLYCODES------------O
ITEM MASTER--------------------- X
PERISHABLE TOOLS--------------O
PRODUCT STRUCTURES----------- X
JIGS AND FIXTURES------------- O
ENGINEERING CHANGES----------O
PRODUCTION PLANNING-----------X
PART GEOMETRY---------------- O
INVENTORY CONTROL-------------X
ASSEMBLY GEOMETRY----------- O
GROUP TECHNOLOGY CELLS----- O
GENERIC BILL OF MATERIAL------X
RAW MATERIAL INVENTORY-------X
ELECTRONIC MAIL-----------------X

A job shop bill of material is a special case of engineering documentation. Often a manufacturing concern is completely customer order oriented and has been operating under a job number system which identifies each purchased and each designed part for the order with the same order number. Implementation of a bill of material seems to take this customized reporting away from management. Using a job shop bill of material can retain the best of both systems.

Once a customer order is received, a generic bill of material for a similar product can be created. This serves the first step of cost and time estimates as well as reserving time in the production schedule, or using the data to determine if a future overload will occur.

As the customer design is created, the generic bill of material is converted to the real one. All these manipulations can create potential errors in the engineering product structures and item master data if proper configuration management is not followed at all times. Thus, if a generic bill of material and its corresponding customer bill of material are maintained in a separate job order file, these errors can be minimized. It should be noted that there is some redundancy in this process, but, in a job shop atmosphere, the advantages may well be greater than this disadvantage.

The customer bill of material can be customized. Several special branches to the bill of material can be added. For example, one branch might show, by part number, all installation drawings, another all engineering changes associated to the order, still another all the customer specifications, and more. This customization can also identify special designs which are not used by or on any other order or design, and actual cost will be isolated.

Use of this method will not seriously affect other systems, such as material requirements planning (MRP), since the order bill of material can be used in a manner identical to those which are exploded in a mass production environment.

1. *Item Master*. As in the case of all bill-of-material formats, the basic data will be obtained from the item master database.

2. *Product Structure*. In a like manner, the level-by-level bills of material will be created by using the part assembly relationships from the product structure database and a generic bill of material for the general product.

3. *Production Planning Data*. Production planning data, such as cost and lead times, can be used to indicate a schedule for the order. It might be important to prepare a lead time chart for the order as well to generate a design and drafting schedule as well.

4. *Inventory Control*. Inventory control data can be itemized for the job order bill of material as the generic bill of material is converted. This may assist in decreasing the lead time of the order as well as mechanically create purchase orders for long lead time items and raw materials.

5. *Generic Bill of Material*. The generic bill of material can become a significant tool in a job order environment.

6. *Raw Material Inventory*. Since raw material inventory is often considered stock and nonstock in a job shop environment, the job order bill of material can serve to identify all raw materials required at an early date so that nonstock materials can be ordered and stock materials reserved.

7. *Electronic Mail*. In a job shop environment, electronic mail can be used to expedite engineering changes release of new designs and conversion of the generic bill of material to a true customer order. Communications are an important phase of this concept.

DESIGN SIMULATIONS

DESIGN RETRIEVAL--------------- X
FAMILY PROCESSES--------------- X
PART FAMILY DATA---------------X
PROCESS POLYCODES------------O
ITEM MASTER--------------------O
PERISHABLE TOOLS------------- O
PRODUCT STRUCTURES--------- O
JIGS AND FIXTURES------------O
ENGINEERING CHANGES--------- O
PRODUCTION PLANNING--------- O
PART GEOMETRY-----------------X
INVENTORY CONTROL----------- O
ASSEMBLY GEOMETRY-------------X
GROUP TECHNOLOGY CELLS-----O
GENERIC BILL OF MATERIAL---- O
RAW MATERIAL INVENTORY----- O
ELECTRONIC MAIL---------------O

One of the important techniques possible, with a combination of design geometry and other data including design analysis algorithms, is the ability to create simulations of individual designs or entire products. This ability seems to be uniquely computer oriented, since the variations and the requirements for visual data are best represented using the computer.

It is quite possible now to create a finished working model of a product on the screen of a terminal, simulate movements, loads, additions of attachments, collecting various modules which make up the finished product, and have it perform as a customer requires. This simulation both of physical components and their actual operation will be a significant advance for marketing since it may be possible to show a potential customer a unique configuration of a product which might not actually exist. A similar example might be that of designing a prototype of a product and testing it in simulations.

A customer, thus informed, might order the configuration shown, which, in turn, automatically releases the necessary bill of materials, creates a schedule, and costs the product. This methodology has been used in rather simple products for years, including pumps and grinding mills. With the addition of pictorial representation, the realism of the simulation is even more helpful.

In the same manner, using design analysis algorithms, simulations can be pictured of design actions and reactions. A notable example of this is use of finite-element analysis of some structure, the results of applying forces to it resulting not only in digital data but pictorial data, which gives the designer a more clear picture of results.

It is clear that to perform these activities, the data of both geometry and bills of material must be integrated correctly. A CIM database which we are describing is a powerful assistance to this endeavor providing it is defined correctly.

1. *Design Retrieval*. The design retrieval system will assist to find those parts and assemblies which have similar algorithms for design analysis. Integration of such algorithms to design assemblies may also be practical.

2. *Family Process*. Family process can be used to simulate the manufacturing process, including flexible manufacturing techniques. A cell of parts, tied to a design analysis program, can create parts which do not exceed the tool package limiations of an NC machine tool.

3. *Part Family Data*. Part family data is also needed if a product simulation is required, followed by estimating costs, lead times, and preparing a bill of material for the customized customer order. The more that design analysis, geometry, and manufacturing data are coordinated, the better and more logical will be the result.

4. *Part Geometry*. Part geometry is mandatory for the types of design models and simulations described here. For example, a family mesh for use in finite-element analysis will expedite the design process. Use of features to prepare the design will create a solid model of the part which can be analyzed rapidly and accurately.

5. *Assembly Geometry*. Assembly geometry means not only simple assemblies, but also the final product. If these products are designed using solid models, simulation in almost any form can be generated. One of the largest models being prepared is that for global weather forecasting. It will soon be possible to perform the same techniques on products, and will certainly be far simpler.

GROUP TECHNOLOGY

DESIGN RETRIEVAL--------------- X
FAMILY PROCESSES--------------- X
PART FAMILY DATA---------------X
PROCESS POLYCODES------------- X
ITEM MASTER--------------------O
PERISHABLE TOOLS---------------X
PRODUCT STRUCTURES----------O
JIGS AND FIXTURES---------------X
ENGINEERING CHANGES----------- X
PRODUCTION PLANNING-----------X
PART GEOMETRY------------------X
INVENTORY CONTROL-------------X
ASSEMBLY GEOMETRY----------- O
GROUP TECHNOLOGY CELLS-------X
GENERIC BILL OF MATERIAL---- O
RAW MATERIAL INVENTORY----- O
ELECTRONIC MAIL---------------O

Group technology is a recent popular concept in the United States. Practiced for many years in Europe, the concept has new names, such as flexible manufacturing and just-in-time (JIT) manufacturing.

The principle of group technology is that of combining parts with similar design or manufacturing characteristics into a batch for economies in their production. The significant savings is in avoidance of set-up time and thus increasing the cutting time of the machine tools over which the parts are routed. Often group technology takes the form of grouping a series of machine tools into a cell in a manner such that all operations required for a series of parts can be performed without moving the parts from one section of a machine shop to another. This saves transit time and the corresponding work-in-process inventory costs.

A special and important case is in the manufacture of assemblies. Some companies have set up a cell of machine tools, or a special NC machine tool to manufacture, in sequence, all parts required for one assembly or product. As the series of parts is completed, the assembly of them follows, and the finished unit may then be shipped. In principle, this not only avoids excessive set-up time but also work in process of the finished parts. Significant saving can be achieved.

Originally, group technology techniques were used to optimize production facilities. In Europe, the cost of carrying excessive inventory was less important than the cost of additional machine tools. Consequently, group technology was proclaimed as a means of saving set-up time for small batches of parts, thus improving the productivity of machine tools and avoiding the requirement to purchase more. Since the principle can work even for fairly high volume parts, this limitation has been ignored in the United States.

NC machine tools are really cells which are used for group technology. However, manufacturing companies do not always utilize them as such. As a result, individual parts are machined on NC tools one at a time. Even though provision is made to prepare the next part for machining on such a tool, excessive set-up time can still be created because the tools in the tool package are identified by part number, not a general operation. When computers are used with NC machine tools, and families of parts are coordinated, the tool path for each part is prepared using a tool location which remains constant. Instant change from one part to another can be maintained if this practice is followed. Cutting productivity has gone from 40% to as much as 95% when these principles are followed.

Care must be taken to develop families of parts correctly in cooperation with engineering. Standard hole sizes, bevel, drilling and tapping, and more are required to minimize the variations of tools required in a tool package.

Design restrictions or limitations must be studied carefully also. For example, keys of varying cross-section and length must be standardized. However, the length of the key may be unimportant if the proper machine tools and cut-off procedures are designed into the cell. Cross-section standardization may be the only variable which should be maintained. Any length of key can be economical.

All parts and all part families are not conducive to group technology. A good design classification system combined with manufacturing polycodes, and an ABC inventory analysis to determine the high usage families of parts, will assist in the analysis of group technology requirements and economies.

1. *Design Retrieval*. The design retrieval system will assist to find those parts and assemblies which have similar algorithms for group technology. In some cases a series of families may be

combined. In others several manufacturing cells may be required for one family of parts.

2. *Family Process.* Family process can be used to simulate the manufacturing process, including flexible manufacturing techniques. Producibility tips, tied to the family process, will standardize tools required for an NC machine tool package.

3. *Part Family Data.* Part family data is also needed to coordinate a group technology cell including estimated part usage and lead times. Tooling available and similar tooling is important also.

4. *Process Polycodes.* Process polycodes will assist in determining which operations on diverse parts can be merged into a cell. When specific machine tools are in short supply, all parts passing over such machine tools may be expedited by grouping them. This, however, may require a polycode. It should be noted that a process or routing, properly prepared, is, in fact, a series of polycodes. Each operation on a process has a work center designation and an operation. If each operation has a standard operation number, the combination of these elements is a built-in polycode which can be used as desired. This can eliminate developing a separate, often costly, additional set of polycodes requiring maintenance.

5. *Perishable Tools.* Perishable tools might be classified in an individual system. Whether or not, a series of perishable tools, standardized by family will assist in creating logical cells. As mentioned previously, all NC tools will show increased productivity if the tool package is controlled. Economies of scale may be had by designating preferred perishable tools to minimize their inventory.

6. *Jigs and Fixtures.* Jigs and fixtures can be standardized also to fit a series of parts when manufactured in a cell. Too often new fixtures are prepared since no reference or retrieval of old ones is possible. If group technology shall work efficiently, parts should be combined so that the same ones will be usable.

7. *Engineering Changes.* Engineering changes must be examined to determine what, if any, changes must be made to a manufacturing cell or, of course, individual parts being manufactured in a cell.

8. *Production Planning.* Production planning is an important factor of group technology planning. Today, many of the parts which pass through a cell are planned on a short-range basis or as-required basis. If group technology is tied to material requirements planning, weekly requirements of all designs can be sorted by a classification system and a family of parts merged together. An economical order quantity calculation may be made, not on individual parts, but families of parts. The combination of these tools will truly enhance requirements planning.

9. *Part Geometry.* Part geometry is mandatory for preparing simulations and family processes for group technology. Variations to each design will determine the practicality of cells. Simulation

of tool paths or other features of the cell will be necessary to make certain the proper family of parts has been selected.

10. *Inventory Control.* Inventory control plays an important part in group technology planning since, in theory, less inventory will be generated. If parts are made for the exact assemblies requiring them, work in process as well as finished inventory should dramatically decrease.

11. *Group Technology Cells.* Group technology cells require careful study before their implementation. Several companies have designated a set of machine tools as a cell and then examined which parts should be processed through it. Of course, this is the reverse of logical planning. Even with careful planning, a set of machine tools placed in a cell may be incorrect. One of the best methods to determine whether the cell is correct is to first study the application, make the decision on which tools should be grouped or combined, and then simulate the activity by tagging the machines, limiting their use to the cell concept, but not moving them physically until the simulation proves to be correct. Implementing group technology is a dramatic example of how an integrated system, or CIM, is required to optimize engineering and manufacturing economies.

ABC INVENTORY

DESIGN RETRIEVAL--------------- X
FAMILY PROCESSES--------------- X
PART FAMILY DATA------------- O
PROCESS POLYCODES------------- X
ITEM MASTER--------------------- X
PERISHABLE TOOLS--------------O
PRODUCT STRUCTURES----------O
JIGS AND FIXTURES-------------O
ENGINEERING CHANGES----------X
PRODUCTION PLANNING-----------X
PART GEOMETRY--------------- O
INVENTORY CONTROL-------------X
ASSEMBLY GEOMETRY----------- O
GROUP TECHNOLOGY CELLS-----O
GENERIC BILL OF MATERIAL------X
RAW MATERIAL INVENTORY-------X
ELECTRONIC MAIL--------------O

The principle behind an ABC inventory analysis is powerful. For perhaps many unknown reasons, this analysis seems to be rarely used.

Certainly it is not publicized sufficiently. Perhaps the reason is that company after company has been so busy gathering data they have failed or had little time to develop methods to analyze the data so arduously obtained.

The ABC analysis is derived from Pareto's law. The general theory notes that any activity may have a growth curve. This growth curve increases at a certain rate which can be expressed mathematically. The ABC inventory analysis takes advantage of the phenomenon. As a very general statement, only 5% of all design usage represents 80% of all dollars expended. Thirty percent of all design usage represents an additional 15%, while the remaining 65% of all design usage represents only 5% of the dollars expended. This relationship is quite useful for a variety of actions.

Since a small number of designs represents a major dollar expenditure, these parts, in inventory, can be controlled more closely than others. On the other hand, the vast majority of parts upon which only 5% of the dollars are expended may become loosely controlled, even floor stock items. High value items are classified A, the next volume B, and the lowest usage C items—thus the ABC inventory designation.

A items are normally purchased or manufactured discretely, B items with some form of economic order quantity, and C items once or twice yearly using a two-bin replenishing system.

In the past and currently many companies take a physical inventory on a yearly basis. This procedure is highly inaccurate and costly because unskilled persons are usually designated the counters, and the plant usually closes during the procedure. It may well be that more errors are introduced into the system than were there before inventory was taken.

A more practical method for auditing inventory is through cycle counting. Here the ABC inventory approach produces dividends. For example, the A class items may be counted every 60 days, the B items every 90 days, and the C items twice yearly. Emphasis is thus placed on the more costly items with a corresponding reduction of labor requirements. As a bonus, this method does not require a plant to close to take the yearly inventory.

The procedures just described are classical and are used by many companies. Additional benefits of this analysis can be made in an integrated system.

1. *Design Retrieval.* A design retrieval system can be used to extract ABC items into families of parts and then take each family to create a family ABC analysis. This provides a means to determine which family of parts should be modernized, made into a cell for group technology, or otherwise examined.

2. *Family Processes.* Since all family processes cannot be prepared at once, and since many of them might not be economical, they can be selected discriminately using the ABC analysis.

3. *Process Polycodes.* If process polycodes are examined in conjunction with the analysis, a machine tool set of requirements, including plant capacities, can be generated.

4. *Item Master.* Item master data may be used in conjunction with the family ABC analysis. For example, are costs similar between parts in a family, do they have similar lead times, and is there a mixture of make and buy items and why?

5. *Engineering Changes.* Engineering changes should not be made carelessly on A items. Since these represent the highest dollar volume and, potentially, the largest dollars in inventory, care should be taken to examine the status of any part before a change is initiated.

6. *Production Planning.* Production planning can be used to create an important report to analyze future production. If a sales forecast or the master schedule is used to generate requirements of all designs for one year, as an example, into the future, and these requirements are totalled by part number then sorted from high to low value usage, and ABC net requirements analysis is generated. If, in addition, all design on order or in inventory are subtracted from this requirement, a descending dollar value of new requirements or excess inventory can be generated.

7. *Inventory Control.* The inventory control records can be audited using the cycle inventory count method based on the ABC inventory analysis. The highest used parts may also be stored in the bins or other locations easiest to reach.

8. *Generic Bill of Materials.* The generic bill of materials for products in a job shop can be used to predict possible ABC analysis by family.

9. *Raw Material Inventory.* Raw material inventory can be forecasted as a result of the ABC inventory analysis of manufactured parts. As another special case, all manufactured parts can be extracted from the requirements planning forecast report, sorted into the ABC analysis format, and the results used to determine tooling, economical order quantities, and more.

PURCHASE ORDERS

DESIGN RETRIEVAL--------------- X
FAMILY PROCESSES--------------- X
PART FAMILY DATA------------- O
PROCESS POLYCODES------------- X
ITEM MASTER--------------------- X
PERISHABLE TOOLS--------------O
PRODUCT STRUCTURES----------O
JIGS AND FIXTURES------------- O
ENGINEERING CHANGES----------- X
PRODUCTION PLANNING-----------X
PART GEOMETRY---------------- O
INVENTORY CONTROL-------------X
ASSEMBLY GEOMETRY----------- O
GROUP TECHNOLOGY CELLS----- O
GENERIC BILL OF MATERIAL------X
RAW MATERIAL INVENTORY-------X
ELECTRONIC MAIL--------------- O

In many companies more than 50% of all designs are purchased. In addition, all raw materials are purchased from an outside vendor, an many designed parts and assemblies are also purchased on a subcontract basis. For some remote reason, purchase order status for these items is often modernized only after all other systems are in place. With this amount of working capital tied up in this process, a good deal more time and effort should be placed on purchasing systems.

MRP has been of great assistance to this process, but on-line systems, combined with an integrated on-line receiving system, create an even better environment for good purchasing practices. Use of a classification system allows all purchased parts to be segregated by buyers and vendors so cost reductions through similar parts and through familiarity of these parts by buyer can be achieved.

1. *Design Retrieval*. A design retrieval system can be used as a buyer's code, a cost comparison analysis, and a vendor analysis. Parts from a particularly costly family of parts can be displayed and competitive vendors allowed to view them to make bids on each and all parts in the family.

2. *Family Processes*. Since all family processes are for manufactured parts, it may not be obvious how these will help. However, the make-buy decision of any part family could determine whether a uniform make or buy decision should be made. If a family of parts is particularly expensive to manufacture, quotations from an outside vendor can be used to compare for a make-buy analysis.

3. *Process Polycodes*. If process polycodes are used to analyze manufactured parts in conjunction with the family processes, they will assist in the make-buy decision.

4. *Item Master.* Item master data may be used in many ways by purchasing. Standard or actual cost comparisons, lead time, tool requirements for vendors, forging and casting part numbers, and more will help in purchased part analysis.

5. *Engineering Changes.* Engineering changes should not be made idly on any purchased part which falls into the A inventory classification, as described previously. In addition, many purchased parts may be based on a yearly or open contract. Forgings and castings are often expensive to change. These variables should be examined carefully when an engineering change is initiated.

6. *Production Planning.* Production planning can be greatly affected by on time delivery of purchased parts. Consequently vendor evaluation of on-time deliveries, quality, and financial status will assist in helping to create high quality vendors and parts. Using material requirements as a base, total dollars committed by vendor and total cash flow weekly or monthly can be forecast. For a company that purchases a significant volume of forgings or castings, it should be known that the volume discounts include set-up time. If the price breaks do not meet production requirements, a costly excess inventory may be created to make use of the discounts. By requesting a new quote which segregates the set-up cost from the cost per item, any quantity of parts may be ordered with an optimum price. One company that requested this information from a large forging supplier found a surprsing amount of savings in the practice.

7. *Inventory Control.* The inventory control records of all purchased parts should be controlled by the ABC inventory analysis. For JIT manufacture, it may be of significant economic benefit for the vendor to store the parts required for instant delivery. In some high volume operations, it has proven economical for the vendor to locate his manufacturing plant near his customers.

8. *Generic Bill of Materials.* The generic bill of materials can be used to simulate the quantity of purchased parts and the costs to assist in determining the type of vendors and the order quantities needed for purchased parts.

9. *Raw Material Inventory.* If raw material inventory items can be standardized and their quantities forecast, mill size orders can be placed, rather than purchasing raw materials from warehouses, or even purchasing just enough raw material to make one or two parts. This practice adds considerable cost to a product. It should be emphasized that mill runs of a raw material, particularly plate or sheet, may be ordered in any width and with any length. The width extras associated with such orders make ordering 60 in. width plate highly attractive. If enough volume of the same or parts allows for a mill purchase, the specific width of plate or sheet will eliminate costly drop-offs. One company practicing this more scientific method for purchasing plate reduced plate sizes 50% and reduced total raw material cost by 10%.

MATERIAL REQUIREMENTS

DESIGN RETRIEVAL--------------O
FAMILY PROCESSES--------------- X
PART FAMILY DATA---------------X
PROCESS POLYCODES------------O
ITEM MASTER--------------------- X
PERISHABLE TOOLS--------------O
PRODUCT STRUCTURES----------- X
JIGS AND FIXTURES-------------O
ENGINEERING CHANGES----------- X
PRODUCTION PLANNING-----------X
PART GEOMETRY---------------- O
INVENTORY CONTROL-------------X
ASSEMBLY GEOMETRY----------- O
GROUP TECHNOLOGY CELLS-------X
GENERIC BILL OF MATERIAL------X
RAW MATERIAL INVENTORY-------X
ELECTRONIC MAIL--------------- O

The definition of material requirements planning (MRP) comes in all shapes, colors, and sizes. To some it is a discipline. To others it is a database, a procedure, or a set of programs. Regardless of this, a general description could be made that it is a material logistics procedure.

MRP is not new. LeTourneau (later known as LeTourneau Westinghouse), implemented MRP in the early 1940s with punched card data. The aircraft industry created special programs called top-down, push-down to create a low-level code for each design usage. (This will be discussed later.)

With the advent of Ramac, an IBM company random access devise, IBM developed several canned programs to implement MRP. Some of these products had acronyms such as IPICS, COPICS, and MAPICS. Since then, an entire industry has developed to create user-oriented software for manufacturing systems, the center focal point being MRP logic.

The principle behind MRP is simple. A bill of material is used to generate requirements of all parts and assemblies by multiplying a sales forecast which has been converted into a master schedule by the corresponding bill of materials. As the requirements are generated, the level of each of them in the bill of material is determined and then backed off by the lead time of each design. This backward loading requirement is characteristic of MRP. All requirements are placed into some time frame, such as months, weeks, days, or even hours for JIT manufacture.

Originally, MRP was computed and regenerated weekly or monthly. This regeneration process was time consuming and inadequate, particularly in a large company, since, by the time the report was analyzed and notes made in the margin, a new report appeared. In addition, no chain from a part requirement to the assembly for which it was needed was tabulated. Thus, if the bill of material was inaccurate, incorrect requirements were prepared.

These problems have been alleviated to some degree by using net change and pegged requirements.

Net change generates new requirements for parts and assemblies only for those products or items on the master schedule which have changed since the past computation. To perform this action, gross requirements must be stored for all the active items. As the new master schedule is used, the results are compared to the old. Only those requiring changes (usually within lead time of the item) are reported. This concept decreases the computer time and computation radically unless the product and the business climate are such that the entire master schedule changes. In that case no real advantage is found.

Pegged requirements simply means the tabulation of item requirements listed by the next level requirement from which it comes. Thus, if the same design is used in several assemblies, all these separate requirements are tabulated in the report and then summarized.

A means to establish the rating of a company in its progress to use MRP has been prepared. A class A user has a bill-of-material accuracy of at least 95% and an inventory accuracy in the same magnitude. An entire list of criteria for becoming a class A user exists.

Implementation of MRP had, and has, a traumatic effect of manufacturing companies. Prior to MRP, order point control of inventories for selected parts was used, and forward loading of material requirements was used, based on the principle that everything should be ordered as soon as possible then expedited. Order point control for the selected items relied on past usage to generate a forecast of future requirements.

Implementation of MRP called for instant new disciplines in finished parts and goods inventory, work in process, and bills of material. New scheduling methods were required for shop loading. This set of large changes simply was not absorbed by many companies. The amount of education, training, and the need for cooperative management was not recognized. Yet, every requirement necessary to implement MRP successfully is necessary to an even greater degree for CIM implementation success.

The original MRP systems were concerned with material logistics only. No consideration was given to shop loading or vendor capacity. Today, feedback from MRP is integrated in some degree with production scheduling and shop floor control. This new configuration is called MRPII.

Before random access storage was devised, MRP operated on sequential bills of material. Since total requirements of identical parts require not only summarization of parts in all bills of materials but also analysis of inventory on a level-by-level basis and the effect of lead times on the date the designs are required. At first these computations were performed in two stages; the first stage created gross requirements for each part is every usage by lead time. These requirements were then sorted and merged together. (No wonder 12 tape drives looking very busy, were required!!)

With random access storage, all these activities were performed at the same time using a low level code. The low level codes, attached to each part and assembly, simply sorted requirements by used-on, but also to the lowest level on which each design appeared. Since the net requirements of one level are the net requirements of the next lower level (backed off by lead time), total requirement, call net requirements, could be computed in one stage.

1. *Family Processes*. The family processes become the source of data to compute manufactured part lead time. While each individual part has its own lead time, the family lead time can be used as an estimate until the design has actually been manufactured.

2. *Part Family Data*. Part family data includes purchased part lead time, tooling requirements, and other data which may be analyzed as a result of generating MRP.

3. *Item Master*. The item master data includes the classification code of the design which can be used to sort part requirements into family part requirements for family planning. The division between make and buy can be determined by a data increment in the item master. All lead times, costs, and inventory status are in the item master. The type of economical order quantity and the purchasing discounts are also available.

4. *Product Structures*. Product structures are used to explode the bill of material on a level-by-level basis to create the data for MRP.

5. *Engineering Changes*. Engineering changes have a direct effect of material requirements since "effectivity" of change is incorporated in the system. For example, a design change may substitute a new part for an old, but the assembly upon which both are used is interchangeable. Only the revision level of the assembly changes. However, there is still a large supply of the old part available, and the change to the new design is not mandatory. With these conditions, the new part will be used on an "effective" date equal to that when the old part has been completely used up. This date is normally determined manually, but some more sophisticated systems calculate it for the user. In any case, the use of effectivity can save money by exhausting obsolete but useable inventory before initiating new designs.

6. *Production Planning*. Production planning was considered a separate function from MRP. A manufacturing facility was loaded to

infinite capacity, and supervisors manually changed shop loads. Today a greater attempt is being made to integrate material requirements with shop capacity, hence MRPII. No matter how well this is performed, conflict will exist. No company can afford machine tools sitting idle waiting for temporary overloads. Companies will always try to meet the schedule requirements of the customer. Those foreign companies who profess to have solved this problem do not always advertise that they have made some of the variables into constants, production volumes, for example. Some algorithms do exist, which eventually take the form of a triangle matrix, that seem to optimize requirements and production. For some relatively simple conditions, linear programming techniques can solve some of the paradoxes. However, this also presumes that the data used is accurate.

One of the key factors in production scheduling is queue time. Queue time is a combination of three factors, the wait time before the next operation is performed, the wait time after the operation is performed, and the transport time from one operation to another. These intervals have rarely been studied scientifically, and have so many conditions causing variations that they seem meaningless. In addition, variations of absenteesism, machine failure, material shortage, scrap, and more all contribute to the chaos of production. Perhaps we will never find an ideal calculating method. We are, however, generating more and more data which will assist in the management of the art called production scheduling. Shop floor control is used to schedule short-range production scheduling.

7. *Inventory Control*. Inventory control maintains the accuracy of parts on hand and on order which are subtracted from total requirements in the MRP calculation. Obviously, inaccurate inventories will create either false part shortages or surpluses. It must be remembered that "stock" and "nonstock" inventories no longer exist. All completed parts and assemblies and all received purchased parts are considered inventory. When parts are taken from inventory using a picking list, the part inventory is reduced, but the work-in-process inventory is increased. These changes used to be transacted in dollars only. Today these transactions are in quantities of parts as well as dollars, and the two must track and be equal.

8. *Group Technology Cells*. Group technology cells can be enhanced with MRP. Once total requirements have been computed on a weekly basis, say, all these requirements may be sorted into an additional array, by family. This consolidated information can become the basis for loading one or more cells for manufacturing with group technology.

9. *Generic Bill of Material*. The generic bill of material, as mentioned previously, can be used alone or intermixed with real products or orders to create material requirements of all parts which have been released to the generic bill of material. It should be

remembered that, in a job shop atmosphere, a new customer order is usually designed from bottom to top, with the final assembly part numbers unknown until, maybe, the product is shipped. If the generic bill of material is used, these unknown assemblies will have phoney part numbers and estimated lead times, but the data can be manipulated in a manner similar to actual bills of material. This takes advantage of the techniques available in MRP logistical systems.

10. *Raw Material Inventory*. Raw material inventory is often not made part of the bill of material. Consequently raw material requirements may be calculated manually or estimated. In either case, excessive inventory or shortages may be the result. When considering a bill of material, think of it as a pyramid. The base of the pyramid in this case is raw material. The result is that the most computations occur at the lowest level, the raw material. Shouldn't this calculation be made mechanically?

PRODUCTION SCHEDULE

DESIGN RETRIEVAL-------------- O
FAMILY PROCESSES--------------- X
PART FAMILY DATA--------------- X
PROCESS POLYCODES------------ O
ITEM MASTER--------------------- X
PERISHABLE TOOLS-------------- O
PRODUCT STRUCTURES----------- X
JIGS AND FIXTURES------------- O
ENGINEERING CHANGES----------- X
PRODUCTION PLANNING----------- X
PART GEOMETRY---------------- O
INVENTORY CONTROL------------- X
ASSEMBLY GEOMETRY----------- O
GROUP TECHNOLOGY CELLS------- X
GENERIC BILL OF MATERIAL------ X
RAW MATERIAL INVENTORY------- X
ELECTRONIC MAIL--------------- O

Production scheduling has been less modernized than most of the other CIM systems. Perhaps it is because of the many imponderables within the manufacturing environment. Poor routing and process information, machine tool capabilities, machine breakdowns, absenteeism, engineering changes, and many more contribute to the chaos and uncertainty of production schedules. Not the least of these problems stem directly from marketing. Many personnel have little or no knowledge of manufacturing and almost no knowledge of the engineering process. It has become a way of life that the delivery date of the customer is more important than any other logic. This philosophy has created all sorts of problems. In many cases production schedules are the results of the loudest voice, not the most practical. When these types of production schedule variables are allowed, they are reflected in the cost of the products, which creates either less customers or customer who can force unrealistic delivery dates on the company. Thus an inverse spiral continues to erode the economics of logical production scheduling.

Obviously, all companies do not face this unrealistic marketing problem. They do face, however, all the others. It is significant that production schedules seem to be more on time when robots, group technology cells, and other modern techniques as the automated factory are implemented. In some cases the advent of these techniques creates some disadvantages, requiring less product variation, and a long-range plan of standardization of machining, assembly, and design practices. When a manufacturing company decides to solve these problems over a long-range development rather than continually attempting to run by short-range profits, the problems will be much more solvable than they are now.

The production scheduling techniques today are really an extension of MRP. If each operation on a manufactured part or assembly is considered, like a part, and the queue and run time for each like lead time, the production schedule shows when to start each operation at each work center which must perform some work on it. To minimize the amount of data requiring storage, a detailed production schedule may be prepared for one month only, for example. All other future months may be summarized to show potential work center overloads on a gross basis.

As mentioned previously, these loads are considered infinite, and only in the short-range planning area are they manipulated, usually manually to level production.

To perform finite loading of a manufacturing plant is very complex. To be optimum, string loading must be considered. For example, the shop load might identify a work center which is overloaded. Moving some parts and their corresponding operations from this work center or changing the completion date may ease this overload. However, previous or subsequent operations from the critical one should be moved also. Otherwise the moved operation will be performed only for the part to remain as work in process overlong. If all operations are thus moved in a string, all other parts in the assemblies for which the part is used should be changed also. Of course this set of movements, their analysis, and the new overloads which such movements might create, and the new overloads which such movements might create will cause a ridiculous (today) amount of calculations with a questionable amount of economic benefits. The problem we are faced with today is to determine at what level of accuracy and economics we should stop the calculation, and which variable it is more important to consider to make accurate.

The production data portrays work loads per work center per operation, and per shift. Included in this listing may be each job priority. Job priorities are often designed with indistinct objectives. For example, the ratio of time remaining before the order is late to the total time required to manufacture the order is called a critical ratio. This does not take into account the difference between a part with one day remaining of a two day total time from a part with two weeks remaining of a four week total time.

The foreman or shop supervisor is usually the recipient of these schedules. Some leeway is given each foreman so he or she can group similar parts as desired to take advantage of reduction of set-up times. Good foremen perform this analysis. However, only local and short-range scheduling optimization can occur.

1. *Family Processes*. Family processes can be used to combine similar parts and manufactured to the production schedule. If the production takes into account family processes, an even more economical or optimum schedule may result.

2. *Part Family Data*. Part family data, in conjunction with a family process, can be used to load the shop in an optimum manner but also predict the machine tools and other work centers necessary to make production scheduling better.

3. *Item Master*. Item master data, such as lead time and economical order quantities go into the calculation of the production schedule.

4. *Product Structure*. Product structures, which are used to create the MRP requirements, use the process and routing data as an extension to create the production schedule.

5. *Engineering Changes*. Engineering changes may require that all parts in process subject to the engineering change be scrapped or reworked. Some changes may make designs obsolete, requiring new parts. Many variables can be created by an engineering change, requiring careful planning in production.

6. *Production Planning*. Production planning is a function which develops the methods and time spans to manufacture parts and assemblies. In the past, a "line of balance" technique was often used. This combined material required with the operation steps needed to create a product. As a manual system it was quite efficient. It seemed to work best in a job shop atmosphere and particularly on products whose main features did not move but to which parts were attached, such as a boiler, an air frame, or a large overhead crane. PERT scheduling and CPM were techniques also devised that resembled line of balance. However, these new techniques were computer oriented and found "critical paths" that

would call attention to those activities which, if not performed on time, would make an entire project late.

Large production boards were also used for a visual schedule. Each main assembly was represented in various colors as a series lines showing start and completion dates. Of course, schedule changes were hard to make, and reporting actual to the original schedule relied on the skills of foremen from past experience. Today, even with all the variables and problems yet unsolved, the production schedules coordinated with MRP develop a cogent and fairly accurate set of schedules and requirements which can be compared to actual results as they occur. Feedback of data is more rapid and accurate for making corrections before catastrophic events can occur.

7. *Inventory Control*. The accuracy of inventory control has become even more important to make certain the production schedule does not show a requirement which cannot be filled in time. Today assembly pick lists are compared by the computer to the on-hand inventory. If parts are short, the assembly order may not be issued. Back orders must be filled in a timely and accurate manner, and expediting of purchased parts is made based on the inventory reports. Thus accuracy, or lack thereof, affects many departments and persons.

8. *Group Technology Cells*. Group technology cells make it somewhat easier to schedule manufacturing. Since the grouping of parts and the tools and set-ups have been predetermined, all one must do is issue the design requirements to the cells and let the preplan take effect. If enough designs were made in this manner, the only problem which might be faced is capacity, certainly not the sequence of manufacture or the sequence of parts in the queue.

9. *Generic Bill of Material*. A generic bill of material can be used to preplan the shop load for a new order long before it has been completely designed. This will help to make certain that no surpises are found later when the design is completed. It also can be considered a means of reserving production capacity to alleviate the strain of rescheduling at short notice.

10. *Raw Material Inventory*. Raw material inventory must be accurate to make certain the production schedule is met. After all, the majority of all in-house designs are piece parts, and completing them on time is based on the availability of their raw material. Accuracy and preplanning for raw material is thus an important requirement.

SHOP FLOOR CONTROL

DESIGN RETRIEVAL--------------O
FAMILY PROCESSES--------------- X
PART FAMILY DATA---------------X
PROCESS POLYCODES------------O
ITEM MASTER--------------------- X
PERISHABLE TOOLS--------------O
PRODUCT STRUCTURES----------- X
JIGS AND FIXTURES-------------O
ENGINEERING CHANGES----------- X
PRODUCTION PLANNING-----------X
PART GEOMETRY---------------- O
INVENTORY CONTROL-------------X
ASSEMBLY GEOMETRY----------- O
GROUP TECHNOLOGY CELLS-------X
GENERIC BILL OF MATERIAL------X
RAW MATERIAL INVENTORY-------X
ELECTRONIC MAIL--------------- O

Someone somewhere in manufacturing a few years ago initiated a verbal policy stating, "never issue to the shop more work than they can absorb in a week." One might say this was the beginning of shop floor control. Shop floor control is a metering device to make certain the manufacturing facilities are loaded but not so much that cavitation sets in. It is surprising how many companies do not follow this procedure. To many, the faster the work orders are issued, th faster the work will be performed. Certainly, the pile of work in front of a machine operator assures him or her of work for tomorrow. This pile, however, may represent routings for parts which could be combined, unnecessary value added to costly raw materials, and excessive work in process generally.

One gear manufacturing company was surveyed, and the results showed that, had all part orders which had been issued been combined in simply identical part combinations, three months of work load representing excess set-up time could have been eliminated.

One of the elements of shop floor control is a short-range production schedule. As weekly requirements are developed, the proper paperwork is issued in the form of shop packets. This packet may contain a routing or process which can double as the work order, a picking list for assemblies, and the engineering drawings. In some cases, punched cards, optical scanning, or bar code paperwork accompanies the packet as move tickets and/or labor reporting instruments.

Since mechanized priority rules are often vague or self-defeating, the shop floor control personnel can manually set priorities for issuing the proper work to the shop.

1. *Family Processes*. Family processes, tied to the coding structure and shop floor control reports, will assist in issuing work orders for similar parts which were not subject to a group technology cell but could be economically manufactured at the same time.

2. *Part Family Data*. Part family data is used in a like manner as in item 1. In addition, the shop floor control personnel must recognize and plan the load on a group technology cell.

3. *Item Master*. Item master data is used to prepare the shop floor control data.

4. *Product Structures*. Product structures are converted into picking lists of assemblies for the shop packet. Coordinated by effectivity, different revision levels of assemblies may be made each with difference parts. Once the data of effectivity is reached, all later picking lists will have a new configuration. In this manner, obsolete, replaced, parts may be used up.

5. *Engineering Changes*. Engineering changes are reflected in the picking lists, but also on the revision level of the routings and process sheets. Thus the latest revision, based on effectivity, should be made into the shop packet.

6. *Production Planning*. Production planning generates the short-range plan, which in turn is used by shop floor personnel to generate and release the proper paperwork.

7. *Inventory Control*. Inventory control data is prechecked in many cases so that the issuance of the work order for an assembly will not suddenly show part shortages. This preposting of inventory assists in expediting lagging part and assembly orders.

8. *Group Technology Cells*. Group technology cells must be integrated into the shop floor control system. The result may be simply a list of part requirements and a group of drawings sent as a group to the work cell, allowing the operators to prepare their own shop floor control.

9. *Generic Bill of Materials*. Generic bill of material is not too effective in shop floor control, which is a shortrange requirement schedule. However, if, suddenly, a part from a generic bill of material with a phoney part number appears on the shop floor control schedule, notice of past due design and/or manufacturing documentation will be sudden.

10. *Raw Material Inventory*. Raw material inventory must be coordinated with shop floor control data so part orders are not issued prematurely, and, if raw material availability is limited.

INVENTORY CONTROL

DESIGN RETRIEVAL--------------O
FAMILY PROCESSES---------------X
PART FAMILY DATA---------------X
PROCESS POLYCODES------------O
ITEM MASTER---------------------X
PERISHABLE TOOLS--------------O
PRODUCT STRUCTURES-----------X
JIGS AND FIXTURES-------------O
ENGINEERING CHANGES-----------X
PRODUCTION PLANNING-----------X
PART GEOMETRY----------------O
INVENTORY CONTROL-------------X
ASSEMBLY GEOMETRY-----------O
GROUP TECHNOLOGY CELLS-------X
GENERIC BILL OF MATERIAL------X
RAW MATERIAL INVENTORY-------X
ELECTRONIC MAIL---------------O

Inventory controls systems were some of the first to get attention in manufacturing. While control of inventory took many forms, the most common was in the utilization of a Kardex system.

Manilla-colored cards with columns to post "on order," "received," and "on hand" were combined with a title which held the part description, sometimes the part number, and the vendor name address. A reorder point and reorder quantity were also tabulated. When, through normal removal from inventory, the part quantity balance reached the reorder point, the amount to be ordered was posted and the entire card sent to purchasing. This procedure was called a traveling requisition.

All sorts of miscellaneous remarks, comments, and instructions were posted to these cards. A red mark meant, "don't order regardless of requirements." A blue mark meant, "has an alternate part number." Often a phone number, an alternate vendor, and an alternate planner were attached. Yet this document was mechanized. The horrible screams of the inventory planner who lost all his history, his notes, and his baby blanket still permeate the air over the United States. In fact, chaos, poor records, and high inaccuracy became the result of mechanization because of little training, misunderstood new procedures, lack of integration into other old systems, and computers which were too small. This should have served as a warning for the future. It didn't.

The Kardex system, once converted to the computer, still was mainly an order point control system. This method simply made note of part usage; it did not document where such usage was generated. As the usage statistics were compiled, they became the source of generating future requirements, a highly inaccurate form of forecasting. Another analysis was that of inventory turns, or the number of times designs, usually by some type of category, were replaced yearly. Each industry had a standard. This measurement was, in reality, only a means to judge how cleverly the inventory personnel had forecasted future usage of designs.

Once material requirements planning was implemented, the requirements of all parts and assemblies were developed as a result of a master schedule for products, orders, and spare part requirements. In conjunction with new style economic order quantity methods such as part period balancing, or with the use of family planning for group technology, the optimum inventory and the inventory turns were generated by the system, not some preestablished inventory turn criteria. This fact was not often recognized by management creating conflict in the amount of inventory suggested by MRP, and the dollars of inventory authorized.

Inventory control systems still are not accurate. People still take parts when they want them without posting the withdrawal. The means to store parts, and a closed inventory area, are still but dreams in some companies. Part of this inaccuracy is due in no small measure to not educating manufacturing personnel. Accurate inventory does require red tape. To the worker, this is unnecessary red tape. Red-blooded American workers destroy red tape whenever possible.

1. *Family Processes*. Family processes assist in creating the economical order quantities for each part in an orderly manner, thus determining the inventory quantities.

2. *Part Family Data*. Part family data, including the actual family code, may be used to store similar parts near one another. It becomes a buyers' code also.

3. *Item Master*. The item master is the repository of data such as cost, ABC inventory status, and lead time, which are used in computation of inventory in MRP.

4. *Product Structures*. Product structures are used in bill-of-material explosions within MRP to generate total requirements of all parts and assemblies. It should be noted that, previously, all parts requirements developed with order point control were developed individually. Using MRP, all part requirements are the direct result of a master schedule and the corresponding product bill of material. Thus requirements may be increased or decreased more rapidly and with a more coordinated set of part and assembly requirements.

5. *Engineering Changes*. Engineering changes are vital to maintaining inventory without excessive scrap or rework. In many cases an engineering change may require these activities. It is mandatory to analyze inventory prior to implementing an engineering change.

6. *Production Planning*. Production planning requires an accurate inventory. MRP is only as good a system as all of its components. Inventory is a significant factor.

7. *Inventory Control*. Inventory control is a means of evaluating capital investment of a manufacturing firm in the performance of creating finished products. In a broad sense, all buildings and capital equipment represent inventory, and the return on investment

is based on all inventory. While it would be nice to believe that, evntually, no inventory will be required once CIM is completely implemented, the fact is, logistics are simply not that concise nor ideal in most industries. Even simple products are not shipped the instant they are completed. Perhaps a telegram is the only product instantly created and shipped, but these may wait in a queue at the receiving end. Thus it behooves us to make inventory systems, efficient, simple, user oriented, and cost effective.

8. *Group Technology Cells*. Group technology cells will, in some cases, be designed to manufacture parts in a sequence such that a completed sequence of parts may be assembled. Thus assembly production planning will eliminate considerable work in process. In addition, the value added to raw material will be decreased by eliminating set-up time. This will be reflected in the cost of parts and their corresponding inventory costs.

9. *Generic Bill of Material*. The generic bill of material will generate requirements for parts in a job order system even when the entire bill of material hasn't been prepared. Thus parts may be ordered and placed into inventory in the proper sequence even as the total design is being documented.

10. *Raw Material Inventory*. Raw material inventory should be based on MRP. On the other hand, the wide varieties of raw materials should be analyzed and controlled so mill production can be purchased. All raw materials must have part numbers, even those purchase to the exact size of the part. If the latter is not performed, no record of expensive material acquisitions will be available to replace them with material which can be used in volume.

PURCHASING DATA

DESIGN RETRIEVAL--------------O
FAMILY PROCESSES---------------X
PART FAMILY DATA---------------X
PROCESS POLYCODES------------O
ITEM MASTER---------------------X
PERISHABLE TOOLS--------------O
PRODUCT STRUCTURES-----------X
JIGS AND FIXTURES-------------O
ENGINEERING CHANGES-----------X
PRODUCTION PLANNING-----------X
PART GEOMETRY----------------O
INVENTORY CONTROL-------------X
ASSEMBLY GEOMETRY-----------O
GROUP TECHNOLOGY CELLS-------X
GENERIC BILL OF MATERIAL------X
RAW MATERIAL INVENTORY-------X
ELECTRONIC MAIL---------------O

An on-line, integrated purchasing system should reduce the redundant initiation of data often required to write a purchase order. If MRP generates requirements, and also calculates quantities by time, many purchase orders could, and should, be generated automatically. Those which cannot should have on-line menus which have all item master data as well as vendor data automatically generated. With proper design, purchasing agents can perform with a greatly reduced amount of clerical effort.

Once the purchase orders are prepared, the on-order status of the part should become an immediate part of the inventory record for instant receiving authorization, if required. The same data should be available at receiving and inspection, and the receiving notice should be written automatically once the material is received.

Within the system, limits on receiving overages should be established so that the receiving department cannot receive excess inventory.

1. *Family Processes.* Family processes are not a significant factor in this system.

2. *Part Family Data.* Part family data may be used to estimate lead time, cost, and potential vendors for new items.

3. *Item Master.* The item master represents the description, lead time, vendor code, last cost, standard cost, and other data needed either to prepare a purchase order automatically, or to generate one manually rapidly.

4. *Product Structures.* The product structures are used to generate MRP and thus the time frame in which purchased parts and raw materials are required.

5. *Engineering Changes.* Engineering changes have a significant effect on purchased parts. In this case, engineering changes may come from the vendor, requiring design analysis of part usage. Purchased parts may be made obsolete by a vendor. Specifications for purchased parts may be changed enough that current vendors cannot match the engineering requirements. If the specifications are not written clearly and concisely, only one vendor may be able to meet the design requirements. This limitation should be eliminated by good engineering documentation and searches by purchasing wherever possible.

6. *Production Planning.* The production planning planning process is really the coordination of four different schedules, assembly operations, manufactured parts, purchased parts, and raw materials. Their integration is vital for concise scheduling and economics of manufacturing. Thus the coordination of MRP and production scheduling is critical.

7. *Inventory Control.* Inventory control accuracy is necessary so that unnecessary parts are not ordered. Since the accuracy of MRP is based not only on parts on hand, but also parts on order,

vendor delivery dates need monitoring and evaluation. Of course, many parts are so bulky or heavy that they are not, literally, moved and stored in inventory. These special cases must be planned.

8. *Group Technology Cells*. Group technology cells may create a new meaning to inventory. Several parts may completed in one batch, for example. Many parts may be made and then assembled, bypassing inventory reporting of the parts themselves.

9. *Generic Bill of Material*. A generic bill of material can generate potential volumes of purchased parts which will be required, including lead times and costs. This may help forecast cash flow, and be particularly important when large quantities of new products and their purchased parts are being considered.

10. *Raw Material Inventory*. Raw material inventory will be better maintained and replenished with part numbers for each raw material, and an attempt to standardize them for mill supply quantities.

COSTED BILL OF MATERIAL

DESIGN RETRIEVAL-------------- O
FAMILY PROCESSES---------------X
PART FAMILY DATA---------------X
PROCESS POLYCODES------------O
ITEM MASTER--------------------- X
PERISHABLE TOOLS--------------O
PRODUCT STRUCTURES----------- X
JIGS AND FIXTURES-------------O
ENGINEERING CHANGES----------- X
PRODUCTION PLANNING-----------X
PART GEOMETRY---------------- O
INVENTORY CONTROL-------------X
ASSEMBLY GEOMETRY----------- O
GROUP TECHNOLOGY CELLS--- X
GENERIC BILL OF MATERIAL------X
RAW MATERIAL INVENTORY-------X
ELECTRONIC MAIL--------------- O

Much has already been implied about a costed bill of material. Realistically, it is a culmination of many different systems which result in a product of order cost. This cost can be based on average costs, last costs (at least on one-of-a-kind designs), or standard costs. In any case, this culmination of data also eliminates the need for a job order costing system which created an accounting-summarized bill of material usable in accounting only. The bill of material allows a more flexible manufacturing system since parts within one order or between orders can be combined for economical manufacturing, not segregated artificially, so the job costing system can be utilized.

1. *Family Processes.* Family processes may be used to obtain an estimated cost for all parts and assemblies for a new customer order, even a new design

2. *Part Family Data.* Part family data, including purchased cost data, for new designs can be used in conjunction with family process data to create an estimated costed bill of material.

3. *Item Master.* Item master data contains material and labor costs for all designs used previously.

4. *Product Structures.* The product structures are used in the cost roll-up procedure to create a costed bill of material. It should be noted that total costs for assemblies are usually not made a part of the database, only labor to assemble. The cost roll-up adds material and labor on a level-by-level basis, including the labor costs for assemblies. If total cost were stored at the assembly level, any adjustment of cost at a lower level would create a requirement for changing all higher level costs, an unnecessary and often costly procedure.

5. *Engineering Changes.* Engineering changes must be carefully reviewed, since all of them could have a significant effect on product costs. In addition, the cost of an engineering change should be estimated before implementing it, since the cost of change may exceed the economic benefits.

6. *Production Planning.* Production planning creates, through the use of an economical order quantity and labor reporting, part and assembly costs.

7. *Inventory Control.* Inventory control records, including part and assembly costs, are used as the database for computing a costed bill of material.

8. *Group Technology Cells.* Group technology cells may create parts far more economically than past practices creating a significantly different costed bill of material for a product.

9. *Generic Bill of Material.* The generic bill of material for a new product may be used to create a costed bill of material. This requires that each design in the product be classified and that the part family have an estimated cost. Even high and low estimated costs will allow comparisons and estimates of value before the product is made.

10. *Raw Material Inventory.* Raw material inventory is a significant part of a product. Thus chaining raw material requirements to the parts which use them makes the creation of a costed bill of material easier, rapid, and more accurate.

QUALITY ASSURANCE

DESIGN RETRIEVAL--------------O
FAMILY PROCESSES---------------X
PART FAMILY DATA---------------X
PROCESS POLYCODES------------O
ITEM MASTER--------------------X
PERISHABLE TOOLS--------------O
PRODUCT STRUCTURES-----------X
JIGS AND FIXTURES-------------O
ENGINEERING CHANGES-----------X
PRODUCTION PLANNING-----------X
PART GEOMETRY----------------O
INVENTORY CONTROL-------------X
ASSEMBLY GEOMETRY-----------O
GROUP TECHNOLOGY CELLS-------X
GENERIC BILL OF MATERIAL------X
RAW MATERIAL INVENTORY-------X
ELECTRONIC MAIL---------------O

We often overlook the benefits to quality that an integrated system will have. For example, a family of similar parts may have many different sets of tolerances. Analysis may find this to be unnecessary. Once a standard is established, proper inspection fixtures, instructions, quality checks of critical dimensions, and more can be maintained. Credibility of inspection is also increased, since specific and important engineering requirements are addressed.

Design analysis emphasizes the parts of the design which might be critical. Such data can become part of the inspection data. The CAD/CAM system develops geometry of complex parts, which, often, inspectors must recompute. This data can now be made available to inspection and avoids conflicts when the recomputed data does not meet the same dimensions of the original.

More and more, automatic inspection equipment will be developed. Use of lasers to measure as well as physical fixtures will improve. Much impetus to this improvement will come once the geometry of a design has been established in the computer and its accuracy is established. Inspection, so necessary to quality and corresponding success in the market place, may be increased with little or no corresponding cost. The future for better inspection technology and the resultant higher quality products is significant once CIM has been implemented.

1. *Family Processes*. Family processes will be used to standardize inspection procedures and obtain a consensus on critical dimensions and operations.

2. *Part Family Data*. Part family data will be used in conjunction with family processes to determine automatic inspection devices which could be created.

3. *Item Master*. Item master data will be used to form families of parts for a coordinated inspection procedure.

4. *Product Structures*. Product structures, once established by engineering and manufacturing, will be the standard so that quality assurance can forecast not only the assemblies which will be made, but the procedures by which they will be tested or the actions with which to simulate their function.

5. *Engineering Changes*. Engineering changes are an important part of the quality assurance process to make certain such changes are incorporated in the inspection procedures. Engineering changes may be initiated which make it impossible to inspect finished parts, make the design dangerous, or are unsuitable for the environment in which the design is used. These items become part of the quality assurance program to keep designs high in quality and operable at a reasonable cost.

6. *Production Planning*. Production planning may be changed if the sequence of operations shows quality problems. Periodic inspections as a design is processed may alleviate costly rework procedures. Quality assurance personnel should be aware of the influence operation sequencing has on quality.

7. *Inventory Control*. Inventory control may be changed due to quality of the designs. Some designs may require deliberate aging, so storage is essential. Some designs deteriorate in storage, requiring rapid use. These conditions must be noted and considered in quality assurance also.

8. *Group Technology Cells*. Group technology cells will change the inspection methods radically. For example, one cell, an NC machine tool, may manufacture critical parts with many critical dimensions. Rather than inspecting each part, the actual operation can be checked once, and then only tool wear is considered. Even this factor might be checked automatically by having a probe, for example, in the tool package. Using the probe, a statistical analysis of the machined dimension can give a warning when tools should be changed or sharpened. The philosophy of automatic inspection must be advanced to eliminate costly and time-consuming manual inspections.

9. *Generic Bill of Material*. A generic bill of material can be used to predict the types of new parts which will be required in a new product, forecasting inspection devices which will be needed.

10. *Raw Material Inventory*. Raw material inventory will require inspection since more and more diverse, exotic, and complex raw materials are being used. Many product liability cases could be eliminated if the primary material for a design were checked before adding labor to make the more costly parts and products.

7
STRUCTURED OUTPUT

Structured output may be considered formal documents used either in the course of business and/or as a formal company document. For example, the detached parts list is actually an extension of the assembly drawing and must be considered so.

The 10 structured outputs are as follows:

Microfilm microfiche
Part geometry
Part drawing
Assembly geometry
Assembly drawing
Detached parts list
Picking lists
Effectivity
Routings
Processes

In the future a "paperless" factory may exist. If it does, these formatted documents will be at least surreptitiously examined through some portable viewing screen. It is true, however, that more and more documentation originally used to transmit data will be transmitted visually through the use of terminals.

The ability to capture data electronically will no doubt spawn inexpensive devices to store and carry large quantities of data for instant display wherever the user may be be. It is realistic to believe these devices will contain all pertinent data, both graphic and written, which is needed by a machinist, a planner, and others. Microwave transmisstion of this data to these devices will keep them constantly updated.

MICROFILM/MICROFICHE

DESIGN RETRIEVAL--------------- X
FAMILY PROCESSES--------------- X
PART FAMILY DATA---------------X
PROCESS POLYCODES------------O
ITEM MASTER--------------------- X
PERISHABLE TOOLS--------------O
PRODUCT STRUCTURES----------- X
JIGS AND FIXTURES-------------O
ENGINEERING CHANGES----------- X
PRODUCTION PLANNING--------- O
PART GEOMETRY------------------X
INVENTORY CONTROL----------- O
ASSEMBLY GEOMETRY-------------X
GROUP TECHNOLOGY CELLS-----O
GENERIC BILL OF MATERIAL---- O
RAW MATERIAL INVENTORY----- O
ELECTRONIC MAIL---------------O

The use of microfilm and/or microfiche to capture and retrain engineering data can be made an important part of a CIM system. Unfortunately, the technology of this media has not seemed to have kept parallel to other methods of capturing data. Rapid retrieval of microfilm data and rapid conversion of data have lagged behind. In the 1960s it was possible to convert a computer-generated drawing directly to microfilm. Some systems probably still allow this rapid conversion of geometric data (certainly it is possible with text data), but they are little advertised and not generally used. Retrieval of a single microfilm from a large file also was available through computer interrogation, but has seemed to become less popular and is certainly not well known. The means to convert drawings to these methods are fairly expensive and time consuming.

Conversion of printed data to microfilm has been used by many installations and is still a popular method for converting large computer printouts to a smaller database. Nothing very exciting has increased the interest in this method fo capturing data.

Conversion from microfilm to microfiche is also difficult. Microfilm is, however, a logical means to reduce a large file of data into a more usable set, and avoid voluminous files of obsolete drawings. Since more and more drawings are being prepared on a CAD/CAM system, one of the easy ways to merge these new designs with all old, and still usable, ones is through a microfilm system. With a classification system, the power of microfiche systems is that the designs can also be sorted by a design retrieval system into families of parts which can be then placed together by families onto the microfiche. Entire families can be scanned and compared at the same time.

If new viewing techniques and new retrieval techniques can be incorporated into these systems, they could become an even more important segment of the CIM technology.

1. *Design Retrieval.* Design retrieval systems can be used to file microfiche by families for comparative retrievals.

2. *Family Processes.* Family processes can be more easily prepared by creating drawings quickly from a family of parts sorted by the microfiche system.

3. *Part Family Data.* Part family data could become a separate microfilm or microfiche database. When scanning the number of document which might be required for one design, their compilation into one source might be enhanced through microfilm and/or microfiche.

4. *Item Master.* Item master data can be microfilmed to avoid the volumes of data prepared as a master file. Where specific data from the item master is required, microfilm is probably not the medium for it.

5. *Product Structures.* Product structures data in the form of a detached parts list becomes a part of the microfilm database for assemblies. Thus the system must be designed to incorporate this document.

6. *Engineering Changes.* Engineering changes have an effect on microfilm and microfiche, since these changes may affect any and all data thus stored. If a microfilm system will be efficient, a relatively inexpensive means to make changes must be created. Of course, copies of microfilm are relatively inexpensive. It is the original cost which needs study for economies.

7. *Part Geometry.* Part geometry will be found on a microfilm system. It is important to remember that old designs and new designs should be stored for easy retrieval and scanning. Otherwise cold designs might be abandoned, creating a substantial cost to recreate them.

8. *Assembly Geometry.* Assembly geometry is required in the same manner as part geometry to avoid creating new ones. In addition, the parts lists for each assembly is required along with the geometry; otherwise it cannot be interpreted.

PART GEOMETRY

DESIGN RETRIEVAL--------------- X
FAMILY PROCESSES--------------- X
PART FAMILY DATA---------------X
PROCESS POLYCODES------------O
ITEM MASTER--------------------- X
PERISHABLE TOOLS--------------O
PRODUCT STRUCTURES----------- X
JIGS AND FIXTURES-------------O
ENGINEERING CHANGES----------- X
PRODUCTION PLANNING--------- O
PART GEOMETRY------------------X
INVENTORY CONTROL----------- O
ASSEMBLY GEOMETRY-------------X
GROUP TECHNOLOGY CELLS----- O
GENERIC BILL OF MATERIAL---- O
RAW MATERIAL INVENTORY----- O
ELECTRONIC MAIL--------------- O

Until recent years, part geometry was represented only by an engineering drawing or, partially, through the generation of a computer-aided design program which created selected dimensions. With the advent of CAD/CAM systems, the geometry of the design is captured in the form of two-dimensional representations, three-dimensional wire frame models, or solids created from predesignated primitives.

Each of these representations is somewhat disarming, for the actual geometry of the part still may be only represented, not defined. Obviously a two-dimensional drawing is not geometry but only a picture of it. The lines and arcs of this type of drawing do, however, have dimensions which represent geometry. A wire frame representation of a part design does not define a circle as a hole, a line as the intersection of two planes, but, again, is a representation of such geometry. Of course, as in the case of the two-dimensional model, the three-dimensional wire frame model contains dimensions which represent geometry.

A solid model of the part design is a true representation of the part geometry. A hole or slot represented through a geometric definition of these physical characteristics. A section passed through such a design will define where the sections, appear accurately, both solid and not, as in a section view through a hole. We can call this portrayal a model of the part design.

As a structured output, the part geometry is not necessarily viewed by the user other than on a terminal.

The geometry definitions may be carefully structured so they can be interpreted in an NC tool path application, a drafting application, a design analysis application, and others. The screen views may show a variety of features, such as color, shading, various views, and more. Dimensions, tolerances, finishes, and sectional views for clarity may be generated also. It should be clear that the part geometry is stored mathematically and all of the visual methods to portray it are just that, portrayals.

1. *Design Retrieval*. A design retrieval system may be used to retrieve the geometry of a part.

2. *Family Processes*. Family processes may be created based on the family geometric characteristics. If the family process is developed for NC tools, the tool path will religiously follow the contours of the part geometry as required.

3. *Part Family Data*. Part family data will not only include the features of the part geometry described in a family, but also the feature variations which might be found in a family.

4. *Item Master*. Item master data may contain critical dimensions of the part as defined and described in a classification system. If a classification system is used, the code number of the family will be in the item master.

5. *Product Structures*. Product structures may be used to identify and collect design features required to make a part design. As more and more features are stored for use in making part geomtries, it may be necessary to create design feature families and collect them through a product structure algorithm.

6. *Engineering Changes*. Engineering changes may create the need for changing a part geometry. Since all portrayals of this geometry rely on the accuracy of the geometry, each part geometry must be examined to determine if the engineering change has affected it.

7. *Part Geometry*. The part geometry is a new series of data increments developed as a result of CAD/CAM systems or from computer-aided design programs. Since this geometry will serve to drive many of the other parts of a CIM system, it is important that this source data be clear, accurate, and consistent. The ability to transfer this data from one computer to another and/or one database to another will enhance the power of CIM, as well as make it more practical and economical for manufacturing companies to implement.

8. *Assembly Geometry*. Assembly geometry will be created as the result of part geometry. Today it seems natural to draw piece parts and then turn around and display almost identical pictures, albeit limited views of the parts, in their assembly. Since the part geometry is well defined, the collection of such parts into an assembly will assist in better defining the assembly, including fits, tolerances, and more. We must be careful to understand associativity when considering assembly geometry. A physical change to a part could cause a change to the geometry of every assembly in which the part is used. Do we wish this to happen? Will we require new

assembly drawings to be prepared? Won't this, possibly, create an unnecessary revision level change to assemblies? How will we control such changes?

PART DRAWING

DESIGN RETRIEVAL--------------- X
FAMILY PROCESSES--------------- X
PART FAMILY DATA---------------X
PROCESS POLYCODES------------O
ITEM MASTER--------------------- X
PERISHABLE TOOLS--------------O
PRODUCT STRUCTURES----------- X
JIGS AND FIXTURES------------- O
ENGINEERING CHANGES----------- X
PRODUCTION PLANNING--------- O
PART GEOMETRY------------------X
INVENTORY CONTROL----------- O
ASSEMBLY GEOMETRY-------------X
GROUP TECHNOLOGY CELLS----- O
GENERIC BILL OF MATERIAL---- O
RAW MATERIAL INVENTORY----- O
ELECTRONIC MAIL--------------- O

The part drawing is the main pictorial representation of a design. In the case of some simple drawings, word descriptions have often been sufficient as an equivalent documentation. Variations to part drawings include tabulated drawings, multipart drawings, assembly drawings which include dimensions for the parts within the assembly, and hand sketches of parts.

Control of part drawings is based on configuration management rules which determine whether the change to a part is interchangeable or not. If not interchangeable, a new part number and a new drawing are required. If interchangeable, the revision level of the old design is changed.

The drawing is created using long-standard customs. Lines and their thickness, section lines, scribe lines and arrowheads, dimension protocol, tolerances, and number of views are all well defined. The result is a special kind of shorthand used to describe, for the esigner, what is to be created. Conversion of the part geometry to this format is of considerable importance.

It has become important to reconsider drawing content since, often, manufacturing data, as opposed to strictly engineering design data, has been intermixed. Not only is the chemistry of the raw material for a fabricated part defined, which is proper, but often the size of the raw material from which it is to be cut, sawed, burned, or sheared. This process data has caused some confusion between engineering and manufacturing, and also redundant activities. A clearer requirement for part drawing content will enhance the speed and economies to be found in an integrated CIM systems.

It should be noted that purchased parts, that is, parts required which are not under the company's design control, have quite different rules than those parts which are under design control by the company. Sufficient specifications or descriptions of the purchased parts should be documented, where possible, to allow their purchase from more than one vendor. Some, but not all, dimensions of such parts may be documented on the drawing. Only dimensions critical to the operation of the design or those necessary to show connection between it and other designs might be necessary.

1. *Design Retrieval.* A design retrieval system will be used to retrieve part drawings. A part number system will be used to identify each part design, and the documents required to fully define the design will be identified with this part number. Each design will be classified into families of parts and the part number and the code number of the family within which each design may be found will serve as a cross-index for retrieval.

2. *Family Processes.* Family processes will be based on specific data shown on the drawing for each design in the family. Both general and specific requirements will be considered.

3. *Part Family Data.* Part family data will be based on a compilation of data requirements as pictured on the part drawings contained in each of them.

4. *Item Master.* An item master record is prepared for each part number, which implies that each part drawing is also represented by such a record. Since a part drawing might really be a series of documents, a table of documents might become a part of this record.

5. *Product Structures.* Product structures are prepared for an assembly or a part which requires raw material. Thus the product structure is a means to collect associated parts into a document commonly called an assembly.

6. *Engineering Changes.* Engineering changes must be examined to determine whether a part drawing has been affected. It is quite possible that an engineering change will not affect the geometry of a part but will change the drawing. Examples are geometric notes, raw material chemistry, hardness, and other such features.

7. *Part Geometry.* Part geometry is represented by the part drawing. Thus the coordination of this data is mandatory. If a company makes changes to a drawing without changing the corresponding geometry, or vice versa, the credibility of the system will become zero rapidly. It should not be overlooked that, as the part design geometry is being prepared, the designer knows much of the data which will become part of the item master database. Thus creating this data during the time when the part design data is created may merge two separate disciplines economically and practically.

8. *Assembly Geometry*. Assembly geometry will be created from part geometry. In one sense, when this is complete, the identity of the parts has been lost. Two types of assemblies can be created, dismantleable, and nondismantleable. Those which are non-dismantleable, such as a weldment, need only assembly dimensions to picture them. Those which might be dismantled may need to retain some or all of the part dimensions, even if they are not pictured on the assembly drawing.

ASSEMBLY GEOMETRY

DESIGN RETRIEVAL--------------- X
FAMILY PROCESSES--------------- X
PART FAMILY DATA---------------X
PROCESS POLYCODES------------O
ITEM MASTER--------------------- X
PERISHABLE TOOLS--------------O
PRODUCT STRUCTURES----------- X
JIGS AND FIXTURES------------- O
ENGINEERING CHANGES----------- X
PRODUCTION PLANNING--------- O
PART GEOMETRY------------------X
INVENTORY CONTROL----------- O
ASSEMBLY GEOMETRY-------------X
GROUP TECHNOLOGY CELLS----- O
GENERIC BILL OF MATERIAL---- O
RAW MATERIAL INVENTORY----- O
ELECTRONIC MAIL--------------- O

Assembly geometry will be stored, like part geometry, but all means to picture it will be nothing more than that, portrayals. As with part geometry, representations of an assembly may be two-dimensional, three-dimensional wire frame, or a solids model.

Several differences, however, may be seen in the stored geometry. If we consider an assembly as a combination of part designs, then each subsequent level upward from such a assembly will consist of the same parts but with additional parts or assemblies added to make the next-level-higher assembly. This geometric assemblage is documented in the bill of material. If we must store the parts and assemblies over and over, much redundant and unnecessary data will be stored.

It is possible that pointers to part designs combined with the geometric transforms or coordinate dimensions for locating one part in relation to some datum line or, at least, to all other parts in an assembly, will be more practical. In this manner, each subsequent level of assembly will consider the level below as a series of piece parts, and the database for such an assembly will contain pointers to them.

As described previously, the pointers and geometric transforms used for an assembly could place each part or subassembly into at least three positions in the geometric model—the final assembly position, the exploded view position, and in a bin surrounding the actual assembly area. In this manner a host of pictorial representations or means to assemble or series of assembly instructions can be prepared. In addition, no redundant data will be stored.

In addition to pointers and transforms, assembly dimensions and/or other requirements to complete each specific assembly must be stored:

1. *Design Retrieval*. A design retrieval system may be used to retrieve assembly geometry. Since classifying assemblies is more difficult than for parts, and since the economies of design retrieval are not so apparent, retrieval of assemblies by function, name, or simply similarity of design may be necessary. As with part designs, each assembly and its geometry will be given a part number. Thus retrieval by part number is always possible.

2. *Family Processes*. Family processes for assemblies can be based on the assembly geometry. However, it is quite probable that only special cases of closely similar fabricated assemblies or assemblies which require considerable machining after assembly will apply.

3. *Part Family Data*. Part family data will be of less importance for assemblies. However, in special cases, part family data can be important to generate.

4. *Item Master*. Item master data will be prepared for assemblies just as for part designs. Since an assembly represents a product structure, it is important to be able to distinguish the type of data being stored for an assembly. Thus a table of documents combined with the methods of locations of storage for the parts may be necessary.

5. *Product Structures*. Product structures might be generated as a by product automatically when an assembly is created. If the part designs have been prepared previously, or parametric drawings of part families are available, creating the assembly geometry may be a matter of using the parts which exist or altering a parametric drawing to a specific design. In both cases, maintaining a record of the parts used is the same as creating a product structure or parts list.

6. *Engineering Changes*. Engineering changes must be examined carefully to determine whether changes in a part design have affected the assemblies upon which it is used. On the other hand, the change to an assembly may affect the interchangeability of the assembly, requiring the issue of a new assembly rather than a revision level change to the assembly. In all cases, close coordination between changes to the assembly geometry, the parts within the assembly geometry, and the parts list and drawing which reflects the assembly configuration is important.

7. *Part Geometry*. Part geometry can be considered segments of the assembly geometry which has been created. If prepared logically, this assemblage of parts which have already been designed may eliminate many of the errors found in an assembly, and may also make design layouts more rapidly and accurately.

8. *Assembly Geometry*. Assembly geometry might be created even more rapidly and accurately if a series of parametric drawings representing part geometries were merged to create an assembly geometry parametric drawing. Properly prepared, the assembly layout, when completed, would develop all piece parts required automatically.

ASSEMBLY DRAWING

DESIGN RETRIEVAL--------------- X
FAMILY PROCESSES--------------- X
PART FAMILY DATA---------------X
PROCESS POLYCODES------------0
ITEM MASTER--------------------- X
PERISHABLE TOOLS------------- 0
PRODUCT STRUCTURES-----------X
JIGS AND FIXTURES-------------0
ENGINEERING CHANGES-----------X
PRODUCTION PLANNING--------- 0
PART GEOMETRY------------------X
INVENTORY CONTROL----------- 0
ASSEMBLY GEOMETRY-------------X
GROUP TECHNOLOGY CELLS-----0
GENERIC BILL OF MATERIAL---- 0
RAW MATERIAL INVENTORY----- 0
ELECTRONIC MAIL---------------0

As in the case of a part design drawing, an assembly drawing is used to represent an engineering design. Of course, this drawing also represents a collation of parts as required by the designer to make the assembly. For the past 20 years or more, significant changes have been made to this document. At one time, tabulated assemblies were used. For this purpose, a list of materials and parts required for the assembly was placed on the drawing, and columns attached to this list for each assembly variation. In this manner a variety of parts, some used in one assembly, some used in another assembly and still others used in all assemblies, were tabulated. This caused sufficient confusion to make even the designers who created this method agree to create one assembly drawing for each new assembly.

At the same time, mechanized bill-of-material systems were developed. The list of material found on the assembly drawing became a "detached" list prepared by the computer. "Find," or "bubble" numbers, found on both documents, serve as cross-references to identify parts itemized on the parts list with their locations on the drawing. Wherever a parts list was not detached, a manufacturing company found itself with two databases, a list of material on the drawing, and, usually, a different list on the computer. This practice has been stopped, more or less, by most companies with successful bill-of-material systems.

Since two documents are now used to represent an assembly, accuracy of both is important. Configuration management systems describe how this will be accomplished. In general, the detached parts list is maintained with a revision level for the assembly documented on the face of the parts list equal to the revision level of the assembly drawing. This practice is often onerous to the designer as, often, several sheets of drawings are required to document an assembly. According to the rules, each of these sheets must show the same revision even though no actual change has been made on one or more of these sheets.

Changes to the content of an assembly, either addition and deletion of parts or changes to quantities of parts, require analysis to determine whether a new assembly design is required with corresponding new drawings and parts list, or a revision level change only is needed. If a revision level change is made, effectivity of the change may need to be analyzed also. We will discuss this principle later.

Before integration of engineering design and manufacturing production is truly integrated, the product structure of a product must be examined. Often the designer creates an assembly of parts without being fully aware of how the parts are actually assembled. This may cause manufacturing to create "phoney" assemblies to reorganize parts according to manufacturing practices. Inevitably, this creates the need for two bill-of-material systems, perhaps even three if accounting needs a version different from these two.

Because structuring has become so important to integrate engineering and manufacturing, industrial engineering and design engineering are becoming more aware of the role each plays in documenting design. Successful companies are allowing industrial engineers to have equal voice with design engineering in the preparation of assembly drawings and their corresponding bills of material. Economies of manufacturing can be significant.

1. *Design Retrieval*. Design retrieval, where applicable, can be used to retrieve assembly drawings in a manner similar to retrieval of part drawings.

2. *Family Processes.* Family processes may be prepared for assembly processes and routings, where applicable also.

3. *Part Family Data.* Part family data will assist in creating standard manufacturing assembly practices, including the division of manufacturing operations which should be considered a portion of the part design rather than those which are made on the finished assembly.

4. *Item Master.* Item master data must be prepared for each assembly in the same manner as that for the part designs. Different data may be necessary for them. A part type code can be used to distinguish parts from assemblies, and a set of part type codes can distinguish one type of assembly from another as "module," "attachment," "kit," and more.

5. *Product Structures.* Three types of product structures exist. A general product structure represents an agreement between industrial engineering and design engineering which describes how the product is assembled. A specific product structure is represented by an assembly drawing, and the parts list represents the documentation of the parts of which the assembly is composed. A generic-set-of-parts list can also be used to create a forecast of a new product before it has been designed. Such parts list can be converted into a generic bill of material.

6. *Engineering Changes.* Engineering changes are particularly important when examining an assembly and its corresponding parts list. Not only is determining interchangeability more complex, but the addition and/or deletion of parts to an assembly creates the need for effectivity of such changes.

7. *Part Geometry.* Part geometry is indirectly represented in the assembly. It should be noted, however, that each assembly drawing should be prepared and dimensioned on the basis that each part within the assembly has been dimensioned previously. Thus the dimensions appearing on the assembly drawing represent only those needed to create the assembly. In a like manner, the material for an assembly is not documented, since materials have been documented for individual parts. It should also be noted that purchased assemblies, such as electric motors and bearings, are considered piece parts within the system. However, such designs must be documented with a company part number and, at least, a specification sufficient to avoid requiring the purchase of such a design from one vendor only.

8. *Assembly Geometry.* Assembly geometry is represented by the assembly drawing. For clarity, many extra views or sections of the assembly may be pictured.

DETACHED PARTS LIST

DESIGN RETRIEVAL--------------O
FAMILY PROCESSES--------------O
PART FAMILY DATA------------- O
PROCESS POLYCODES------------O
ITEM MASTER--------------------- X
PERISHABLE TOOLS--------------O
PRODUCT STRUCTURES----------- X
JIGS AND FIXTURES------------- O
ENGINEERING CHANGES----------- X
PRODUCTION PLANNING--------- O
PART GEOMETRY---------------- O
INVENTORY CONTROL----------- O
ASSEMBLY GEOMETRY-------------X
GROUP TECHNOLOGY CELLS----- O
GENERIC BILL OF MATERIAL---- O
RAW MATERIAL INVENTORY----- O
ELECTRONIC MAIL--------------- O

We have referred to the use of a detached parts list several times. As previously stated, the detached parts list replaces the manually prepared list of materials usually found on an assembly drawing. Elimination of the old material list is important to maintain accuracy of a bill of material, since one or another of the two lists will always be out of synchronization. In fact, detaching the list of material and making it a part of the computer system makes it possible to achieve more accuracy of data as so many other people now use it, and it is critical to many other systems within CIM.

The power of the parts list data, when residing in the computer, is often overlooked. One parts list represents one section only of a total bill of material for a product. (It is often called a single level bill of material.) The computer can be so programmed that it is possible to "chain" upward or downward through the bill of material. A parts list connects two levels of the bill of material hierarchy. Thus subsequent higher or lower levels are connected one to another. By inverting the data, a where-used structure can be prepared from the lowest level raw material to the final products. By exploding the bill of material downward, the total parts, assemblies, and raw materials required for a product may be compiled. In this manner, a summarized bill of material, an indented bill of material, and a costed bill of material are prepared. Separate assembly costs can be prepared also.

The engineering parts list should be considered the official document which represents a finished segment of a product. It should also be considered an engineering document which is under engineering change control. Mixing manufacturing, inventory control, or accounting data on this rather formal document creates unnecessary recopying of the document when any of such data is changed. Engineering-controlled data should be displayed only. If a combination report or other data is required, it should be prepared separately. Remember that all engineering parts lists should be maintained on microfilm with their corresponding assembly drawing.

1. *Item Master*. Item master data is merged with product structure data to generate a parts lists. Incidentally, this ability to merge data from the item master with that from the product structure eliminates considerable preparation of redundant data in an engineering department. For example, how many times, in a manual system, does a designer or drafter write the same part descriptions in preparing parts lists? This combining of data also reduces errors in documentation.

2. *Product Structures*. Product structures are the vehicles to create the proper association of parts into an assembly, and the parts list (converted to a picking list), combined with the assembly drawing, become a portion of the shop packet issued to manufacturing.

3. *Engineering Changes*. Engineering changes are reflected on the parts list. Since an engineering change can be made immediately in engineering but may not be implemented in manufacturing until the effectivity date is reached, the parts list will reflect current status. However, in engineering, it is possible to interrogate the files of assembly data and determine all configurations of an assembly from its original release to the present. This maintenance of history is also an important feature of the bill-of-material system.

4. *Assembly Geometry*. Assembly geometry, represented by the assembly drawing, combined with the parts list, composes the complete documentation of the assembly. It should be emphasized that, in a true part-number-oriented system, more than one assembly sheet may be needed to fully document a complicated assembly. Each of these sheets must have the same part number and revision level, and is differentiated by sheet number or page number. For clarity, a weldment, for example, may need to be documented in two stages, as a weldment only, and as a machined weldment. While it is possible to document these two steps with one part number and two sheets, it is often more practical to create one line parts lists and create two part numbers for these manufacturing stages.

PICKING LISTS

DESIGN RETRIEVAL--------------O
FAMILY PROCESSES----------------X
PART FAMILY DATA------------- O
PROCESS POLYCODES------------O
ITEM MASTER--------------------- X
PERISHABLE TOOLS--------------O
PRODUCT STRUCTURES----------- X
JIGS AND FIXTURES------------- O
ENGINEERING CHANGES----------- X
PRODUCTION PLANNING-----------X
PART GEOMETRY---------------- O
INVENTORY CONTROL-------------X
ASSEMBLY GEOMETRY----------- O
GROUP TECHNOLOGY CELLS-------X
GENERIC BILL OF MATERIAL---- O
RAW MATERIAL INVENTORY-------X

The picking list is a special form of the engineering parts list. It is often not understood that the same database is used to prepare the picking list that is used to prepare the engineering parts list. The main difference is content.

When material requirements planning schedules assembly requirements, and when production scheduling and shop floor control determine a shop packet is to be issued for the assembly, the picking list is prepared. Two methods for its issue exist. It is possible to prepost or postpost inventory prior to issuing the picking list. When preposting, the part requirements for the picking list are examined to determine whether sufficient inventory is available to meet the requirement. Often these parts are then allocated, although it is possible to allocate them before this operation also. Depending upon the rules, part shortages may or may not delay the picking list issue.

The postposting of inventory assumes that parts are available, and the interrogation of inventory does not occur.

When the picking list is issued, a good concept is to place the bin location for each part on the picking list. If the picking list is sorted by this number, a more rapid picking of parts will occur. Some parts may be so large they are not found in regular inventory, and some parts are considered floor stock. These conditions can be duly noted on the picking list also.

The picking list represents the current configuration of the assembly being manufactured. This configuration may not be the latest documented by engineering. For example, an assembly picking list may show revision a, but the engineering parts list may have been changed to revision B. This difference is based on the principal of effectivity which notes that the parts in revision A are to be used up before changing to revision B. More will be said about this in discussion effectivity.

1. *Family Processes.* Family processes will become a portion of the picking list in the future if group technology is used to manufacture assemblies. It may be that a combined picking list will be used to pick parts for a series of assemblies to be manufactured as a family.

2. *Item Master.* Item master data is used to prepare the picking list in a manner similar to that of preparing an engineering parts list. In addition, manufacturing data is included.

3. *Product Structures.* Product structures are used to determine the parts listed on the picking list. The effectivity of changes and the corresponding dates and revision level are examined before the picking list is prepared.

4. *Engineering Changes.* Engineering changes may create several different picking list stets of parts, depending on the revision levels and dates of changes.

5. *Production Planning.* Production planning activates the picking list based on lead time and MRP data. As part of the shop packet, a process and/or routing is also issued to instruct the assembler.

6. *Inventory Control.* Inventory control data is used to determine whether the parts are available before the picking list is issued. Often a set of back orders or expedite messages are prepared as a result of this examination. The picking list may or may not be issued if there are part shortages, depending on the particular rules set forth previously.

7. *Group Technology Cells.* Group technology cells for assemblies will create the need for a combination form of picking lists, or, at least, the issuance of several picking lists as a set.

8. *Raw Material Inventory.* Raw material inventory is also interrogated before the issuance of a picking list for piece parts. This may not take the form of a picking list, but, rather, the issuance of a process or routing, the first operation describing the part number and quantity of raw material needed to make the part. If the part is a forging or casting, it may be that a single line picking list will be issued since the rough casting and/or forging should have a part number other than the finished part.

EFFECTIVITY

DESIGN RETRIEVAL-------------- O
FAMILY PROCESSES--------------- X
PART FAMILY DATA------------- O
PROCESS POLYCODES------------ O
ITEM MASTER--------------------- X
PERISHABLE TOOLS-------------- O
PRODUCT STRUCTURES----------- X
JIGS AND FIXTURES------------- O
ENGINEERING CHANGES----------- X
PRODUCTION PLANNING----------- X
PART GEOMETRY---------------- O
INVENTORY CONTROL------------- X
ASSEMBLY GEOMETRY----------- O
GROUP TECHNOLOGY CELLS------ X
GENERIC BILL OF MATERIAL---- O
RAW MATERIAL INVENTORY------- X
ELECTRONIC MAIL--------------- O

We have mentioned effectivity several times as portions of the bill of material have been discussed. Simple in concept, the use and definition of effectivity sometimes seems complex. Let us look at the reasons for effectivity and how it is used.

When engineering prepares a bill of material, the components in each assembly are defined. These components represent the product configuration at that moment in time. Inevitably, for whatever reason, changes are made to this configuration later. Some of these changes represent replacing parts or assemblies with new parts or assemblies having new part numbers, since they are not interchangeable. Whether the bill of material system is manual or mechanized, these changes are sent to manufacturing in the form of new parts lists and/or new bills of material. If the data thus created is used in a mechanized system, and the new designs replace the old in the documentation, the old design will suddenly seem to be unneeded, while the new design may need to be expedited to replace the old. This condition, where engineering seems to make an instant change with no regard to lead times or inventory, causes much trauma in manufacturing. Consequently, manufacturing has often maintained their own set of bills of materials, one showing the current configuration, another showing the latest engineering configuration. This, too, causes problems since engineering and manufacturing documents never seem to match. The problems created by two bills of material was solved using effectivity.

It must be remembered that the rules of interchangeability must be followed correctly for effectivity to work. If the replacement of one part in an assembly creates an assembly which is noninterchangeable with the old, the new configuration must be documented with a new part number. This investigation must continue from the point of change up to the finished product part number. With this understood, how is effectivity used?

Once a part in an assembly is replaced with a new part in the assembly, for example, the effect of this change must be examined. If the old part has usages other than within the assembly being changed, and the other assemblies upon which it is used do not also need to be changed, stock of the old part will be used in other assemblies and the effective date of change can be made immediate, or when the lead time for the new part registers that they can be received into inventory.

If the old part has no use in any assembly, and the part is in inventory, or the new part requires tooling and/or has a long lead time, the day when all these conditions are met, exhausting old inventory, receiving new parts, and so on, is called the effectivity date. No system is yet so accurate that the date can be determined with absolute precision. Nevertheless, a rough date can be established.

If the change is mandatory, the effectivity date may be instant. However, this may cause an assembly stoppage, or create assemblies with parts missing.

There are those who suggest that effectivity of change loses identity of assemblies which have different parts since all assemblies are stocked and documented by part number only. Thus an assembly with revision A with one set of parts will be stocked with revision B having a different set of parts. It is here again we must emphasize that the rules of interchangeability must be maintained. Otherwise chaos does, in fact, exist.

Still another condition exists. Some of the obsolete parts may be serviceable parts. In fact, several parts may be changed in an assembly and the assembly is interchangeable, but the sum of the serviceable parts may have changed drastically, with no one-to-one relationship. This creates, usually, a condition of noninterchangeability and requires a new assembly part number. In fact, such changes and configuration control rules may force a change of all part numbers in subsequent assembly levels up to the finished product.

Engineering changes are expensive. Thus the service level of products has slowly risen until it is often more expensive, or impossible, to buy parts of a product, rather than buying a whole new product, when the simplest of parts fails within it.

1. *Family Processes*. Family processes may be affected by effectivity, since the assembly process may change when old parts are replaced with new. For example, a weldment of two strips of steel might be replaced with an angle of the same dimensions.

2. *Item Master*. Item master for the assembly will usually reflect the latest revision level of the assembly. If more than one revision level for the item master of the assembly need be examined, the face of the assembly drawing usually has a brief history of the changes, and the engineering change log records them also.

3. *Product Structures*. Product structures maintain all the old configurations as well as the latest. Provision in the structure must be made to show parts which are "in" currently, and "out" and the dates when these events occurred. Since it is possible to take a part design out of an assembly today, replace it with a new one, and then replace this replacement again with the old, a record of each of these transactions must be maintained to completely document the history of the assembly configuration. Four pieces of data on the product structure are used to control changes, "in" and "out," date or serial number of change, the engineering change number of the document authorizing the change, and the revision level of the assembly when the change was made.

4. *Engineering Changes*. Engineering changes are used to document parts and/or assembly replacements due to design changes. The engineering change notice is used as the instrument to determine cost of change, inventory quantities, disposition of parts in inventory, and order, or in process. It is possible that the cost of using up obsolete parts exceeds their value and instant change is more practical. In any case, the parts list will usually show parts "in" and "out", date of change, and the engineering change number which created the change. This allows tracking of change history later. There are some who insist that engineering changes must be made by serial number, not date. Under carefully controlled conditions, this may be possible, but it is a far more complex method. If part or lot traceability is required, this may be the only acceptable means to control changes. It is also much more expensive and time consuming.

5. *Production Planning*. The production planning system, when properly designed, makes note of the effective dates and issues the proper picking lists at the appropriate time. This is performed in conjunction with the inventory control function.

6. *Inventory Control*. Inventory control records of parts which have been replaced in a nonmandatory change are monitored to determine when an exhaust stock condition has been reached. This monitoring, in conjunction with production planning, determines when a later revision of an assembly may be created. In complex changes where several parts are being changed, manual monitoring is probably necessary, and no exact balance of stocking for all parts are probably possible. If, however, all parts are extremely expensive, it may be economical to balance all inventories.

7. *Group Technology Cells*. Group technology cells will be an important manufacturing activity that may sometimes minimize engineering changes. If families of parts are manufactured, less inventory of parts may result, making effective dates almost instantaneous.

8. *Raw Material Inventory*. A raw material inventory status is an important feature of effectivity, since design changes may make raw material obsolete also. If high volumes of a raw material are stored, care again must be taken to determine an effectivity date which does not obsolete an expensive raw material prematurely.

PROCESSES

DESIGN RETRIEVAL--------------- X
FAMILY PROCESSES--------------- X
PART FAMILY DATA---------------X
PROCESS POLYCODES------------- X
ITEM MASTER--------------------- X
PERISHABLE TOOLS---------------X
PRODUCT STRUCTURES-----------X
JIGS AND FIXTURES-------------- X
ENGINEERING CHANGES-----------X
PRODUCTION PLANNING-----------X
PART GEOMETRY------------------X
INVENTORY CONTROL-------------X
ASSEMBLY GEOMETRY-------------X
GROUP TECHNOLOGY CELLS-------X
GENERIC BILL OF MATERIAL------X
RAW MATERIAL INVENTORY-------X
ELECTRONIC MAIL--------------- O

Nothing emphasizes the importance of a CIM system more than to scan the database segments influencing or influenced by processes. It was stated previously that the process and routing data is to manufacturing what the item master and product structure are to engineering. The process describes the "how" of the design, while the drawing describes the "what."

Much has been said about generative process planning and variant process planning. Each method may be practical under certain circumstances. Regardless of the method used, a significant set of data needs to be known to prepare an economical, practical, and effective process. A cursory examination of processes created for similar parts in a manufacturing company prepared without the benefit of part family data shows radical variations in the process sequence, the process times, the work centers utilized, the tooling prepared, the set-up and run times estimated, in fact, in almost any facet of the process.

A design retrieval system or a set of polycodes which can bring together similar operations or part families will significantly improve preparation of processes. It will eliminate the whims of various process planners and remove reliance upon their memories also.

Preparation of a family process will also tend to standardize the manufacture of parts, making the process sequence and cost more reliable.

More and more NC tools are replacing one or more machine tools. This replacement, together with combinations of machine tools in cells for group technology, will create the need to reexamine processes. Indeed, at least one company has purchased special machine tolls which will make all parts of an individual assembly in sequence with no set-up time, eliminating work in process of parts and expediting the shipping of finished assemblies. This type of new technology creates a completely new methodology in the preparation of processes.

There are many roadblocks to increasing the productivity of a manufacturing operation. Diversity of processes which haven't been itemized is but one of them. Tools have been designed for individual parts, jigs and fixtures are specialized, little communication between industrial engineering and design engineering to alter designs for more economical manufacturing, nonstandard holes, taps, boring, keyways, and other operations, all these have contributed to costly manufacturing processes. Even incorrect product structures have created illogical assemblies for which it is difficult to plan a process.

A process is a combination of how to manufacture a design with an abbreviated routing of the design through a manufacturing department. Many companies did, and still do, manufacture with no written instructions. Slowly, but surely, however, this verbal process is being replaced. Old journeymen who knew all the processes, master foremen, and others have retired. The labor market is being replaced with less dedicated workers, and, too, the designs have become more complex and critical. The complexity of design combined with competitive pressures have forced more documentation of manufacturing practices. The next step must be cooperative communication with design engineers to make operations more standard.

Today some unnecessary redundancy is found. A process for parts to be manufactured on an NC tool may be made, only to find the NC tool part programmer creating an identical or similar set of instructions. Often the process is too detailed. The accuracy and clarity of a process is greatly dependent upon the writing skills of the planner, who may have had little writing education.

1. *Design Retrieval*. A design retrieval system must be used to bring together part families for analysis to study differences in processes between them and to prepare for standardization of the process.

2. *Family Processes*. Family processes will be the major tool in reducing cost of manufacture and to increase more standard methods. It will be found that, due to the retrieval system, one family may require more than one process, and also that many families of parts can be defined with one process. In both cases, standards can be set more easily and master processes created.

Producibility tips and design and drafting standards can be created for each family, and new designs, when released, can be compared to this family data rather than examined on an individual basis. In fact, all new designs should be sorted by families and worked on as a family to eliminate the learning curve now required each time an individual part is examined for a process.

3. *Part Family Data.* Part family data will be the basis and the repository of all standards and practices needed to create consistent and economical processes.

4. *Process Polycodes.* Process polycodes may be needed to identify important manufacturing operations common to parts in a variety of dissimilar families. Depending upon the product being manufactured, polycodes may be more important than part families. However, it is felt that a combination of families with polycodes is the best of both requirements and should be carefully examined as a unit with specific objectives in mind before implementing either.

5. *Item Master.* Item master data is the source of at least some of the initial data for the process. To keep the item master and the process plan database equivalent, and to avoid creating redundant data, the two should be considered close relatives.

6. *Perishable Tools.* Perishable tools are often described on a process in a variety of ways. All perishable tools should have part numbers, and an inventory record of them maintained. The variety of perishable tools should be catalogued and segregated by preferred and nonpreferred. As a part of the producibility tips, a record of the preferred should be available to designers and drafters. As a special and important part of this identification, tool packages on NC tools should be identified and all part families, which are to be manufactured on such tools, examined to make certain the tool variety required on new designs does not exceed the capacity of the tool package.

7. *Product Structures.* Product structures should reflect actual assemblies if consistent process plans are prepared for them. Assembly processes, if not on an actual assembly line, are often inexact and inaccurate. If they are to be improved, the same consistency required for piece parts will be needed.

8. *Jigs and Fixtures.* Jigs and fixtures should be maintained in inventory. The part numbers of all jigs and fixtures should also be tied to part families so that investigation of their use on similar parts can be made.

9. *Engineering Changes.* Engineering changes can alter tools and jigs and fixtures, as well as the process. In fact, as more

similar parts are processed together as a group, and identical tooling is utilized, changes may become even more significant, requiring closer examination.

10. *Production Planning*. Production planning may change the method of preparing processes. For example, a cell utilized for group technology may require a general rather than a specific process. It may also create two stage processes, one segment for the cell and the other for operations performed outside the cell. In some instances, a process may be eliminated since common operations will be performed in the cell and the learning curve and instruction may not be required.

11. *Part Geometry*. Part geometry will be passed on directly to the NC part programmer. In fact, combining the part geometry with the process may mean direct generation of the tool path and the process. As a parametric drawing is used to create a new design, the dimensions of the part which are identified in the process can be entered into the geometry of the design and the corresponding instruction of the process as well as in the tool path algorithm.

12. *Inventory Control*. Inventory control quantities are based on an economical order quantity which contrasts cost of set-up to inventory carrying cost. If set-up times become very small, and/or a series of parts are made together, less inventory will be economical or required. Work-in-process inventories will have to become less. This will tend to tighten up queue times, and make the entire process more rapid and close knit. The logistics of this phenomenon must be examined and planned.

13. *Assembly Geometry*. Assembly geometry for those similar assemblies in a family will follow the same principles and require the same consideration as part families. If robots are considered part of the process, consistent assembly components will assist in this activity.

14. *Group Technology Cells*. Group technology cells will, as mentioned previously, greatly change or affect the methods by which processes are created. Group technology will be impractical or not economical if consistency of parts both in general and with specific operations cannot be prepared. Close cooperation between industrial engineering and design engineering is required.

15. *Raw Material Inventory*. Raw material inventory may be reduced if consistent manufacturing operations are considered. For example, some similar parts today are sawed separately from raw materials, others are cut from a large piece of raw material on a turret lathe or engine lathe. This creates inconsistencies in the type and sizes of raw materials required and increases the proliferation of raw materials needed.

ROUTINGS

DESIGN RETRIEVAL--------------- X
FAMILY PROCESSES--------------- X
PART FAMILY DATA---------------X
PROCESS POLYCODES------------- X
ITEM MASTER--------------------- X
PERISHABLE TOOLS---------------X
PRODUCT STRUCTURES----------- X
JIGS AND FIXTURES---------------X
ENGINEERING CHANGES----------- X
PRODUCTION PLANNING-----------X
PART GEOMETRY------------------X
INVENTORY CONTROL-------------X
ASSEMBLY GEOMETRY-------------X
GROUP TECHNOLOGY CELLS-------X
GENERIC BILL OF MATERIAL------X
RAW MATERIAL INVENTORY-------X
ELECTRONIC MAIL--------------- O

A routing can be described as a shortened version of the process, or an abstract. It is often combined with production data and issued as a work order which authorizes the manufacture of a part or assembly. The quantity required and the work order number are displayed on the routing. Labor is charged to the work order number, and the quantity of designs are made as recorded. The routing is also a part of the shop packet, accompanying a drawing and a parts list where required. In some companies, a punched card or a bar code listing accompanies the shop packet for a data collection system.

Contents of the routing include the part number of the design, the sequence of operations, the work centers to which the part must go, a brief description of the operation to be performed, the run time per design, the set-up time, if any, per operation, and the part numbers of the tools required for each operation. Queue times between operations may be shown also.

1. *Design Retrieval*. A design retrieval system may be used to combine similar routings for shop consideration in the manufacture of parts. This can be considered an informal group technology compilation, allowing a foremen or shop planner to make decisions on the sequence in which parts will be manufactured.

2. *Family Processes*. Family processes will make routings more consistent. The consistency will allow more rapid preparation of a routing for a new design.

3. *Part Family Data*. Part family data will assist in making routings for similar parts more consistent, making it easier for machine operators to understand what operation is to be performed.

4. *Process Polycodes*. Process polycodes have been discussed several times. However, it should be noted that a unique set of polycodes is available in the routing system if consistent routings are prepared. Let us first consider that the operation number is standard for an operation. Thus 20 is always drill, 23 always saw, 43 always mill, for example. When the routing is prepared, standard phrases and operation numbers are used. A sequence number must then be added to the routing to maintain the sequence of operations. Each work center should be numbered. However, these numbers should be coded to describe what operations are possible at that work center. Once a routing has been prepared, each operation and work center can be considered a polycode. Thus all part which have the operation "mill" can be identified. The machine tools which can "mill" are also identified. By simulating MRP, the total hours of milling required and the number of parts requiring milling examined to determine whether machines are overloaded or underloaded, the plant capacity, and the need for changing some routings and processes to more evenly distribute shop loads.

5. *Item Master*. Item master data should be made a part of the routing when it is prepared to avoid generating redundant data. As a matter of fact, if the process and routing are prepared concurrently, the routing is but an extraction of data from the process.

6. *Perishable Tools*. Perishable tools to manufacture a part should be identified on the routing and in the routing database to be able to evaluate total perishable tools required and the number of nonstandard tools which have been designated.

7. *Product Structures*. Product structures are not directly used in a routing, but if the product structure is correct, the routing should reflect the actual assembly operations.

8. *Jigs and Fixtures*. Jigs and fixtures should be identified on the routing and in the routing database. If it is critical, issuance of the routing in the form of a work order can also trigger a tool requirement notification so the proper jigs and fixtures can be moved to the proper work center on time.

9. *Engineering Changes*. Engineering changes will also affect the routing. However, a routing and process may change without any engineering change. Consequently two revision levels may be required on the routing, the engineering revision level, and process and routing revision level.

10. *Production Planning*. Production planning determines when the shop packet should be prepared and issued. The routing should be issued each time the part is manufactured. Otherwise an incorrect routing may be stored at some work centers, which could create an incorrect part. Special work orders and routings may be needed for scrap and rework, and will be a special consideration of industrial engineering.

11. *Part Geometry*. Part geometry in the form of a drawing will accompany the routing. In the case of parts made on NC tools, the picture of the tool path may be a necessary document also.

12. *Inventory Control*. Inventory control algorithms, such as the economical order quantity, determine the number of parts which are manufactured using a routing. The set-up time is a factor in determining this quantity.

13. *Assembly Geometry*. Assembly geometry in the form of a drawing and a parts list form a shop packet with the routing.

14. *Group Technology Cells*. Group technology cells may require a family, not a part, routing. Exceptions to the family routing may have to be identified and accentuated if a series of parts are routed to a cell of machine tools.

15. *Raw Material Inventory*. Raw material inventory, and the methods used to make designs from it, will often determine how the first operation of a routing is generated. The part may require a specific piece of raw material purchased specifically for it. The part may be sawed from a raw material. A series of parts may be sawed from a raw material to be cut into separate parts later. The part may require a "first piece" extra because it must be held firmly as the pieces are cut from it. These are but a few of the variations which may be required on routings of parts made from raw material.

QUALITY

DESIGN RETRIEVAL--------------- X
FAMILY PROCESSES--------------- X
PART FAMILY DATA---------------X
PROCESS POLYCODES------------- X
ITEM MASTER--------------------- X
PERISHABLE TOOLS--------------- X
PRODUCT STRUCTURES----------- X
JIGS AND FIXTURES---------------X
ENGINEERING CHANGES----------- X
PRODUCTION PLANNING-----------X
PART GEOMETRY------------------X
INVENTORY CONTROL-------------X
ASSEMBLY GEOMETRY-------------X
GROUP TECHNOLOGY CELLS-------X
GENERIC BILL OF MATERIAL------X
RAW MATERIAL INVENTORY-------X
ELECTRONIC MAIL-----------------X

Nothing permeates a manufacturing system more than quality, or lack thereof. While described under structured output, its presence is felt in every input, in every output, and in the database accuracy.

Of course, a CIM system merely reflects data handled in engineering and manufacturing. It does not create quality, only reflects it. If inventory accuracy is not high, excessive material will be purchased or manufactured. If the bill-of-material data is not accurate, the quantities of parts needed will be transmitted inaccurately.

Consistency of data and how it is used will enhance quality. For example, a family routing, once created, need be examined and checked only for changes, not in its entirety. Quality control principles were first created in World War II due to lack of adequate labor. Inspection accuracy was based on X bar R charts, prepared to statistically examine critical dimensions of a design. This management by exception enhanced and controlled quality. In the same manner, examination of differences from well-established data will enhance preparation of planning and action documents.

The reusability of parts, assemblies, tools, jigs, and fixtures, and the implementation of standard tolerances, routing sequences, and more, in the form of producibility tips will assist the inspector looking for the real problems, not waste time on the continuous differences. Realistic dimensioning and tolerances appearing in the same place in the same manner create standard communications for better understanding between engineering, manufacturing, and inspection.

These are but some of the benefits of CIM. Use of robots and NC machine tools are also vehicles to create manufacturing repeatability. Combining accurate data with high quality manufacturing equipment and consistent standards will increase accuracy, communications, and, finally, quality.

8
CIM SYSTEMS WITH DESIGN RETRIEVAL

CLASSIFICATION REQUIREMENTS

Since the first steam engine or rifle or Colt revolver came off the assembly line, proliferation of unnecessary design has existed. During the past 50 years, this proliferation has wasted millions of design and drafting hours. Parts already in existence have been drawn and redrawn. Like looking for the proverbial needle in the haystack, finding the one specific part required in a drawing file with 100,000 parts has been almost impossible, or, at best, too time consuming. Some enlightened companies, like Caterpillar Tractor Co., Control Data Corporation, New York Air Brake, Morgan Construction Company, and Cincinnati Millicron, began to understand this waste and implemented design retrieval systems to avoid such duplication. An example of the savings potential is as follows for one company:

100,000 current drawings on file
7000 new designs issued yearly
$400.00 to $6000.00 savings in cost per design
10% retrieval rate with classification
$300,000 estimated savings yearly

These savings are based on finding parts exactly identical to those required by the designer. Finding similar parts, altering existing parts to make them useful for two applications, or using an old design to "copy" for a new design saves an additional 18% of the cost for initiating a new design. Thus, through classification systems, up to 28% of the former design cost may be eliminated. This type of data repeats itself over and over from company to company.

WHAT IS CLASSIFICATION?

What is classification? What is its value? What are the types of classification? How should it be used? These questions will be answered in this section.

Figure 1 shows a progressive flow of the importance and pervasiveness of classification. Originally, two objectives for implementing a retrieval system existed, design retrieval and standardization of operations and processes in the machine shop. Many companies designed their own systems of retrieval, which generally fell into the categories of name coding, block coding, random coding, picture coding, multiple codes, or decision tree coding. Most coding systems have been developed by well meaning amateurs so they have, more often than not, failed to become the powerful tools needed in CIM, much less for any other project for which they were originally intended. Professionally designed classification systems almost exclusively utilize either a classification system utilizing a decision tree or multiple codes.

Again, looking at Figure 1, notice that the new reasons for using a design retrieval system include CAD/CAM, and a large database from which one of several designs must be retrieved. In addition, a family of parts and associated data in engineering and manufacturing are significant modern requirements.

When used properly, a design retrieval system becomes a pervasive influence in all areas of CIM, including production and inventory analysis, group technology, and manufacturing resource planning.

Classification is defined as:

> A system in which individual items or features are grouped together by virtue of their similarities, and then separated by their essential differences.
>
> Multiple codes divide designs into a series of logical but unrelated characteristics. Decision tree coding separates, through a sequence of multiple but sequential and dependent characteristics the designs into "families" of parts based on their features or function.

These classification systems are also called more technically a polycode, and a monocode.

CLASSIFICATION POPULARITY

While many respectable companies implemented a classification system, until recently they have not been implemented as a generally accepted tool. Here are the reasons:

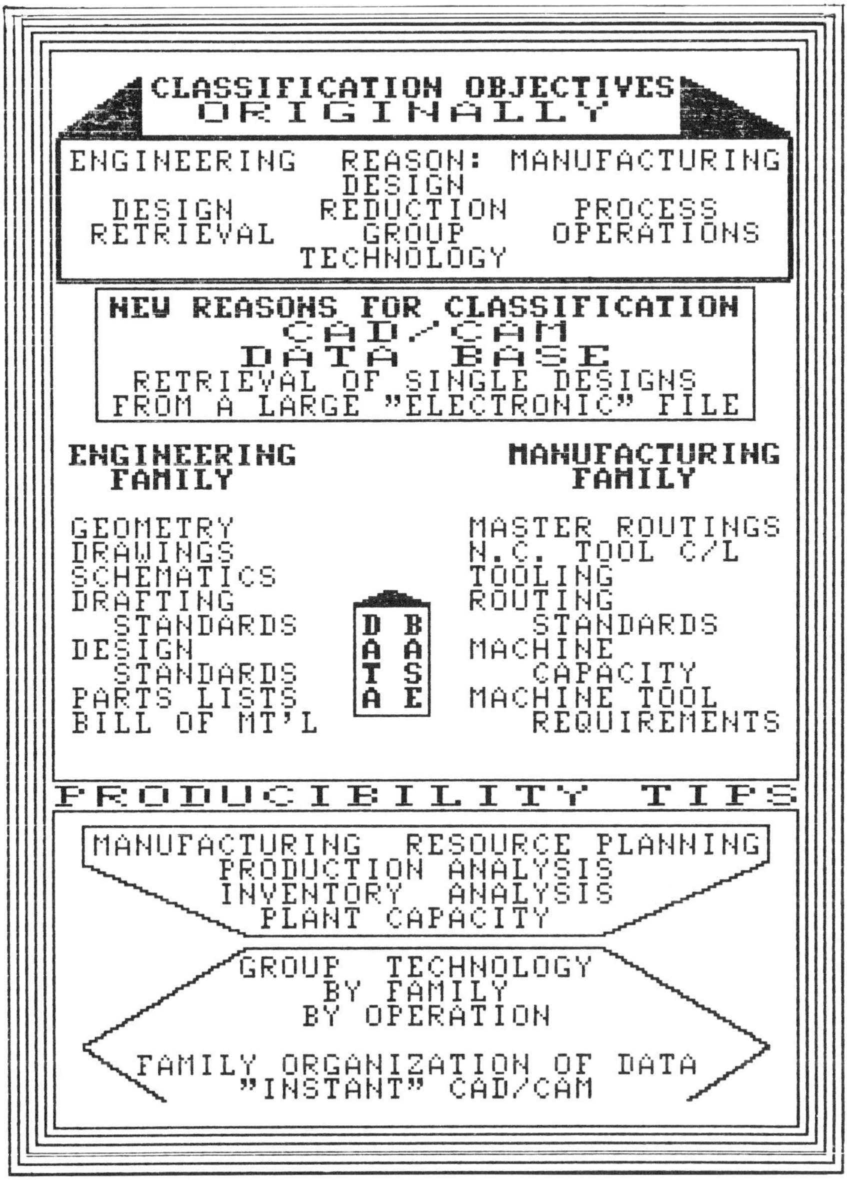

FIGURE 1 The objectives of classification have broadened and become more important with the implementation of a CIM system.

1. Most modern systems have been sold from the bottom up. Only when general management has understood the value of a classification system has it been implemented successfully.
2. Most manufacturing personnel have given up trying to show engineering how duplicate parts exist and/or how parts could be made more economically. With no feedback from manufacturing, there seems to engineering to be no cause, therefore no effect.
3. For the better part of 10 years, at least, management of companies have aimed for profit based on short range goals. Classification systems are not immediate pay-off projects (unless CAD/CAM is considered).
4. Most professional classification systems are designed by consulting firms who have little marketing experience, and a natural reticence to divulge too much of their system.
5. The consulting firms implementing classification systems have little capital to invest in promotion and depend upon word-of-mouth publicity.
6. Engineering departments and their management have often strongly rejected classification systems as a means of finding old designs. In their minds, ability to "create" becomes restricted. In addition, most amateur systems make it quite difficult for a designer to retrieve easily.
7. A retrieval system, when it works, is colorless. It just makes money with no fanfare.

Will retrieval systems become more successful and popular? There have been many signs in the past few years that classification systems are being studied with much more interest. More and more companies realize they must clean up their present drawing files and also set up a system where they can find new designs. Several reasons exist for this renewed interest:

Shortage of good designers
Fewer mass produced products
Conservation of materials
Recognition of part families
Group technology
CAD/CAM systems
Foreign competition

NEW REASONS FOR CLASSIFICATION

This brings us to the second box in Figure 1, "New Reasons for Classification." Today, drawings are physically filed in large cabinets. Anyone can browse through them to find parts and assembly drawings. (How would you like to "browse" through drawing files containing 80,000 drawings in A, B, C, D, and E sizes?)

There is growing recognition that this practice will cease, unless each new design made on a CAD/CAM system, after being plotted, is filed in the same manner. More modern and inexpensive filing methods are now available as microfilm and microfiche. Soon it will be possible to convert new CAD/CAM designs directly to one of these media economically.

If each old design cannot be directly scanned as before, some faster means must be discovered to find them. A classification system is the solution. If, tied to the classification system for each family of parts, a catalog of them with critical dimensions and other features, such as thread sizes and material, is made available, rapid retrieval becomes a reality.

Thus two conditions exist. A large database requires a retrieval system, and a retrieval system allows a large database to exist.

PROFESSIONAL CODING SYSTEMS

As implied earlier, two different types of professional classification systems are prevalent. Each type has different strengths and weaknesses and performs different functions in manufacturing.

A multiple code has the ability to take specific features of a design and emphasize them or separate them from all others. This makes the system ideal for a manufacturing application, particularly in applying group technology methods. An example of features which might be highlighted in a multiple code is shown in Figure 2. Each of the distinguishing features or attributes assists the manufacturing planner to determine the type or types of machine tools required to manufacture a part. Note, however, the penalties of this method of retrieval. Each attribute requires a separate code attached to it. Thus the 21 features listed in Figure 2 require at least a 21-digit code, best separated with dashes. If all of the different types of threads, for example, need to be coded within the feature threads, an additional one or two digits may be needed for this attribute. Thus the code number may total many more than 21 digits. Notice also, that retrieving similar shaped parts, or parts with the same functions, would be quite difficult with these features. A clever designer might select the proper series of features and obtain a small family of parts, but the amount of time needed to decide how to do it can be excessive, and there is no guarantee that all the parts with similar shapes or functions can thus be located.

Figure 3 is the picture of a code page from a decision tree code system. At the top right-hand corner, the number 3XXXX appears.

ENVELOPE
- BATCH SIZE
- MAXIMUM DIAMETER
- LENGTH
- TURNING
- SINGLE DIA.
- MULTI–DIA.
- DIAMETERS
- SINGLE
- MULTIPLE

EXTERNAL
- CHAMFERS
- RADII
- ANNULAR GROOVES
- TAPERS
- KNURLS
- THREADS
- DRILLING AND BORING
- CENTER HOLES
- LONGITUDINAL HOLES
- TRANSVERSE HOLES

INTERNAL
- RADII INTERNAL
- ANNULAR GROOVES
- THREADS
- MILLING. SLOTTING
- SPLINES

SPECIAL
- KEYWAYS
- HEAT TREATMENT
- FORMING
- ANCILLARY DATA

FIGURE 2 Some examples of general polycodes may illustrate the types which can be added to a design retrieval monocode system.

By definition, all parts on this page are designed parts. Starting from the left top, and proceeding to the right bottom of the page, decisions are made to determine a family of parts. The dotted lines between statements are like commas, the solid lines like periods. Let us follow one path

The part we will look for is classified by shape not function.
The shape is regular, not irregular.
The shape is axial, not other.
The axial part is round not other.
The round part is straight not bent.

(Having reached a specific set of boxes, other decisions appear across the top of them and beneath the sequential numbers.)

The round straight part has a multiple not a single diameter. The multiple diameter is at one end, not between ends.

We have now found a family of parts, 305XX. An additional page takes this family and divides each box shown on this page

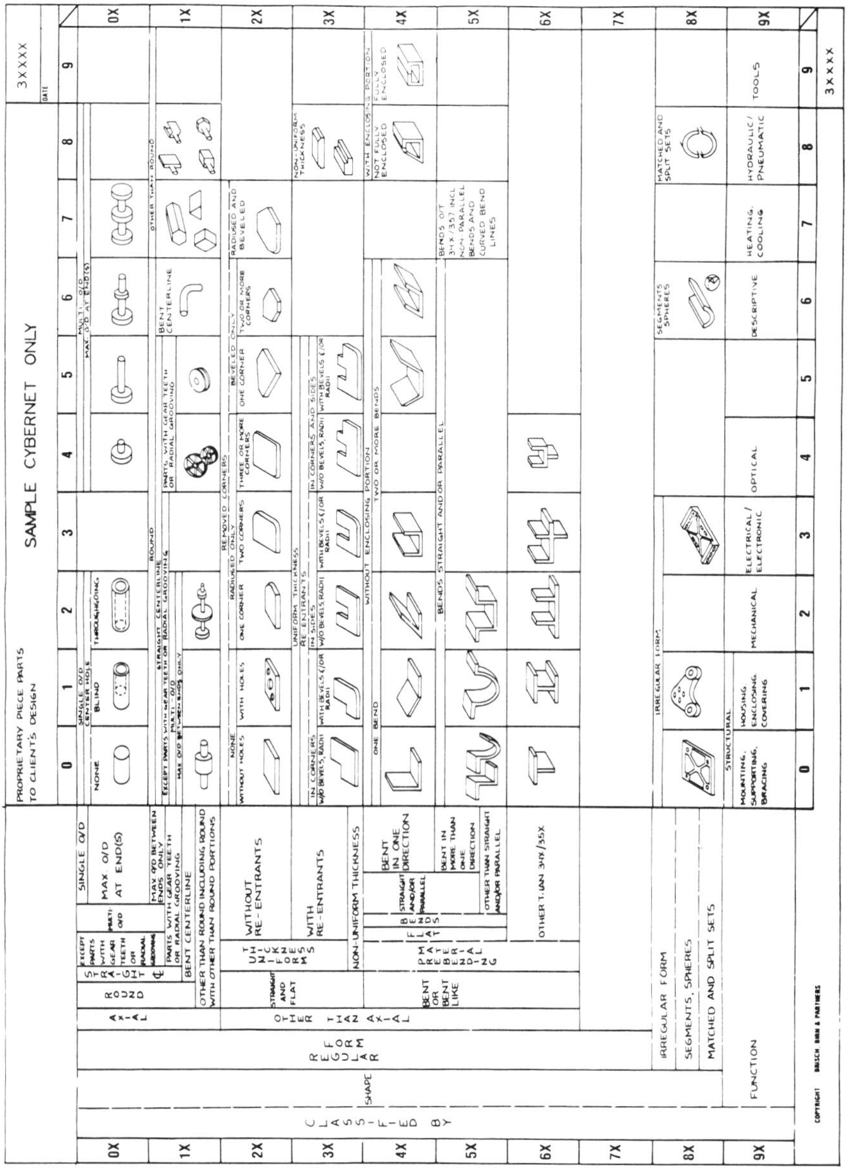

FIGURE 3 The code page shown here will be valuable both to the designer to retrieve parts and from which to create parametric part drawings, but also for manufacturing to create master routings and all other manufacturing data needed for a family of parts. (Courtesy of Brisch Birn and Partners, Fort Lauderdale, Fla.)

into an additional 100 boxes. This provides the potential of 10,000 families of parts.

Figure 4 shows a family tree diagram which is used to generate code pages such as the one pictured. Along the right edge of the family tree appears the quantity of drawings to be found under each segment. This characteristic of a decision tree system is important, since it allows a means to distribute designs evenly through the system and provide also for expansion.

The retriever is "led" to his retrieval efforts. Once he knows the rules, the system does much of the thinking for him. However, the system is not designed to find a specific attribute throughout all the families. Threads on a part, for example, may or may not always be distinguished one family from another. Thus the ability to isolate a series of features is probably impossible.

It seems clear that, with the lack of popularity of classification systems in engineering, many more multiple coding systems might be installed and utilized by manufacturing. Such is not the case, however, because group technology has not been popular in the United States. This, too, is changing, and many more systems are being implemented.

In Europe and Africa the predominant classification systems are decision tree systems, even though group technology is more important in Europe. Why has this occurred? It has been found that tying a multiple coding system to a decision tree coding system makes the best and most powerful combination. Group technology works well with both families of parts and specific features. Figure 5 shows a set of similar part families brought together as a "super" family and is called a composite, making an even more effective tool. All of them may be machined on the same equipment with the same set-ups. Figure 6, on the other hand, shows a variety of dissimilar parts, each, however, with identical features, as boring, and drilling, and bolt circles. Imagine the power of the combination.

COMBINED CLASSIFICATION SYSTEM

The decision tree system retrieves a family and specific parts from a large file today. It will be far more efficient and responsive when used to interrogate a large electronic file. But how about the multiple codes? Figure 7 shows a picture of multiple codes attached to a series of part families which were first classified by a decision tree system. On the left side of the page, the features desired have been tabulated. Some additional characteristics of the parts have been tabulated across the top of the page as nonmetals, metals which are ferrous or nonferrous, and so on. The parts covered by this page are all thin, short, round parts, where the length is less than the diameter. The external characteristics highlighted for these parts are flats, radial slots and/or grooves, gear and/or

```
******************************************************************************
******************************************************************************
*****                           HIERARCHIAL                               ****
*****                              TREE                                   ****
*****                    AXIAL STRAIGHT CENTER LINEE                      ****
*****                    ROUND SINGLE O. D. PORTION                       ****
******************************************************************************
*****CENTER HOLE                                                 TOTAL 1004 ****
***** I  NONE                                                          1004 ****
***** I    I  WITHOUT THREAD OR SPLINES                                1004 ****
***** I    I    I  PLAIN                                                216 ****
***** I    I    I    I  HOLES                                           216 ****
***** I    I    I    I    I  NONE                                           ****
***** I    I    I    I    I  TRANSVERSE ONLY                             90 ****
***** I    I    I    I    I    I  BLIND                                  34 ****
***** I    I    I    I    I    I    I  ONE                               12 ****
***** I    I    I    I    I    I    I  TWO                               22 ****
***** I    I    I    I    I    I    I  THREE OR MORE                        ****
***** I    I    I    I    I    I  THROUGH GOING                          56 ****
***** I    I    I    I    I    I    I  ONE                               56 ****
***** I    I    I    I    I    I    I  TWO                                  ****
***** I    I    I    I    I    I    I  THREE OR MORE                      0 ****
***** I    I    I    I    I  LONGITUDINAL ONLY                           27 ****
***** I    I    I    I    I    I  BLIND                                  25 ****
***** I    I    I    I    I    I    I  ONE                               12 ****
***** I    I    I    I    I    I    I  TWO                                  ****
***** I    I    I    I    I    I    I  THREE OR MORE                     13 ****
***** I    I    I    I    I    I  THROUGH GOING                           2 ****
***** I    I    I    I    I    I    I  ONE                                  ****
***** I    I    I    I    I    I    I  TWO                                1 ****
***** I    I    I    I    I    I    I  THREE OR MORE                      1 ****
***** I    I    I    I    I  TRANSVERS AND LONGITUDINAL                  99 ****
***** I    I    I    I    I    I  BLIND ONLY                              1 ****
***** I    I    I    I    I    I  THROUGH GOING                          98 ****
***** I    I    I    I    I    I  BLIND AND THROUGHGOING                    ****
***** I    I    I  OTHER THAN PLAIN                                     788 ****
***** I    I    I    I  WITH OUT PROTRUSIONS                            788 ****
***** I    I    I    I    I  WITH OUT HOLES                               0 ****
***** I    I    I    I    I    I  WITH FLATS ONLY                         0 ****
***** I    I    I    I    I    I    I  BETWEEN ENDS ONLY                    ****
***** I    I    I    I    I    I    I    I  ONE                            ****
***** I    I    I    I    I    I    I    I  TWO OR MORE                    ****
***** I    I    I    I    I    I    I  AT END(S) ONLY                       ****
***** I    I    I    I    I    I  WITH SLOTS ONLY                         0 ****
***** I    I    I    I    I    I    I  ONE END ONLY                         ****
***** I    I    I    I    I    I    I  BOTH ENDS                            ****
***** I    I    I    I    I    I  WITH GROOVES ONLY                         ****
***** I    I    I    I    I    I  WITH KEYWAYS/ SEATS ONLY                  ****
***** I    I    I    I    I  WITH HOLES                                 756 ****
***** I    I    I    I    I    I  TRANSVERSE ONLY                        35 ****
***** I    I    I    I    I    I    I  WITH FLATS ONLY                   23 ****
***** I    I    I    I    I    I    I    I  BETWEEN ENDS ONLY           23 ****
***** I    I    I    I    I    I    I    I    I  ONE                      ****
***** I    I    I    I    I    I    I    I    I  TWO                   23 ****
***** I    I    I    I    I    I    I    I  AT ENDS ONLY                  ****
***** I    I    I    I    I    I    I    I  AT ENDS AND BETWEEN ENDS      ****
***** I    I    I    I    I    I    I  WITH SLOTS ONLY                   12 ****
***** I    I    I    I    I    I    I    I  AT ONE END                  12 ****
***** I    I    I    I    I    I    I    I  AT BOTH ENDS                  ****
***** I    I    I    I    I    I    I  WITH GROOVES ONLY                    ****
***** I    I    I    I    I    I    I  WITH KEYWAYS/SEATS ONLY              ****
***** I    I    I    I    I    I  LONGITUDINAL ONLY                      67 ****
******************************************************************************
```

FIGURE 4   This chart is an example of a family or decision tree created as a classification system is designed.

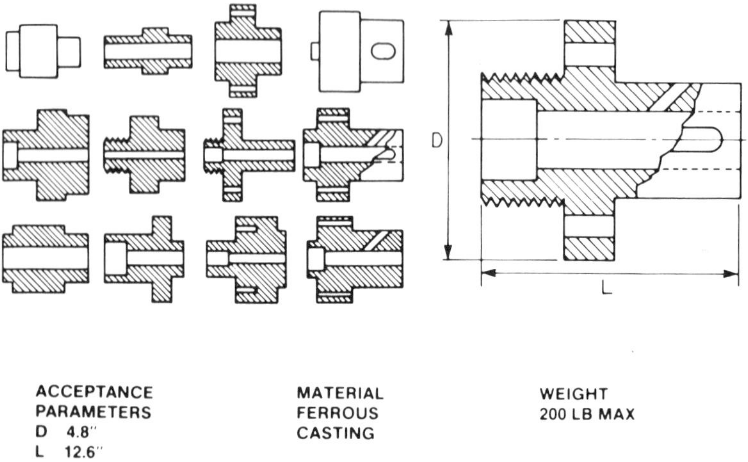

FIGURE 5 Utilizing group technology, parts may be grouped together using parameters as diameter (D), length (L), material, and weight. All parts together make a composite and may be routed over the same machine tools, and together, increasing the amount of parts made in one batch.

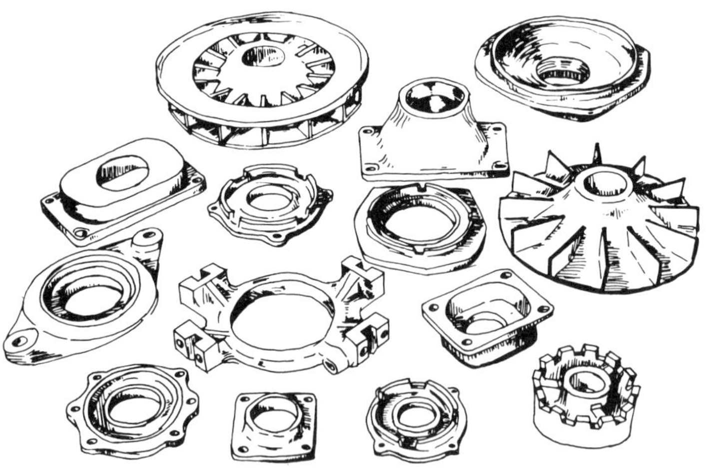

FIGURE 6 Another (and the original) concept of group technology is to bring together parts, whether similar or not, with the same operations to be performed. The parts in this picture have identical boring and drilling operations. Therefore, minimum setup time is required.

BRISCH, BIRN & PARTNERS for: POLYCODES FOR AXIAL PARTS WITH GEAR &/OR SPLINED PORTIONS INCLUDING RACKS

			0	1	2	3	4	5	6	7	8	9
MATERIAL	CHEMISTRY			NON-FERROUS METALS		NON-STEELS		FERROUS METALS	STEELS / ALLOYS INCL. STAINLESS & TOOL MACHINEABILITY			
			NON-METALLIC 00	ALUMINUM & 01	COPPER & 02	CAST IRON DUCTILE IRON 03	SEMI-STEELS 04	PLAIN CARBON 05	>.90 06	≤.90 >.70 07	<.70 08	MATERIALS O/T 00/08 09
	FORM/SIZE		TO DRAWING SPECIFICATION BLANK			ROUND ONLY DIAMETER		NOT TO DRAWING SPECIFICATION AXIAL O/T ROUND SIZE				
			CASTING 10	FORGING 11	O/T 10/11 12	≤2" 13	>2" ≤5" 14	>5" 15	≤2" 16	>2" ≤5" 17	>5" 18	O/T AXIALFOR RACKS 19
	HEAT TREATMENT			SELECTIVELY		DRAW ONLY O/T 22 TO		HARDEN (AND)	CASE ONLY			
			NONE 20	ONLY 21	DRAW 22	≤35 Rc 23	>35 Rc 24	CYANIDE DRAW 25	STRESS RELVE NITRIDE 26	CARBURIZE DRAW 27	ANNEAL DRAW 28	O/T21/28 INCL. COMBINATIONS, DOUBLE DRAW 29
METHOD OF MANUFACTURE OF FEATURES	GEAR TEETH	EXTERNAL		W/O FINISH MACHINING			MACHINED / WITH FINISH MACHINING / SHAVE INCL. CROWN SHAVE			O/T SHAVE INCL. GRIND, LAP, BURNISH		
			NOT MACHINED 30	HOB 31	SHAPE 32	MILL 33	HOB 34	SHAPE 35	MILL 36	HOB 37	SHAPE 38	MILL 39
		O/T 3X, INCL. INTERNAL & END(S)		O/T ENDS, FACES ONLY / W/O FINISH MACHINING			MACHINED / WITH FINISH MACHINING			ENDS, FACES ONLY		
			NOT MACHINED 40	HOB 41	SHAPE 42	BROACH 43	HOB 44	SHAPE 45	BROACH 46	SHAPE 47	O/T SHAPE 48	O/T 41/48 INCL. COMBINATIONS 49
	SPLINES, SERRATIONS, KEYWAYS			MACHINED / SPLINES INCLUDING KEYWAYS / MALE / W/O FINISH MACHINING				FEMALE / W/O FINISH MACHINING				
			NOT MACHINED 50	HOB 51	SHAPE 52	MILL 53	51/53 WITH FINISH MACHINING 54	HOB 55	SHAPE 56	BROACH 57	55/57 WITH FINISH MACHINING 58	O/T SPLINES KEYWAYS 59
	O/T 3X/5X	CENTER HOLE		MACHINED / BLIND / W/O THREADS		WITH THREADS		THROUGHGOING / W/O THREADS			FULLY THREADED	
			NONE/NOT MACHINED 60	DRILLED(&) REAMED 61	DRILLED, BORED, (&) REAMED 62	DRILLED (&) TAPPED 63	DRILLED (&) BORED 64	DRILLED (&) REAMED 65	DRILLED, BORED, (&) REAMED 66	PARTIALLY THREADED 67	DRILLED (&) TAPPED 68	DRILLED (&) BORED 69
		O/T CENTER HOLE		ROUND HOLES ONLY / LONGITUDNAL ONLY		TRANSVERSE ONLY				O/T 70/75		
			NONE/NOT MACHINED 70	DRILLED 71	DRILLED (&) TAPPED 72	DRILLED 73	DRILLED (&) TAPPED 74	COMBINATIONS OF 71/73 75	HOLES O/T ROUND 76	FLATS ONLY 77	71/75 +76 78	71/75 +77 79
MAXIMUM SURFACE TO NEXT NEAREST STANDARD			NONE OR CLEAN UP ONLY 2000 √ 80	ROUGH 1000 √ 81	SEMI-ROUGH 500 √ 82	MEDIUM 250 √ 83	SEMI-FINE 125 √ 84	FINE 63 √ 85	SMOOTH 32 √ 86	GROUND 16 √ 87	POLISH 8 √ 88	SUPER FINISH 4 √ 89
LOT SIZE PARTS/ORDER			1 OFF 90	2 OFF 91	>2 ≤5 92	>5 ≤9 93	>9 ≤15 94	>15 ≤20 95	>20 ≤30 96	>30 ≤50 97	>50 ≤100 98	>100 99

EXHIBIT B

FIGURE 7 The classification chart shows an example of multiple codes for which can be attached to specific part families to enhance the total classification effort. (Courtesy of Brisch Birn and Partners, Fort Lauderdale Fla.)

sprocket teeth, keyways and serrations. Each of these characteristics can be further distinguished by machining characteristics, such as turning, threading, facing, milling, grinding, combinations, and hobbing. Further, by having first separated all the population of parts into a manageable set of families, the number of features requiring coding are less and can be handled and absorbed by the user better. Due to the organization of the features on the page, fewer codes are required to classify the features. We thus have the best of both worlds.

Figure 8 is a pictorial summary of a combination classification system. The first two pages are used to identify a family of closely related parts whose family code is 32413. A multiple code, designed for this type of part and other families closely similar, is found on an additional page. Selecting additional features as necessary completes the process, like the "tail" of a dog. The multiple codes merge with the decision tree.

CIM AND CLASSIFICATION

We have now described the principles of classification and recognized how designs may be divided into either families with similar shapes

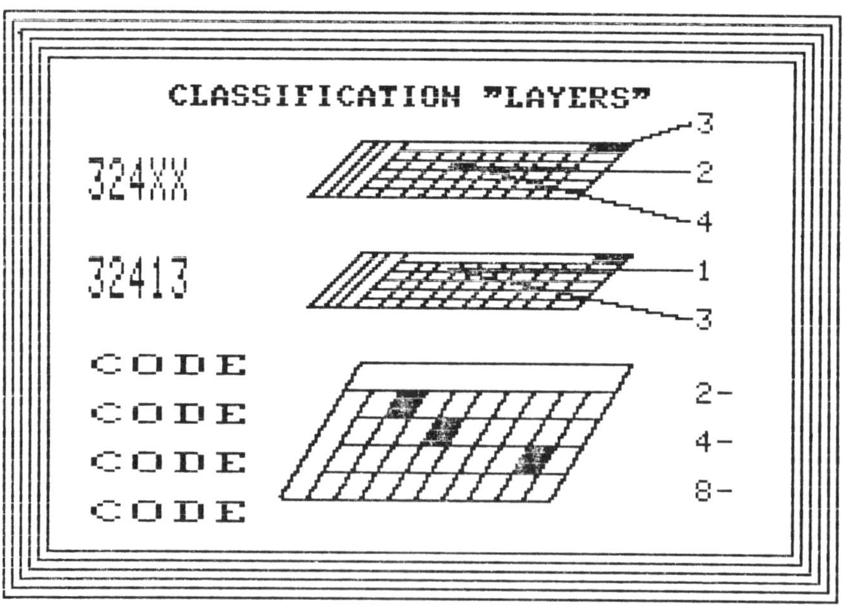

FIGURE 8 The illustration summarizes the use of a monocode for design retrieval combined with polycodes to create a powerful set of retrieval criteria.

or features with the same characteristics. Let us see how a family of parts adds significant value to a CIM system.

Engineering Requirements

Figure 1 shows a list of engineering family attributes needed to enhance a CIM system:

Solid geometry
Surface geometry
Electronic data
Mechanical designs
Schematics
Design analysis
Parts lists
Bills of material

If this data is presumed to be in a database and available to engineering, the following activities can be improved as described.

Drafting

If a company has 60,000 current drawings on file which are divided into 2000 families, each family will have approximately 30 parts. Since they are closely similar, data for one part applies generically to any other. Thus the family drawing, tied to a graphics program, can be utilized to make the drawing of a new part in the family with minimum drafting time or rearrangement of data. A design analysis program for gears, for example, will apply to all parts in the same family.

Design and Drafting Standards

Today, drafting and design standard books gather dust on the shelf of the designer and drafter. CIM systems will make them come alive. Now the specific standards applied to a family of parts will become more customized and applicable and can be pictured graphically in the form of dimensions, arrow heads, dimensions of fillets, and standard welding symbols. This type of information is impossible to manage on a part-by-part basis. Past experience has proven it.

Design Retrieval

Engineers will no longer criticize classification as a means to decrease creativity, and will, instead, hail it as a means for shortening their work and making it easier. Retrieval of identical parts will no longer be the motivation of classification. Retrieval of information will become dominant.

Figure 9 shows the picture of typical family drawings. Figure 10 shows the catalog for this family. The family drawing has generic dimensions which correspond to the columns of data in the catalog. In addition, special characteristics are listed, in this case material specification, revision level, and, of course, the part numbers of specific parts.

Figure 11 shows some other family drawings, each with their own family number. It seems to be a rare design so complex it can"t be represented pictorially in a family. (This seems to be the case for at least 80% of all drawings.) (Another significant possibility worthy of exploration is to store the family geometry as a master file, and store the differences only in the part file. When the geometry is displayed for a specific part, the two are combined. This could reduce the size of the database for parts considerably.)

It is thus more practical and easier to work with 2000 families than 60,000 individual parts.

Design Inversion

A classification system will also assist in changing the entire concept of design. Today designers make general and detailed layouts of new designs. Drafters then take them and extract from them the detailed parts necessary to conform to the design. Many times the drafter's interpretation is faulty during this transfer of information. Also, the drafter often traces parts of the layout a piece at a time to draw the detailed parts. There is much duplication of effort.

Tomorrow, the designer will select appropriate parts through the classification system and align them on the screen of the terminal and "build" the design or layout. Where parts don't fit or meet the requirements, designers will stop, utilize the proper graphics programs to prepare new part designs, and continue with the design. In essence, designers are using real "building blocks" as they design. The speed, accuracy, economics, and satisfaction in this technique will radically increase the productivity of design.

Parts List

The transmission of design data takes the form not only of drawings, but paperwork in the form of bills of material and parts lists. As an assembly is designed; the parts required for it are tabulated. If this data is extracted as text when the drawing is made, it must then be sent to the business systems utilizing the data. This pass-through of data is clear and straightforward. It may be the desire of engineering, however, to retain the data for use by designers when engineering changes and/or new designs are prepared. This will cause some conflict since which segment of the database it should be stored in must be determined. The chart in the book shows them

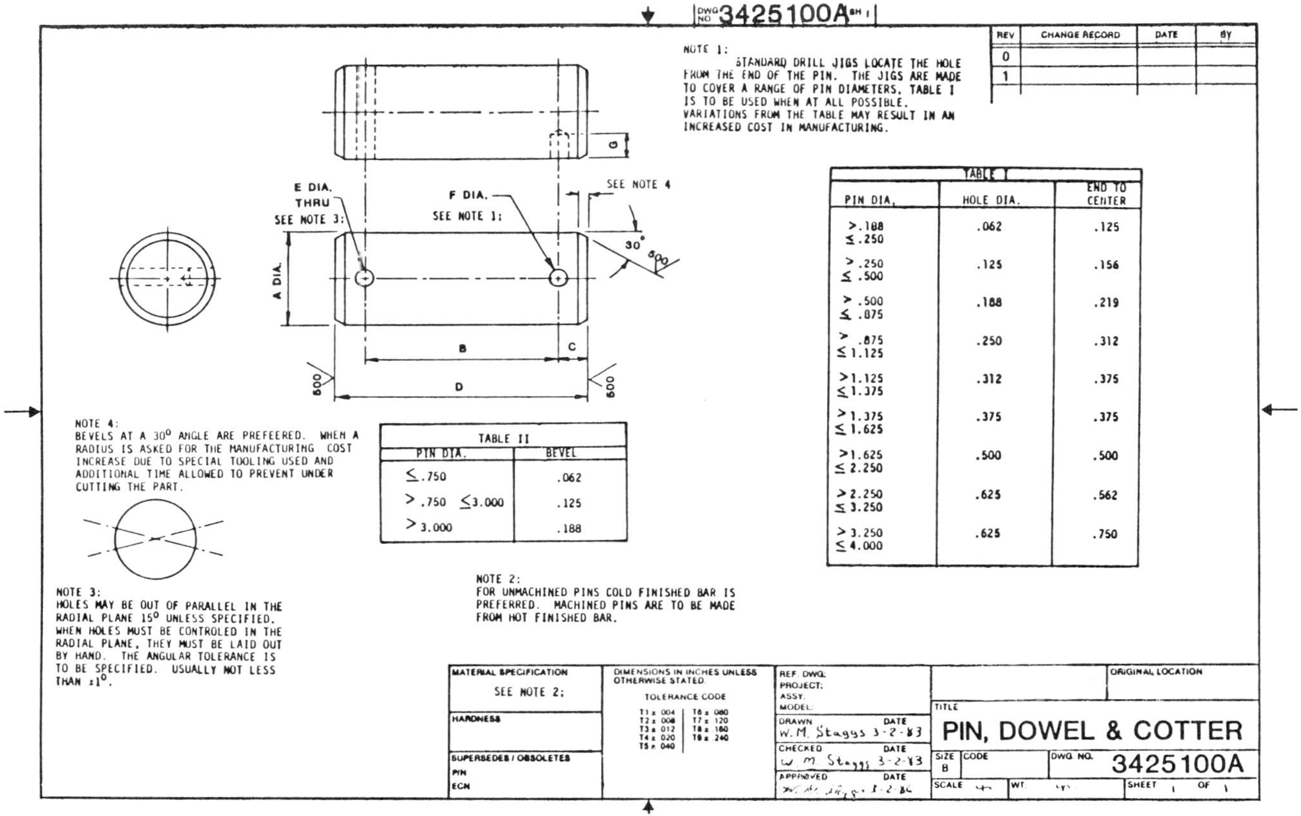

FIGURE 9 A typical family drawing for family 34251 with specific producibility tips and letter dimensions for use with a catalog of parts is an example of Engineering and Manufacturing integration using classification.

ENGINEERING CLASSIFICATION MASTER CATALOG

DATE 08/13/86

PAGE NO. 1

AXIAL, ROUND, SINGLE O.D.
OTHER THAN THIN (LENGTH IS GREATER THAN 1/4 O.D.).
WITHOUT THREAD ON O.D., GROOVES, KEYWAYS, TAPER PORTIONS.
WITH ROUND TRANSVESE HOLES ONLY.
ONE BLIND AND ONE THRU ONLY.

FAMILY - 34251

SECTOR RANGE 001 THRU 899 ALL HOLES WITHOUT THREAD ONLY.

FAMILY DRAWING 3425100A

SEC	PIN DIA A	CTR TO CTR HOLES B	BLIND HOLE END TO CENTER C	LENGTH D	DIA THRU E	DIA. F	X DEEP G	OTHER DATA	MATL SPEC	PREF PART	PART NUMBER	REV LEV	SEC
001	.750	4.812	.219	5.250	.188	.250 X	.375		A017		450489	00	001
145	.750	24.900	.375	25.750	.188	.250 X	.375		A017		448934	00	145
120	.750	51.100	.375	51.900	.188	.250 X	.375		A017		448929	00	120
119	.750	68.100	.375	68.900	.188	.250 X	.375		A017		448935	00	119
002	1.000	23.250	.312	24.000	.266	.312 X	.375		A017		436107	00	002
224	1.249	7.969	.500	9.000	.312	.375 X	.375	CHROME PLATE	A017		1036026	01	224
003	1.250	4.875	.500	5.750	.312	.375 X	.375		A017		402070	00	003
197	1.250	5.188	.500	6.188	.312	.375 X	.375		A017		401907	00	197
112	1.250	5.750	.500	6.750	.312	.375 X	.375		A017		400113	00	112
168	1.250	8.000	.375	8.750	.312	.375 X	.375		A017		440728	00	168
184	1.250	10.000	.375	10.750	.312	.375 X	.375	SUP.BY 440728	A017		416478	01	184
075	1.250	11.250	.562	12.375	.312	.375 X	.375		A054		1054483	00	075
154	1.250	13.062	.500	14.062	.312	.375 X	.375		A017		L 6869	00	154
004	1.250	13.562	.500	14.562	.312	.375 X	.375		A017		412840	01	004
092	1.250	15.188	.500	16.188	.312	.375 X	.375		A017		409492	00	092
222	1.250	15.438	.500	16.438	.312	.375 X	.375		A017		506687	00	222
005	1.250	16.406	.500	17.406	.312	.375 X	.375		A017		430251	00	005
160	1.250	20.750	.500	21.750	.312	.375 X	.375		A017		464095	00	160
006	1.250	21.469	.500	22.469	.312	.375 X	.375		A017		1101001	00	006
181	1.250	23.438	.312	24.062	.312	.375 X	.375	SUP.BY 436674	A017		413587	02	181
007	1.250	26.625	.500	27.688	.312	.375 X	.375		A017		436012	02	007
202	1.250	27.000	.500	28.000	.312	.375 X	.375		A017		454960	01	202
066	1.250	29.358	.500	30.358	.312	.375 X	.375		A017		1052820	00	066

(A)

ENGINEERING CLASSIFICATION MASTER CATALOG

DATE 08/13/86

PAGE NO. 2

AXIAL, ROUND, SINGLE O.D.
OTHER THAN THIN (LENGTH IS GREATER THAN 1/4 O.D.).
WITHOUT THREAD ON O.D., GROOVES, KEYWAYS, TAPER PORTIONS.
WITH ROUND TRANSVESE HOLES ONLY.
ONE BLIND AND ONE THRU ONLY.

FAMILY - 34251

SECTOR RANGE 001 THRU 899 ALL HOLES WITHOUT THREAD ONLY. FAMILY DRAWING 3425100A

SEC	PIN DIA A	CTR TO CTR HOLES B	BLIND HOLE END TO CENTER C	LENGTH D	DIA THRU E	DIA. X DEEP F G	OTHER DATA	MATL SPEC	PREF PART	PART NUMBER	REV LEV	SEC
008	1.250	30.500	.500	31.500	.312	.375 X .375		A017		406838	00	008
171	1.250	31.062	.500	32.062	.312	.375 X .375		A017		442998	00	171
125	1.250	33.625	.500	34.625	.312	.375 X .375		A017		409493	00	125
009	1.250	38.250	.500	39.250	.312	.375 X .375		A017		1102402	00	009
188	1.250	39.000	.500	40.000	.312	.375 X .375		A017		1100471	00	188
221	1.250	39.375	.531	39.375	.312	.375 X .375		A017		1104030	02	221
081	1.250	42.590	.500	43.594	.312	.375 X .375		A017		438161	00	081
010	1.250	42.625	.500	43.625	.312	.375 X .375		A017		1035485	00	010
110	1.250	43.000	.500	44.000	.312	.375 X .375		A017		410962	00	110
011	1.250	47.125	.500	48.125	.312	.375 X .375		A017		422209	00	011
250	1.250	48.625	.500	49.625	.312	.375 X .375		A017		1076515	00	250
013	1.250	54.031	.500	55.031	.312	.375 X .375		A017		443281	00	013
246	1.250	54.219	.500	55.219	.312	.375 X .375		A017		1076699	00	246
012	1.250	72.875	.500	73.875	.312	.375 X .375		A017		419460	00	012
116	1.250	88.875	.500	89.875	.312	.375 X .375		A017		431618	00	116
098	1.374	4.562	.625	5.820	.375	.500 X .375	CHROME PLATE	A027		449229	01	098
162	1.375	5.312	.625	6.578	.375	.500 X .375	CHROMIUM PLATED	A027		449263	01	162
080	1.500	4.125	.625	5.375	.375	.500 X .375		A017		404669	00	080
014	1.500	4.385	.625	5.635	.375	.500 X .375		A017		438185	00	014
129	1.500	4.750	.625	6.000	.375	.375 X .375	255-302 BHN	A028		442772	00	129
015	1.500	4.750	.625	6.000	.500	.500 X .375		A017		426421	00	015

FAMILY - 34251

(B)

FIGURE 10 The catalog of parts shown here for family 34251 shows the proliferation of parts found when they were classified.

ENGINEERING CLASSIFICATION MASTER CATALOG DATE 08/13/86

PAGE NO. 3

AXIAL, ROUND, SINGLE O.D.
OTHER THAN THIN (LENGTH IS GREATER THAN 1/4 O.D.).
WITHOUT THREAD ON O.D., GROOVES, KEYWAYS, TAPER PORTIONS.
WITH ROUND TRANSVESE HOLES ONLY.
ONE BLIND AND ONE THRU ONLY.

FAMILY - 34251

SECTOR RANGE 001 THRU 899 ALL HOLES WITHOUT THREAD ONLY. FAMILY DRAWING 3425100A

SEC	PIN DIA A	CTR TO CTR HOLES B	BLIND HOLE END TO CENTER C	LENGTH D	DIA THRU E	DIA. F	X	DEEP G	OTHER DATA	MATL SPEC	PREF PART	PART NUMBER	REV LEV	SEC
152	1.500	5.000	.625	6.250	.375	.484	X	.375		A017	M	6115	00	152
177	1.500	5.500	.625	6.750	.375	.500	X	.375		A017		437444	00	177
183	1.500	5.500	.625	6.750	.500	.500	X	.375		A017		426420	02	183
157	1.500	6.750	.625	8.000	.375	.500	X	.375		A017	L	6631	00	157
016	1.500	7.375	.625	8.625	.375	.500	X	.375		A017		419458	00	016
186	1.500	7.563	.625	8.812	.375	.500	X	.375		A017		412259	01	186
017	1.500	8.750	.625	10.000	.375	.500	X	.375		A017		417827	00	017
018	1.500	9.375	.625	10.625	.375	.484	X	.375		A017		1250342	02	018
019	1.500	11.188	.625	12.438	.375	.484	X	.375		A017		412355	02	019
020	1.500	11.500	.500	12.500	.375	.500	X	.375		A017		402178	00	020
239	1.500	18.500	.625	19.750	.375	.500	X	.375		A017		1071661	00	239
108	1.501	23.750	.625	25.000	.375	.500	X	.375		A017		430700	00	108
223	1.750	3.250	.625	4.500	.500	.500	X	.500		A017		501620	00	223
153	1.750	3.500	.625	4.750	.500	.500	X	.500		A017		512787	00	153
086	1.750	3.500	.750	5.000	.500	.500	X	.500	HARD CHOMIUM PLATE	A054		1108568	00	086
187	1.750	4.000	.625	5.250	.500	.500	X	.500		A017		404668	00	187
207	1.750	12.125	.625	13.375	.500	.500	X	.375		A017		1046339	00	207
156	1.770	4.375	.625	5.625	.500	.484	X	.500		A036	F	6149	01	156
078	1.771	11.250	.562	12.375	.312	.375	X	.375		A054		1054490	00	078
077	1.771	11.312	.625	12.562	.500	.500	X	.500		A054		458364	00	077
191	1.808	3.625	.500	4.625	.375	.359	X	.375		A036	L	4282	01	191

FAMILY - 34251

(C)

FIGURE 10 (Continued)

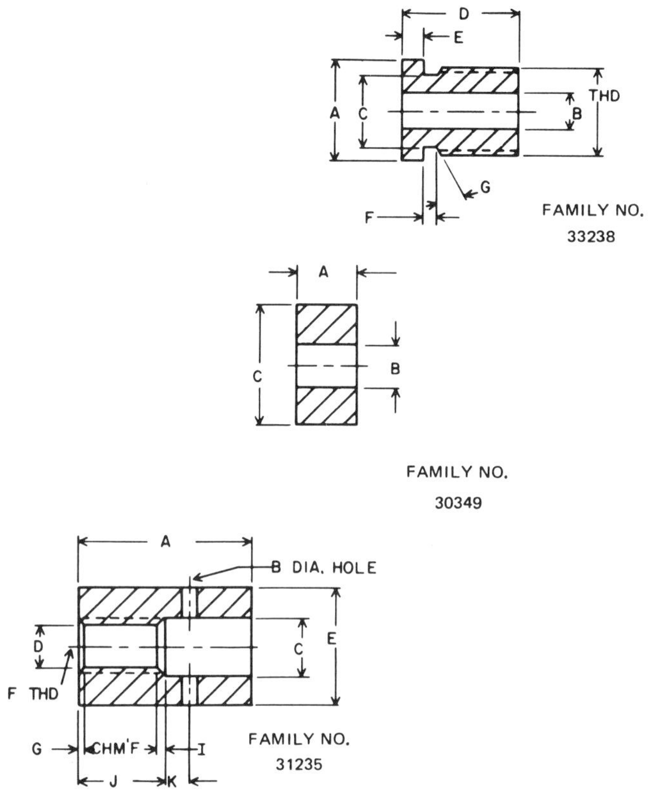

FIGURE 11 These three drawings represent families of similar parts utilizing shape and form. Each family has a number as 33238, 30349, etc. The drawing has generic alpha dimensions which can be replaced with actual dimensions when required. (Courtesy of Control Data Corporation, Minneapolis, Minnesota.)

as separate, but compatible, segments. The catalog of data for each family of parts will be maintained on the CIM system, and no conflict should exist for this data storage.

Manufacturing Requirements

Figure 1 also shows a list of family attributes required by manufacturing:

Master routings
Numerically coded (NC) tool cutter center line data

Tooling requirements
Routing standard time data
Machine capacities
Machine tool requirements
Manufacturing standards

When, in conjunction with the data prepared by engineering, manufacturing personnel utilize this additional data, manufacturing systems will also be improved as suggested below.

The Planning Process

Newly released designs enter the manufacturing cycle at random. The first requirement of industrial engineering is the preparation of a process for making the part, and a routing which abbreviates the process and shows machine set-up and run times. Tooling must be designed or old tooling found. These activities are time consuming and tedious. Imagine the industrial engineer who receives a large gear design and, after completing the process and tooling requirements for it, must next plan a large welded assembly. The learning curve for each new design he or she receives decreases his or her productivity and accuracy.

Master Routings

Supposing, however, the planner has a classification system. As a new design is released, the planner enters the CIM database and finds a family master routing, the tool path for the part, if parts in the family have operations on NC tools, the robotic program for the family if robots are used for assembly or on a dedicated machine line, and standard time data for the family. In most cases, the work planners must perform is an extrapolation of data they already possess, and they do not route the part inefficiently over the wrong machines.

A routing system is necessary for the manufacturing business system. This interface can be made quite clear to the user. First, the family master routings must be a portion of the CIM database to take advantage of the geometry, family attributes, and producibility tips. Once a detail routing for a new part is completed, however, the data can be transferred to the business system database and used in conjunction with all other routings. This breakoff between the two systems is clean and unambiguous.

Tooling

If all the tooling used on any of the parts in the family have been recorded, the possibility of using drill fixtures, clamps, and welding fixtures which already exist is excellent. Minor modifications may also be possible. Even when new tooling is required, other tool drawings can be found and marked up quickly. Manufacturing standards will be maintained on an active basis similar to the engineering standards.

One of the most significant and untapped areas of cost savings is in tool design. Thousands of obsolete tools are kept in storage. Retrieval, and even documentation of their existence, is crude. Classification of tools will become an important part of manufacturing in the future, and, tied to CIM, make the design of tools more economical and rapid using a CIM system.

Figure 12 portrays the amount of time required at the one company to design new tools, whether internally, or purchased from vendor. It takes about eight months to acquire new tooling. This long lead time must, and can, be shortened with a CIM system. A retrieval system designed for tooling will eliminate much of the present redundancy.

Producibility Tips

How will the "/" be erased from CAD/CAM and made CIM? Producibility tips, organized through the use of a classification system, will go a long way. Figure 1 pictures them as a sort of amalgamation of engineering and manufacturing data. What are they? Each type of design, whether a piece part or an assembly, can be manufactured more economically if certain standards are met. These standards are the summation of working knowledge of the shop, the machine tools available, the tolerances which can be held, and, in essence, the common sense of design and manufacturing.

The integration of design knowledge and the practical knowledge of manufacturing will emphasize productivity, eliminate much redesign, optimize production facilities, and coordinate all other activities. A CIM system, with its ability to maintain a large database, coupled with a classification system which can be used to concentrate the producibility tips into manageable segments, will become one of the main advantages of the system.

It is often thought that producibility tips are required for machined parts only. Figure 13 shows examples of such tips for fabricated parts. Any and all types of manufacturing have their own economies. Capturing them will be the key to productivity.

Perhaps we should explain briefly why these tips are necessary. Many years ago, the design function was much closer to the actual manufacturing area. It was not unusual to place new engineers in the shop and give them "hands on" experience. It was also not unusual to find workers promoted from the shop into the drafting department. These activities have almost ceased, in some part because unions have voiced disapproval of an "extra" work force.

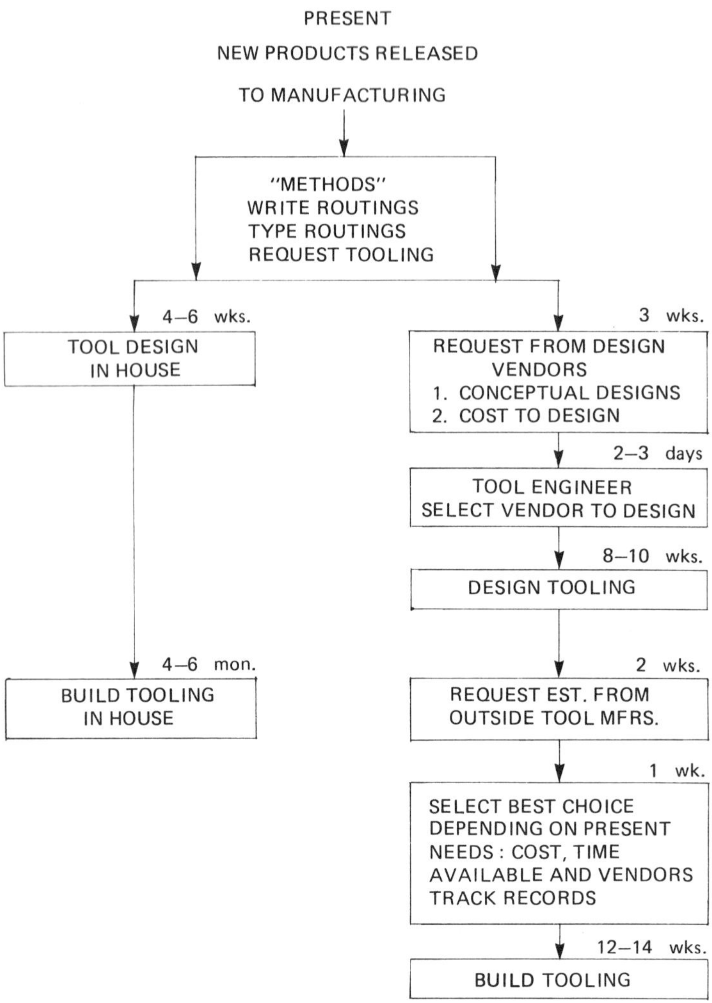

FIGURE 12 The time and cost to prepare tooling for new designs may be an incentive to re-use old designs when practical.

Thus engineers have become functional illiterates in shop problems and practices, particularly in the mechanical products area.

Integrated Database

A database is central to the requirements of engineering and manufacturing. Combining all these requirements with the capability of

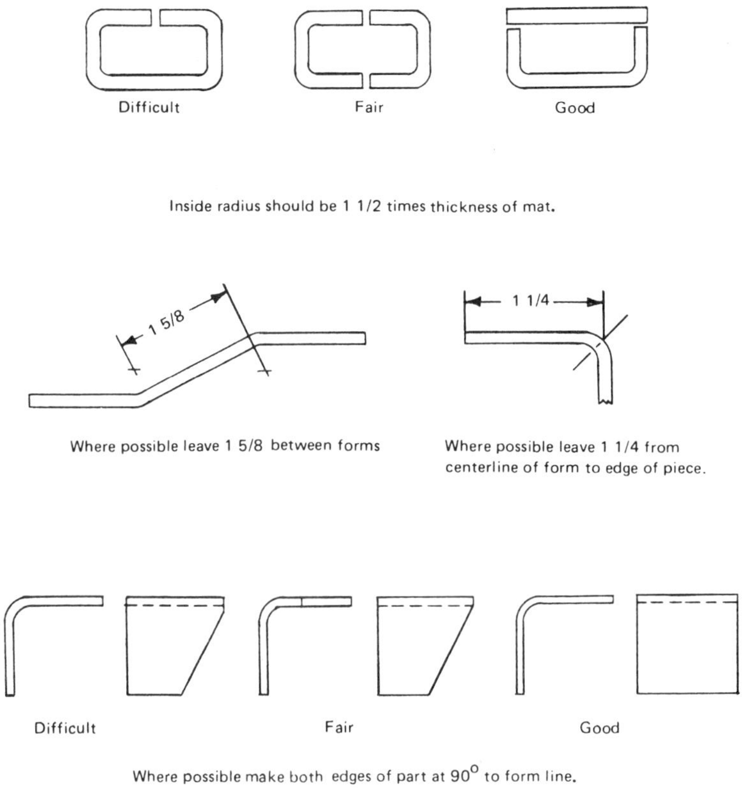

FIGURE 13 Every machine tool has its own set of producibility tip information. For example, designs requiring press brake operations will be more economical if the above rules are followed.

rapid retrieval gives a CIM user a great deal of power, as indicated in Figure 14. Entry into the database either by part family number, or part number allows the user to find part geometry, model images, drawings, analysis results, text data as the item master and product structure for the bill of material, and master routings for manufacturing. Thus classification is truly the glue or paste that holds the data together.

Manufacturing Business Systems

Figure 1 next shows a block which includes manufacturing resource planning, production analysis, inventory analysis, and plant capacity. It has been fairly clear until now how a CIM system interfaces with

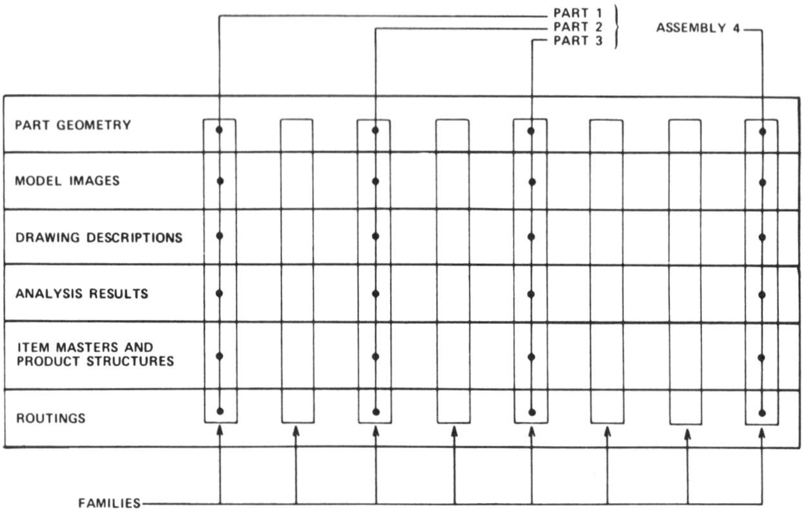

FIGURE 14 Typical database is shown here. It is filled with part geometry, images, analysis results and much more. A good database should provide the means to easily interrogate it using several keys. Here is shown entry either by family number to obtain family data, or by part number to obtain specific part data. (Courtesy of Control Data Corporation, Minneapolis, Minn.)

manufacturing. Process planning, tool design, NC machine tools, and machine capacities have almost always been related in CIM. The next block in Figure 1, however, lists manufacturing resource planning (MRPII) (the latest buzz words for material requirements planning—MRP), production analysis, inventory control, and plant capacity. Where do they and classification fit?

MANUFACTURING RESOURCE PLANNING

MRP is a tool used to forecast all part and assembly requirements based on lead times for each design, a product forecast, and the bills of material for the products. This manipulation of data in a data processing environment is not a feature of CIM. However, the result of MRP will be enhanced by it. Once the requirements for each design are computed, they may appear in a series of "time buckets" as shown below, in Figure 15. Requirements are usually extended to 52 weeks or more. Each weekly requirement not only takes production time but also requires machine set-ups. These set-ups are costly and are not productive time. If they can be

--->SEPARATE REQUIREMENTS WEEKLY---->

PART NUMBER	REQUIREMENTS WEEK 1	WEEK 2	WEEK 3	WEEK 4->
65513	100	100	50	50
65527	60	65	47	67
76623	55	12	10	45
89913	200	150	150	50

---ETC--->>>>

FIGURE 15 Normal results of MRP; part requirements in time buckets.

reduced by combining several weeks of requirements, the parts will be made more economically. However, the parts made ahead of time create a temporary surplus inventory which costs from 25% to 40% of their value during the period they are in excess. Using well-defined formulas generally defined under the name "economic order quantity," the parts are indeed often combined, as shown in Figure 16. The decrease in set-up time pays for the additional cost to carry the excess parts.

--->COMBINED REQUIREMENTS WEEKLY---->

PART NUMBER	REQUIREMENTS WEEK 1	WEEK 2	WEEK 3	WEEK 4->
65513	200		100	
65527	239			
76623	67		55	
89913	350		200	

---ETC--->>>>

FIGURE 16 Recomputed MRP using economic order quantities (EOQ).

The key part classification plays in an MRP system is shown in Figure 17. Here we see, extracted from the MRP data, the requirements for a family of parts. A completely different analysis of requirements may now be made. Rather than combining weekly requirements and storing excess inventory, all the part requirements for one week may be combined and manufactured in one batch. For example, in week 1, the combined requirements of 20, 10, 100, 20, for the four parts will make a total batch size of 150, with identical or closely similar tools and set-ups. Not only are set-up times decreased or eliminated, but excess inventory is not generated.

This analysis does not create a conflict within CIM. As in many other examples of correspondence between two segments of the database, the classification code is "passed on" from one segment to another and utilized as desired.

Production Analysis

Once the manufacturing resource planning system develops material requirements by time period, the business system must next be used to analyze whether sufficient labor and machine resources are available to manufacture the necessary parts. Utilizing the detailed routings combined with work center data provides information for

```
 COMBINED VERTICAL REQUIREMENTS 32413
 ------------------------------------

   PART    COMBINED REQUIREMENTS

FNUMBER WEEK 1 WEEK 2 WEEK 3 WEEK 4->
A       I      I      I      I      I
M32458  I   20 I   15 I   10 I    5 I
I       I------I------I------I------I
L       I      I      I      I      I
Y43496  I   10 I    5 I   25 I    15I
        I------I------I------I------I
O       I      I      I      I      I
F65513  I  100 I  100 I   50 I   50 I
        I------I------I------I------I
P       I      I      I      I      I
A72246  I   20 I   20 I   30 I   25 I
R       I---------------------------I
T          ^      ^      ^      ^
S 32413    ^      ^      ^      ^
           ^      ^      ^      ^
          COMBINE VERTICALLY
                 ---ETC--->>>>
```

FIGURE 17 MRP results coordinated into a family of parts plan.

machine and general shop loads. The analysis does not require interrogation of a CIM database. However, if a multiple code has been implemented, it may be used in conjunction with shop loading to determine general and specific loads on each machine and work center. Based on such an analysis, machine capacity, new machine tool requirements, and thus general plant capacity can be developed.

Multiple code usage is still in the infancy stages. Most companies have had a difficult time generating a representative algorithm which reflects actual shop conditions. Thus they have concentrated on stabilizing data and general reports without paying too much attention to the analysis described here. However, the initiation of a CIM system will make users more aware of the data available, so analysis will become an important part of manufacturing planning. Figure 18 shows one analysis of a company's production. Since over 39% of all parts are manufactured in lots of one unit, there is much to be gained by combining these parts into batches. Figure 17 pictures how parts might be batched together by family. Parts may also be batched together by operation. Thus the use of a multiple code, as mentioned previously, plays a part in the analysis.

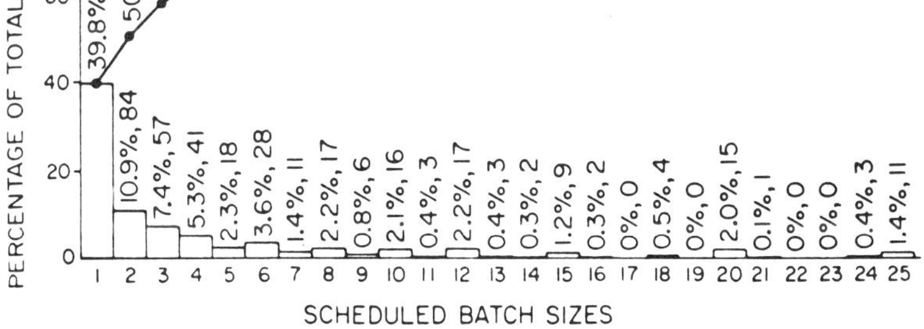

FIGURE 18 Most people today think of manufacturing as a large-scale mass production system. The figures shown here disprove that idea. Scheduled batch sized for from 1 to 17 parts in manufacturing cover 80% of all production in this example. Almost 40% of all parts are built one at a time. (Courtesy of Control Data Corporation, Minneapolis, Minn.)

Inventory Analysis

Another important value of a classification system is in the evaluation of inventory. Figure 19 shows a tabulation of parts sorted from high dollar usage to low. Multiplying cost of each part by its usage creates the list. Figure 20 shows how this tabulation is segregated into A, B, and C segments, based on the percentage of total dollars expended. Figure 21 shows how this data may be further analyzed using a classification system. The illustration shows all parts segregated by dollar usage and by part family. In addition, total requirements for the parts are segregated. With this data, parts which might be combined and mass produced, machine loading requirements and machine types, parts for group technology, parts with identical machining operations, even plant capacity, may be analyzed. Thus a classification system has a direct impact on the management of inventories and the production plan.

Group Technology

Figure 1 next shows group technology as an outgrowth of analysis and planning. Requirements for group technology are accurate routings, standard operations, a classification system, probably both multiple and decision tree codes, and good analysis of production and inventory.

There are many incorrect definitions of group technology. The original and basic concept of group technology follows:

> Group technology is the collection of parts having similar form, shape, function, or operations, such that they may be manufactured—in small batches—in a similar manner utilizing identical tooling or machine tools, resulting in the economies of high volume manufacturing.
> Collection by group is defined as a technique where a set of parts have similar operations which may be combined and performed in a specialized machine shop (multiple code).
> Collection by type is defined as a technique utilized for parts with similar form or shape and having the same material and routing, are combined in one batch (decision tree code).

Emphasis was first placed on small batch manufacture. The reason is based on the fact that group technology was introduced to Europe and Russia first, where mass production as understood

Item No.	Classification	Annual Units	Unit Cost	Annual Usage	Cumulative $ Sales	%
683246	37712-004	51.553	317	77.102	1.652.385	5.0
703611	36115-002	98.406	470	46.251	3.304.769	10.0
965527	36223-001	6.768	4.876	33.001	4.957.154	15.0
332016	36112-002	4.250	7.369	31.318	5.254.533	15.9
894642	34117-002	44.560	675	30.078	5.618.107	17.0
532231	34220-001	8.680	3.286	28.522	5.882.489	17.8
202646	34220-002	27.581	930	25.650	6.609.538	20.0
162678	36620-001	3.428	5.900	20.228	7.600.969	23.0
198155	36711-004	52.765	368	19.993	8.261.923	25.0
928234	36777-001	1.105	14.676	16.217	9.914.307	33.0
856549	35002-001	23.908	640	15.001	10.443.070	31.6
903432	35002-010	2.690	5.475	14.723	11.004.881	33.3
910211	35007-019	11.378	980	11.150	13.219.076	40.0
567809	31001-001	244.690	045	11.011	13.252.124	40.1
933986	30110-002	22.224	450	10.001	14.276.602	43.2
910932	30349-002	7.391	1.054	7.790	16.523.345	50.0
262011	30776-001	2.089	3.540	7.396	17.184.799	52.0
525105	31876-001	56.304	115	6.475	18.209.277	55.1
786399	31877-001	9.984	556	5.551	19.828.614	60.0
584348	33238-005	3.756	1.234	4.635	21.414.903	64.8
376242	32100-001	21.683	205	4.445	21.844.523	66.1
563453	32101-001	23.796	181	4.307	22.042.809	66.7
579996	32102-002	33.743	113	3.813	23.133.393	70.0
612172	32103-002	7.239	490	3.547	23.662.146	71.6
201808	32104-003	3.571	840	3.000	25.050.149	75.8
998699	32105-003	14.774	190	2.807	25.413.674	76.9
662163	33238-002	1.500	1.650	2.475	26.433.152	80.0
237495	31235-009	1.212	1.876	2.274	26.834.724	81.2
350198	31295-001	9.967	209	2.083	27.429.583	81.7
264353	31296-001	1.138	1.720	1.957	26.683.107	82.0
782245	33455-001	3.509	450	1.579	29.015.872	86.0
463812	33456-001	243	5.729	1.391	29.445.492	87.0
217486	33238-001	1.042	1.256	1.309	29.742.921	87.1
590427	33567-001	2.337	475	1.110	30.403.875	87.9
260184	33568-001	2.857	250	1.020	30.536.066	89.0
621958	33569-001	15.360	050	768	31.395.306	90.0
206874	36789-001	3.494	176	615	31.891.021	95.0
741326	33333-001	1.904	282	537	32.122.855	97.2
377241	33334-001	2.842	120	341	32.618.070	98.7
977369	33335-001	2.439	123	300	32.717.213	99.0
661393	33336-001	2.670	103	275	32.783.308	99.2
261853	33337-001	3.750	048	180	32.915.499	99.6
2605	33338-001	198	505	100	32.998.118	99.85
6562	33339-001	210	143	30	33.034.471	99.96
613216	33310-001	500	620	0	33.034.781	100.0
3742	33311-001	1000	20	0	33.034.981	100.0

FIGURE 19 The tabulation shows a listing of all parts used during the year in sequential order from highest dollar volume to the lowest dollar volume usage. (Courtesy of Control Data Corporation, Minneapolis, Minn.)

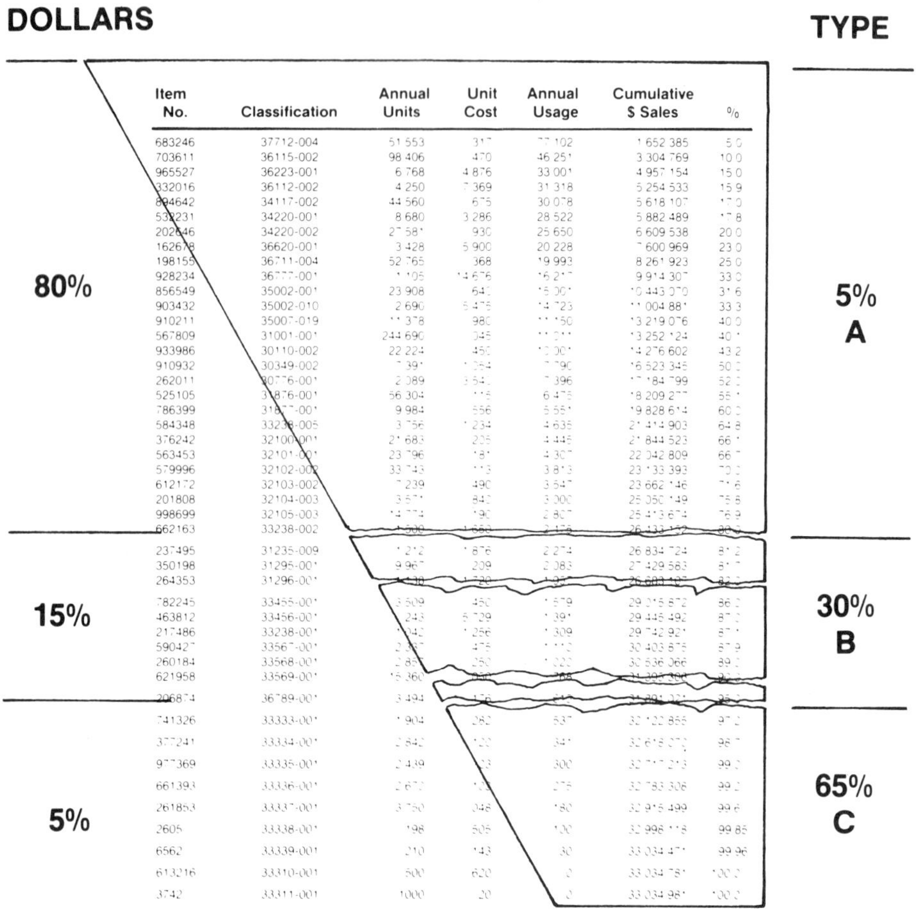

Item No.	Classification	Annual Units	Unit Cost	Annual Usage	Cumulative $ Sales	%
683246	37712-004	51 553	317	77 102	1 652 385	5 0
703611	36115-002	98 406	470	46 251	3 304 769	10 0
965527	36223-001	6 768	4 876	33 001	4 957 154	15 0
332016	36112-002	4 250	7 369	31 318	5 254 533	15 9
894642	34117-002	44 560	675	30 078	5 618 107	17 0
532231	34220-001	8 680	3 286	28 522	5 882 489	17 8
202646	34220-002	27 581	930	25 650	6 609 538	20 0
162678	36620-001	3 428	5 900	20 228	7 600 969	23 0
198155	36711-004	52 765	368	19 993	8 261 923	25 0
928234	36777-001	1 105	14 676	16 217	9 914 307	33 0
856549	35002-001	23 908	641	15 001	10 443 070	31 6
903432	35002-010	2 690	5 475	14 723	11 004 881	33 3
910211	35007-019	11 378	980	11 150	13 219 076	40 0
567809	31001-001	244 690	045	11 011	13 252 124	40 1
933986	30110-002	22 224	450	10 001	14 276 602	43 2
910932	30349-002	7 391	1 054	7 790	16 523 345	50 0
262011	30776-001	2 089	3 541	7 396	17 184 799	52 0
525105	31876-001	56 304	115	6 475	18 209 277	55 1
786399	31877-001	9 984	556	5 551	19 828 614	60 0
584348	33238-005	3 756	1 234	4 635	21 414 903	64 8
376242	32100-001	21 683	205	4 445	21 844 523	66 1
563453	32101-001	23 796	181	4 307	22 042 809	66 7
579996	32102-002	33 743	113	3 813	23 133 393	70 0
612172	32103-002	7 239	490	3 547	23 662 146	71 6
201808	32104-003	3 571	840	3 000	25 050 149	75 6
998699	32105-003	14 774	190	2 807	25 413 674	76 9
662163	33238-002	1 500	1 650	2 475	26 133 152	80 0
237495	31235-009	1 212	1 876	2 274	26 834 724	81 2
350198	31295-001	9 967	209	2 083	27 429 563	81 7
264353	31296-001	1 138	1 720	1 957	[illegible]	[illegible]
782245	33455-001	3 509	450	1 579	29 015 872	86 0
463812	33456-001	243	5 729	1 391	29 445 492	87 0
217486	33238-001	1 042	1 256	1 309	29 742 921	87 1
590427	33567-001	2 387	475	1 112	30 403 875	87 9
260184	33568-001	2 857	250	1 020	30 536 066	89 0
621958	33569-001	15 360	[illegible]	768	[illegible]	[illegible]
206874	36789-001	3 494	[illegible]	[illegible]	[illegible]	[illegible]
741326	33333-001	1 904	282	537	32 122 855	97 0
377241	33334-001	2 842	120	341	32 618 072	98 7
977369	33335-001	2 439	123	300	32 717 213	99 0
661393	33336-001	2 670	103	275	32 783 308	99 2
261853	33337-001	3 750	048	180	32 915 499	99 6
2605	33338-001	198	505	100	32 998 118	99 85
6562	33339-001	210	143	30	33 034 471	99 96
613216	33310-001	500	620	0	33 034 781	100 0
3742	33311-001	1000	20	0	33 034 981	100 0

FIGURE 20 When analyzed, the usage listing shows roughly 80% of all dollars are spent on only 5% of the parts, 15% of all dollars on 30%, and 5% of all dollars on 65%. (These percentages may vary from company to company.) This distribution is called an ABC inventory analysis. (Courtesy of Control Data Corporation, Minneapolis, Minn.)

in the United States was unknown, and the cost of carrying excess inventory not significant. Thus group technology is now being altered to fit the unique needs of this country. The original concept would make less of a contribution to productivity. Notice, again, the words "small batches." By definition, parts in the small batch could be those with very small dollar usage. Thus such parts would fall into the C inventory classification structure, which repre-

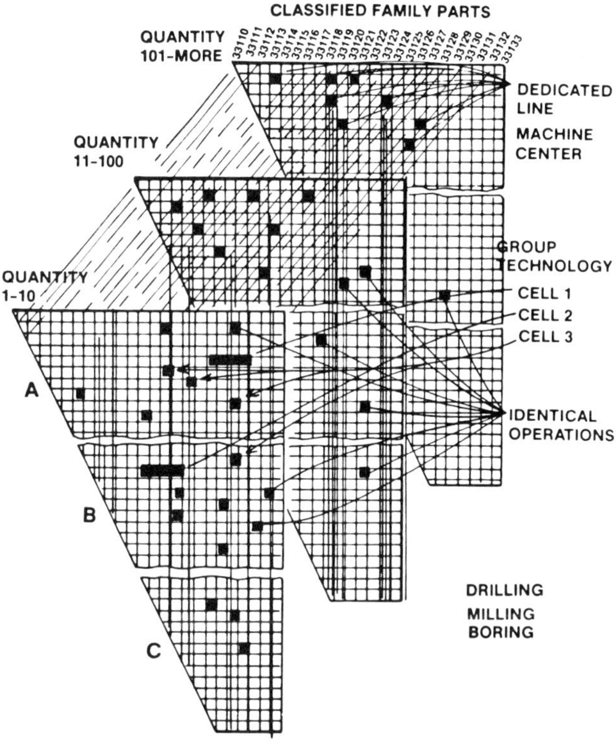

FIGURE 21 The production analysis will show those parts which can and should be massed produced, those which may be manufactured using group technology of cells, and those utilizing group technology for identical operations. If each part has a manufacturing polycode tied to it, providing machining characteristics, machine capacity and requirement may also be measured. (Courtesy of Control Data Corporation, Minneapolis, Minn.)

sents only 5% of the total value of parts a company manufactures. To fully implement group technology successfully in the United States will require the manipulation of a large database in combination with both a decision tree and a multiple code. Thus a CIM system will become the major instrument to develop accurate data so it may be combined with a production control and analysis sytem.

NC machine Tools

Parts programming will be greatly enhanced by a classification system and CIM. Not only will similarly shaped parts be available so a programmer can "copy" tool paths, it is probably practical to retain a tool path by family in a manner similar to storing a family drawing. By interactively making changes to such a family tool path, new ones may be made quite rapidly and accurately. A family tool path is to manufacturing what a family drawing si to a designer.

SUMMARY

Figure 22 is an attempt to summarize the data we have prepared previously. Entering a CIM database by using the classification system, a family of parts can be obtained. A family drawing, a master routing, the producibility tips, design and drafting standards, and manufacturing standards are available to make a quality design. Family multiple codes and the NC machine tool paths are available for manufacturing. Tied to the family is a catalog to investigate old designs for applicability. In addition, the family database is tied to files of specific drawings and their geometry, design analysis

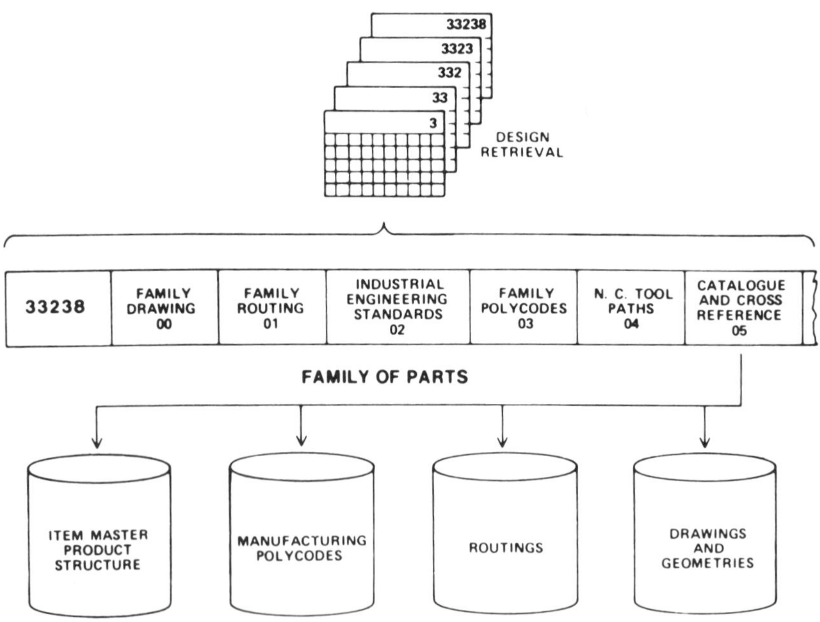

FIGURE 22 Here is shown a summary of the family database reached through a classification system. Once in a family, specific parts and associated data may be obtained. The sequence of events allows an orderly storage of data in logical arrays. (Courtesy of Control Data Corporation, Minneapolis, Minn.)

formulas, and specific data files known as routings, the item masters, and product structures (bills of material).

If we plan ahead while implementing a design retrieval system, then parametric drawings, tool paths, and other CIM data can be prepared by family. When implementing CIM, "instant CAD/CAM" will result by having the ability to go into production of designs quickly and economically.

Let us also summarize how the flow of data will appear within CIM.

Figures 23 and 24 show the flow. Figure 23 begins this flow in marketing. On the left side is shown the product database. On the right side the family database is shown.

1. When a customer specification is reviewed, functional requirements are defined in terms of functional devices segregated by family. Using the family data to determine the capacities, size, and other characteristics, the user will identify any existing devices which appear to satisfy the customer. If a match is found, the task is complete and the customer satisfied.

2. If an existing product is not satisfactory, then a new design must be created. The designer will use the terminal to prepare a three-dimensional layout for the new design based on the family design standards and analysis. In so doing, the part and/or assembly classification system will be used to find alternate parts which might fit the new configuration.

3. Next a "rough cut," or generic, bill of material for the new product will be prepared, consisting of part numbers to identify existing parts and family numbers to identify new components. A preliminary cost estimate can now be prepared for the new device or product based on the documented costs for the existing parts, and on the family cost, meaning the average or accepted cost of all the parts in that family.

4. In a similar manner, the bill of material for the new assembly layout will be used to estimate the manufacturing lead time and delivery date, based on the part family lead times, and family routings, lead times for new parts and actual lead times for existing parts. The MRP system will be used to predict the earliest possible delivery date based on current shop load.

5. Once these tasks are complete, the customer proposal can be documented, including two-dimensional layouts and isometric drawings, family functional schematics, and standard documentation, and an accurate cost and delivery schedule.

Figure 24 shows the details necessary to prepare new designs for parts and assemblies:

1. The designer begins with a concept of the new design. Based on this conceived geometry, he identifies the family of parts similar to it. By studying the family catalog, which lists the parts

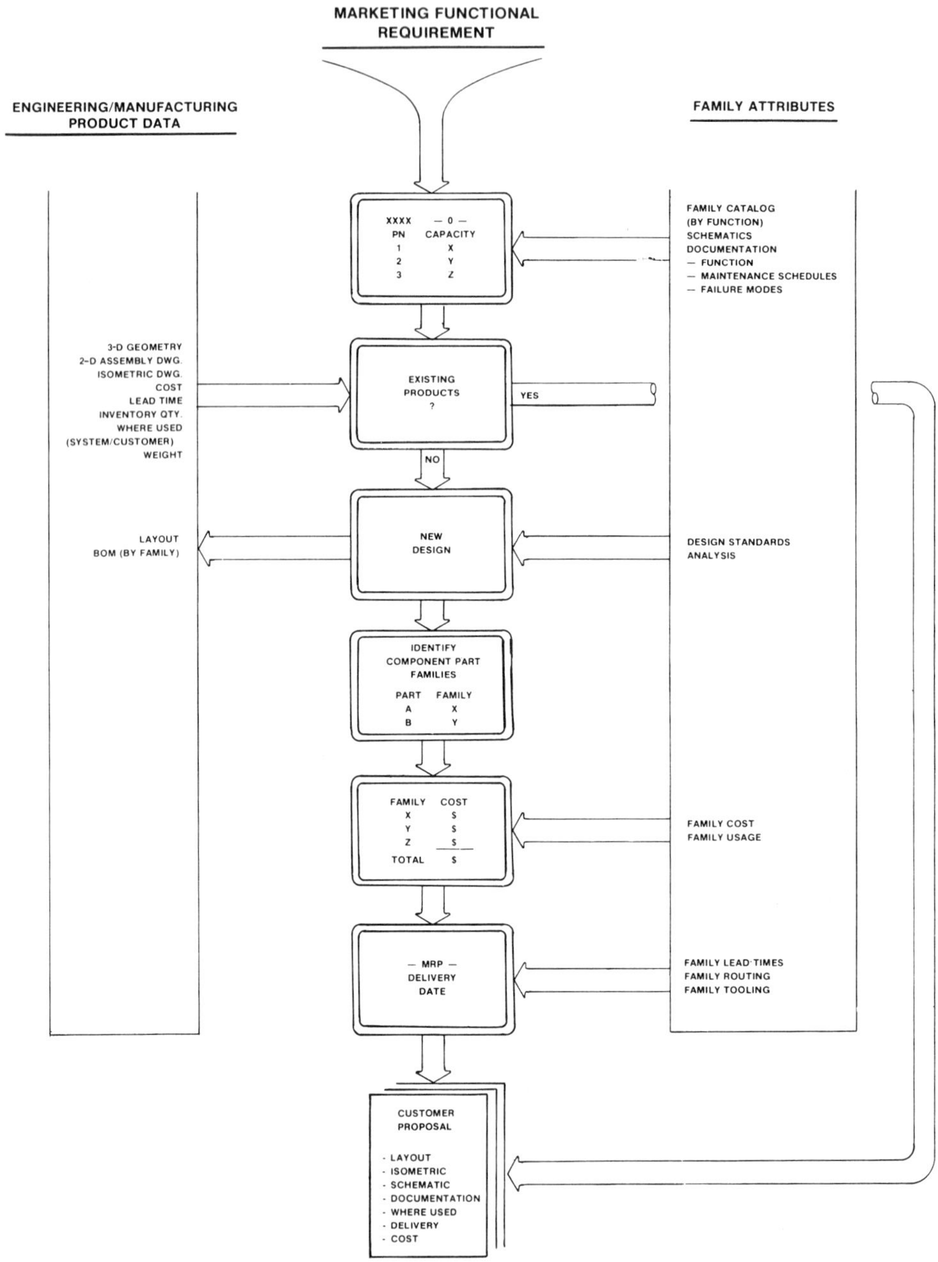

FIGURE 23 Integration of CAD/CAM systems in marketing and engineering require that engineering documentation have integrity and an orderly flow.

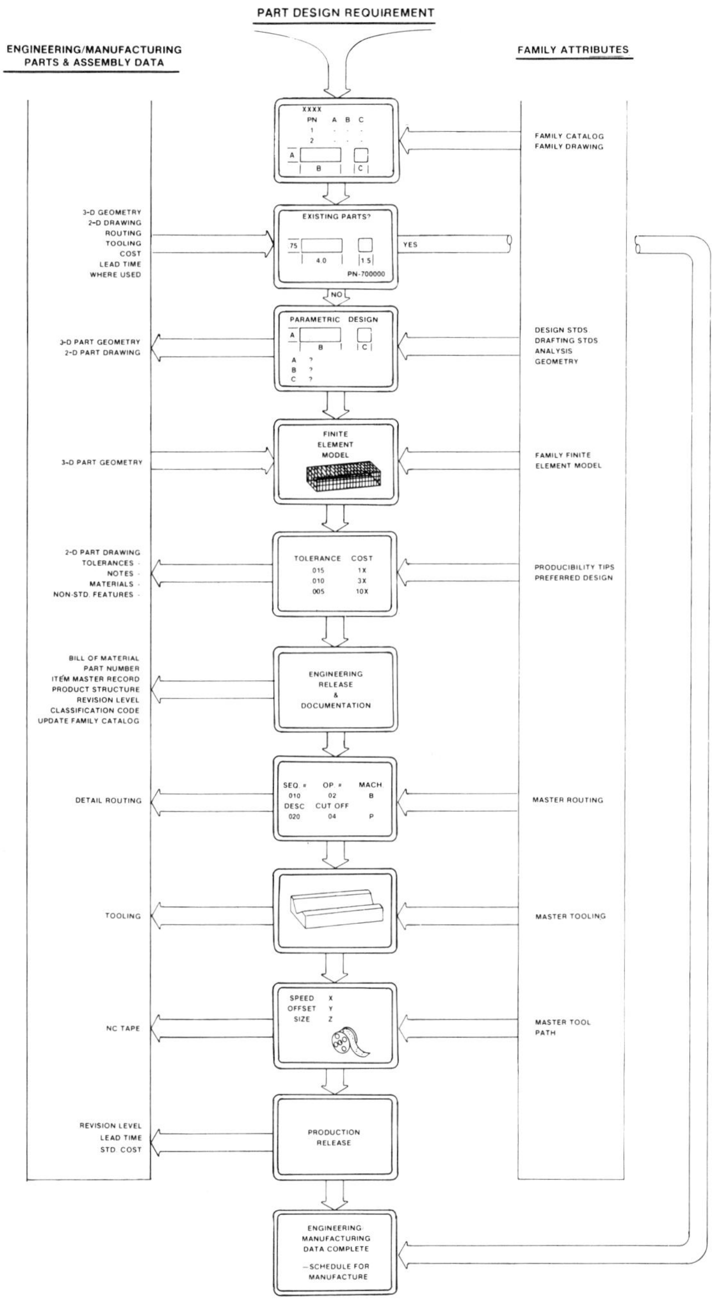

FIGURE 24 Detailed design requires an accurate database, a retrieval system, and accurate engineering data.

in that family, as well as their characteristic dimensions, he may also look at the family drawing to make sure he is in the proper family. Once identified, the drawings of the existing parts may be viewed. If an existing part satisfies the design requirements, the designer will specify it, and the task is complete.

2. If a new part is required, the designer will use the parametric design program for the family to design the new part by interactively specifying dimensions or performance. The parametric design program will use these inputs and the family drawing and geometry to create the part drawing and geometry without further interaction. In some cases, the inputs to the parametric design program may come from Fortran or other analysis programs which calculate the dimensions required to achieve a specified performance. Further, the resulting geometry may be used to parametrically create a finite-element model for that part from the family finite element mesh model.

3. Since it is not practical or necessary to write parametric design programs for all families, in many cases the designer will use the interactive graphics tools to create the geometry. However, standardization will be achieved by the design and drafting standards made a portion of the family data.

4. The resulting two-dimensional part drawing is completed interactively by the user, who adds any necessary nonstandard, nonpreferred features such as special tolerances, materials, and process notes. Family producibility tips and preferred design features will be prepared by industrial engineering in cooperation with design engineering. The drawing will be made using a graphics program which accepts data from the keyboard and enters it into the family drawing, replacing the generic alpha dimensions.

5. When the design is completed, the part will be assigned a part number, and the catalog of family parts updated. The bill of material and parts lists will be released to the system.

6. Once the designs have been released, industrial engineering will make use of the family master routing to make a detailed routing for the new part. The routing data will then be transferred to the business system database if separate.

7. The tooling required will be examined using the data for old tools tied to the family of parts. If no old tooling is discovered, new tooling will be prepared or old altered.

8. To create NC tapes from geometry, the master family tool paths will be utilized. An interactive, parametric program, or input trace, which will extract the part geometry information from the three-dimensional model, will be used to automatically create the tool path for that part based on the master tool path.

9. To release the part, industrial engineering new enters the remaining item master data, such as lead times and standard cost.

Figure 25 pictures the total family data set. Separated into families, the universe of parts can be analyzed to find the following:

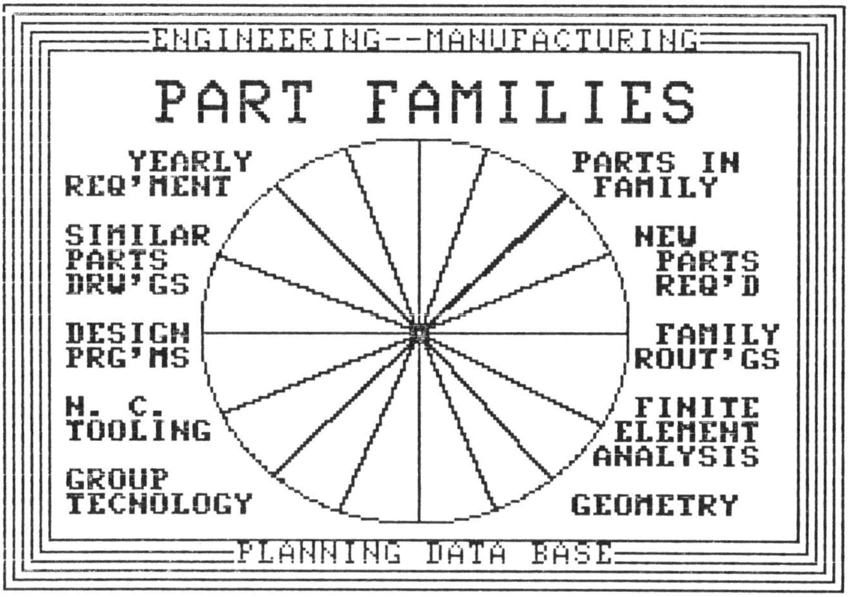

FIGURE 25 Analysis by part family using engineering and manufacturing data combined will allow analysis for many important comparisons including, by family, total parts, total new parts, total yearly requirements, design programs available, NC tool paths, family routings, finite element meshes, and family geometry.

Total yearly requirements
Total number of parts in the family
Families with part generic drawings
New designs released yearly
Families with parametric design programs
Master routings
NC tool paths
Finite element modeling geometry
Family geometry
Parts associated for group technology

We have thus achieved the ability to create a master planning database.

We have attempted to encapsulate the uses and advantages of classification systems. We have attempted to show both their strengths and their weaknesses. The significant benefits to a CIM system have been emphasized.

9
PROCESS PLANNING

PAUL F. MUCHNICKI

Todays manufacturing engineer is faced with the challenge of providing process plans that are concise, timely, flexible, ensure product quality, and provide for a competitive edge. Engineers traditionally pull from historical files and formulate new processes based upon previous successes. The basic obstacles that commonly confront the engineer are:

1. Time
2. Engineering changes and notification
3. Access to appropriate data
4. Knowledge of tools available

Today's engineer requires access to intelligent data in a timely fashion to formulate process plans. Many companies are pursuing the generative process plan approach (the ability to automatically computer generate process plans with minimal decision-making input). This approach, which I believe is appropriate, requires some fundamental building blocks to succeed. Some of these building blocks are:

1. Knowledge base of existing capability
2. Part family determination
3. Equipment facility organization
4. Family tooling analysis
5. Routing and process optimization
6. Risk analysis
7. Business integration techniques and constraints

The creation of a data base which can be utilized to not only develop new process plans but optimize existing ones is critical. The following suggests a methodology and rationale used to create a data base specific for a particular manufacturing industry and generic in nature to accommodate others.

Over the last 30 years, engineering and manufacturing departments have remained autonomous in their execution of duties. Individual charters, each duly justified, allowed for duplication of drawing files, bills of material, and parts lists. Productivity was measured on the average number of engineering changes released per part. As departmental productivity eroded, costs to manufacture climbed. Executives and departmental managers began to realize that if the company were to survive, something had to change.

With the advent of computerization and in particular, CAD/CAM, managers began to understand that the capture of quality data that can be continually used was vital to their overall success. Managers further realized that the overall objective of any organization was to produce low-cost, quality parts in a timely manner. This meant doing things right the first time, improving communications, and working under an organizational plan.

CAD/CAM has enabled manufacturing facilities to consolidate numerous files of information into a central data base. The control database is the key to the reduction of change, particularly in process planning. To begin with, CAD/CAM encompasses numerous activities (see Figs. 1–3) most of which directly impact the process plan. The reorganization of these three areas, CAD, CAM, and MRP into a data flow chart (see Fig. 4) conceptually details the flow of data in the manufacturing facility. The data that will be utilized to develop the process plan, although residing in the central database, will come from the individual activity files shown on this flow chart.

The database, however, does not come as a predeveloped module. The structure is there for filing and retrieval but the collection, formation, and input of data is up to the individual. The problem then is how to formulate and input data to enable the creation of a process plan.

Let us analyze our objectives and determine how process plans should be generated. There have been numerous discussions held on the best process planning approach—that being variant or generative. There are many pros and cons for each method. Certainly time and money to implement are the leading contenders. Fortunately, there does exist a commonality between the two methods that merits discussion—the database.

The creation of a database to enable the generation of process plans could evolve into the variant or generative definition.

The objectives in the development of a process planning database would be to achieve:

1. Optimization
2. Higher throughput

CAD
-Design Analysis
-Kinematics -Prototypes
-Modeling -Structures
-Classification & Coding
-Assembly Design
-Configuration Control
-Value Analysis
-Design Retrieval
-Drafting
-Catalog Development
-Product Structure
-Parts List
-Release Notice
-Solids Modeling
-Finite-Element Analysis
-Systems Design
-Tolerancing
-Standardization of Parts, Design, & Materials
-Material Matrix Analysis
-New Product Introduction

FIGURE 1

3. Work-in-process reductions
4. Higher profitability
5. Training reductions
6. Better utilization of assets
7. Automation
8. Higher and quicker successes

The tangible derivatives of a data-based process planning system would be reductions in:

CAM	
Computer-Aided Engineering	Computer-Aided Manufacturing
-Parts are ordered	-Parts Handling
-CRP	-CNC Lathe
-Process Plans	-CNC Mill
-Factory Planning	-CNC Grind
-Simulation	-EDM, ECM
-Production Planning	-In-process Inspection (Coordinate) (Visual) (Robot) (Lazer)
-Layout	-Assembly
-Quality Assurance	-Testing
-WIP Analysis	-Labor Reporting
-Classification & Coding	-Statistical Quality Control
-Family Principles	-Quality Costs
-Numerical Control	-Robotics
-Computer Layout	-Auto Storage & Retrieval
-Tool Design Fixtures Jigs Dies	-Tool Control & Inventory -Adaptive Controls -DNC (Direct Numerical Control)
-Producibility Tips	-Tooling Packages
-Cost Reductions	-Tool Setups
-Line Balancing	-Bar Coding & Inventory Control
-Maintenance Schedule	-Near Net Shape
-Group Technology Principles (Generative)	
-Make vs. Buy	

FIGURE 2

MRP
-Master Scheduling
-MRP (Rough Cut Capacity)
-CRP (Capacity Requirements Planning)
-BOM (Bill of Materials)
-Routing
-Resource Requirements Planning
-Inventory Management
-Cycle Counting
-Purchasing/Receiving
-Subcontract Operations
-Shop Floor Control
-Finance/Costing Control
-Standards Development
-Forecasting
-ABC Analysis
-Master Catalogs
-Where Used
-Work Center Reports
-Tool Where Used Inventory
-Vendor Performance Reports
-Supply/Demands Report
-Engineering Revision Inquiry
-Variance Reports
-EOQ (Economic Order Quantity)
-Scrap/Rework/Trends

FIGURE 3

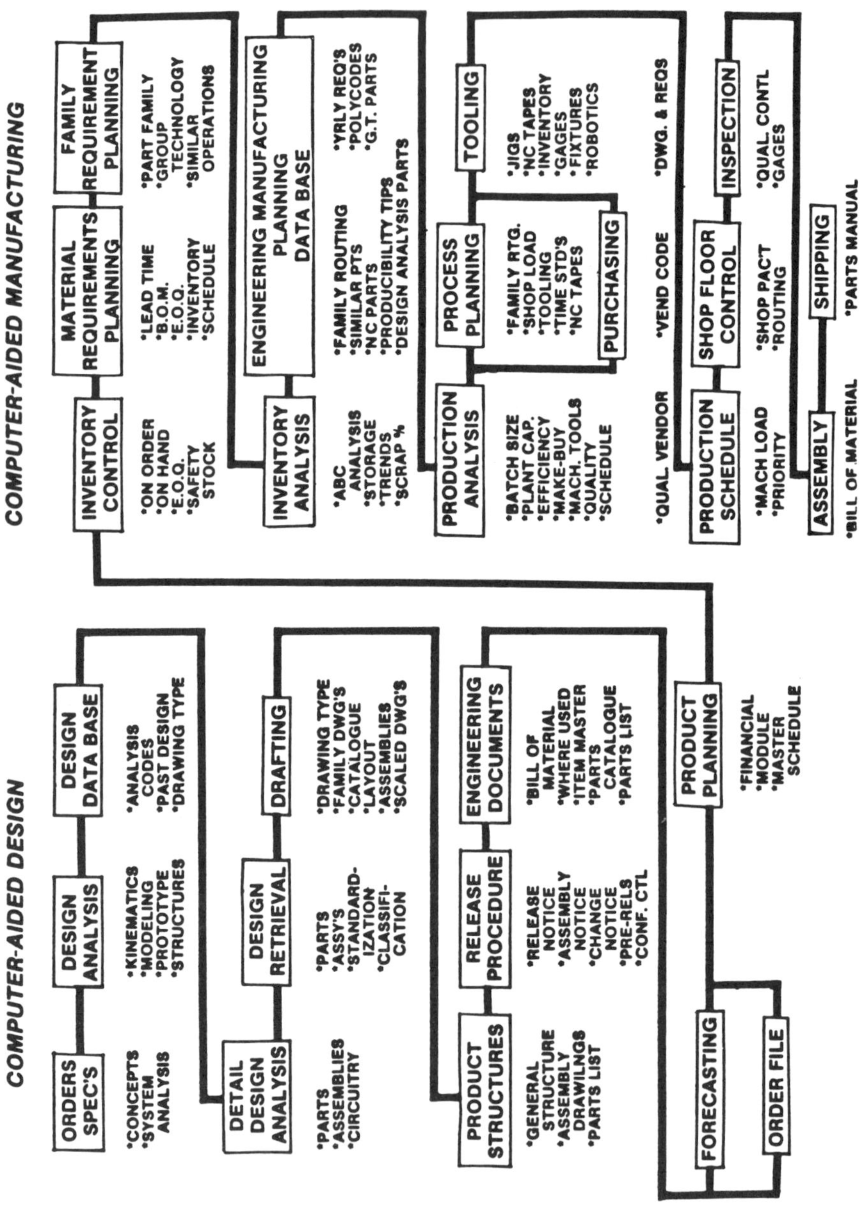

FIGURE 4 Data flow chart.

1. Tooling inventory
2. Scrap and rework
3. Setup and teardown
4. Duplication of effort
5. Paperwork

A preliminary work effort might require that the engineer first resolve some of the following questions.

1. Has anyone ever processed a similar part before?
2. Who is the design engineer?
3. How many of these parts will we have to make?
4. Has there been a cost established?
5. Do I need hard or soft tools?
6. What machines do I rout the part on?
7. Will there be sufficient capacity hours?
8. Do I process the part across a numerical control or conventional machine?
9. What will this part go into?
10. What are the datum surfaces to define geometry for NC?
11. Do I make the part out of bar stock, casting, forging?
12. In what lot sizes will the part be produced in?
13. How do I inspect the part?
14. What inspection tools will be needed?
15. When are the parts needed?
16. Is there perishable tooling available for the types of geomtry required?
17. Are there alternate processes needed such as plating, EDM, grinding?

To effectively answer these and other planning-type questions the engineer will have to interview different people, research many files, examine in detail the piece part drawing, and be educated on how to implement all these tasks. These activities are by no means a science and the eventual result is usually something less than perfect.

This methodology does not optimize the engineers' time and generally does not provide for a successful plan. Once the written plan is completed execution on the shop floor or "proveout" is required to optimize the process plan.

If the engineer was assigned to process a new part with a medium degree of difficulty, this proveout phase could take up to six months to produce an acceptable part. Even worse, the process might never be optimized and probably will be changed with each new lot of parts or materials.

Figure 5 reflects the linear mode required to produce a specific part. Without assuming any risk, each phase should be completed in sequence.

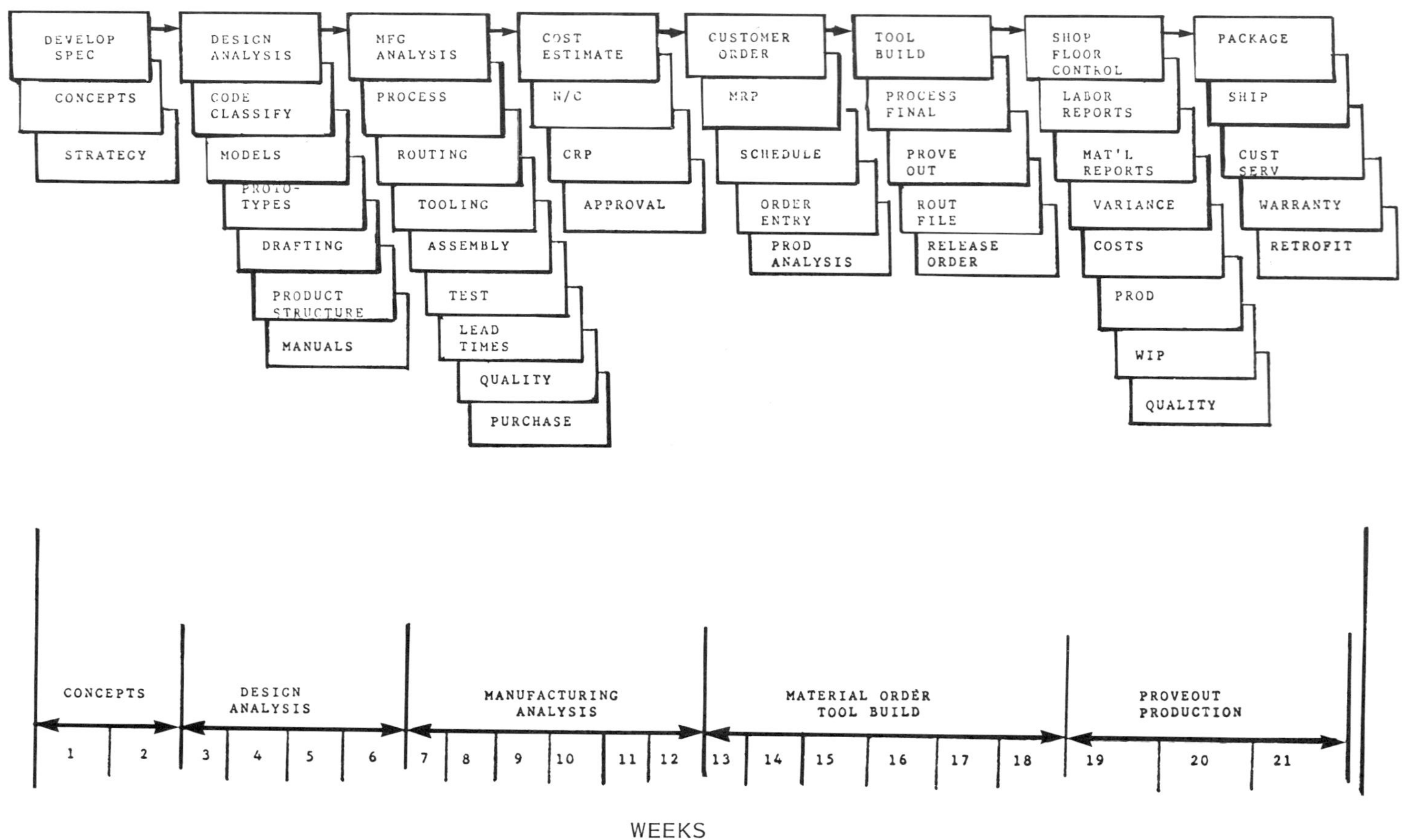

FIGURE 5 Linear mode for specific part production.

The introduction of engineering changes only creates costly delay. The manufacturing engineer needs to have "fingertip access" to locate and retrieve data to answer those process-related questions from a manufacturing database.

GEOMETRICAL ENTITIES

The development of a manufacturing database actually begins in the design engineering department. The classification and coding of piece parts, assemblies, and raw materials into families is the first step. Although the method of data retrieval is variable, the categorizing of known piece parts into similar groups based on function and geometrical structure is the foundation for process planning.

Again, the primary objective of a manufacturing facility is to produce piece parts. The manufacturing engineer (ME) is concerned about part production while the design engineer (DE) is concerned with part creation.

The DE will place the tightest tolerance on a piece of geometry (entity) to ensure functionality, but if the existing manufacturing capability cannot economically produce the entity—is the tolerance justified?

MEs produce process plans to create parts from raw material. The ME systematically breaks the part down into geometrical entities defining the steps, machine, and tools required. By defining the entity, the ME can optimize locally and then string entities together to produce a part. The specific responsibility of an ME is to provide process plans that create geometric entities culminating into piece parts.

The ME will analyze each geometric entity and determine the process function and tooling definition required to produce the entity. The process function, bore, ream, mill, turn, etc., will help the ME determine the best machine center to use. The tooling definition will determine if standard or special purpose tools are required.

MACHINE CENTER TOOL PACKAGES

Since machine centers are function oriented and there exist tools to form each specific geometric entity, then the machine center can be utilized to create geometric entities and not specifically piece parts.

Consider the following two types of machine centers and their functions.

Mill	Lathe
Bore	Turn
Ream	Bore
Drill	Ream
Tap	Thread
Mill	Face
Countersink	Drill
Profile	Trace
Pocket	
Interprelate	

Each piece part had previously been defined and classified into a family. The ME has examined each geometrical entity and assigned a function and tool required to produce it.

The ME could now classify machine centers not by families of parts, but rather by similar geometrical entities. Tooling packages

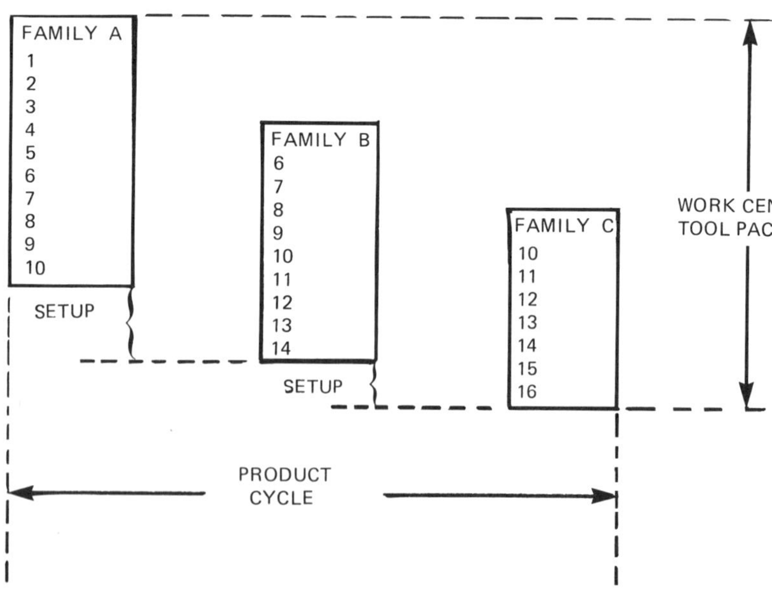

FIGURE 6

can be developed for machine centers and parts, exclusive of shape similarity. Parts can now be routed to machine centers based upon the requirements of their individual geometric entities.

Once the entity analysis is completed, the engineer can form a composite of entities, functions, and tools that define the piece part and design a tool package number to the part. By grouping parts based on process functions and tooling definition, a composite work-center tool package number can be assigned. The unique piece part tool package number can be listed in an open MRP field file and parts subsequently scheduled across work centers based on requirements, function, and tooling similarity.

Figure 6 suggests that parts can be scheduled by knowing the differences in set-up/tear down of tool packages. In the same product requirements cycle, it would be highly more productive to schedule parts 1–10 (family A), 6–14 (family B) and 10–16 (family C) in that order rather than 2, 16,8,13,15,10, etc.

In a captive environment where each machine center has limitless tool storage, the restraint to maintain productivity is now the fixturing required to hold the part.

Fixtures and holding devices can be classified into specific machine center functions based upon the stringing of process elements.

Two examples are:

Mill	Lathe
Serrated vise	2-Jaw chuck
Special vise	3-Jaw chuck
Clamps/nuts/bolts	Centrifugal chuck
Rotary table	Box chuck
Indexer	Special
Building blocks	
Multiple	
Hydraulic/pneumatic	

Each device can be assigned a specific code based upon function. The ME can assign to each processed part the code requirements to secure it within the work space.

The combination of the tool and fixture code within the MRP field of data will allow for day to day scheduling of parts across specific machine centers or groups. Additionally, alternate process planning can be easily identified.

To establish tool packages the manufacturing engineering department will need to undertake a coding and classification scheme similar to design engineering. Optimization to reduce inventory and the introduction of standardization can now be realized. One step to achieve optimization requires the establishment of producibility tips.

PRODUCIBILITY TIPS

The proper creation of the design engineering database is critical to the efforts of the ME. The classification and coding (C&C) efforts will enable the sorting, identifying, and retrieval of piece part data. The similarity of data will reduce redundant efforts by all users. Through (C&C) techniques, the database can be constructed so that associativity of elements and attributes about piece parts can be used to reduce redundancy. The DE and ME efforts must be harmonized so that quantitative and qualitative data can be entered into the database.

Within each manufacturing organization there exist numerous historical records of piece parts, associated processes, costs, and tooling. The extraction of certain pieces of data can be formalized into a working tool to aid the engineer in day-to-day decisions. By establishing boundary conditions on piece part families, the engineer can examine accepted tolerances, finishes, processes, costs, etc., and then assign manufacturing-recommended attainable conditions. These boundaries are called producibility tips.

CREATION OF PRODUCIBILITY TIPS

Each family of parts must be analyzed to establish these tips. Priorities can be based upon production requirements, cost reduction projects, or new product introduction.

A task force consisting of members from the manufacturing, engineering, and quality departments as a minimum should be formed to conduct each analysis session.

The basic process to develop this part of the database would be as follows. For each family of parts construct a matrix that would list all part members down the vertical column and all part attributes across the horizontal. Attributes can be technical data, cost information, preferred process number, quality specification, tooling package, tolerances, material and so on (Fig. 7).

Place within each window the appropriate attribute value. These values can be manually extracted from drawings, process sheets, specifications, or any qualified source.

Parts	GEOMETRY									
	Surface finish	Concentricity	Parallelism	Tolerance	DIM 'A'	DIM 'B'	DIM 'C'	DIM 'D'	DIM 'E'	DIM 'F'
A	125	0.005	0.002	±0.005	5	5	.5	–	–	–
B	32	0.004	0.002	±0.002	5	3	1	–	–	–
C	125	0.005	0.003	±0.005	4	3	1	–	–	–
D	250	0.005	0.003	±0.005	8	6	3	–	–	–
E	125	0.003	0.003	±0.001	3	4	2	–	–	–
F	125	0.005	0.003	±0.003	6	6	2	–	–	–
Composite	32	0.003	0.002	±0.001	8	6	3	–	–	–

FIGURE 7 Attributes and specifications.

Once the matrix is completed, the task team must carefully analyze the matrix to establish a value field for each attribute. Areas to evaluate might be a wide variation in surface finish from 8 micro to 64 micro. The 8 micro probably requires an additional grinding operation resulting in higher costs. Historical review might also discover that the 8 micro was assigned to a particular lot of parts to resolve a problem and that all subsequent lots were manufactured to the 8 micro. Agreement on boundary conditions for each attribute must be achieved.

It is important for the design engineer to have a value-added sense of the cost to manufacture based on a decision to assign tolerances. Examination of the individual costs among similar parts might lead to a value engineering activity. Once established, these costs, along with process standards can be used to develop cost to manufacturing charts that the engineer can reference. Figure 8 reflects the cost to manufacture due to surface finish tolerance. It

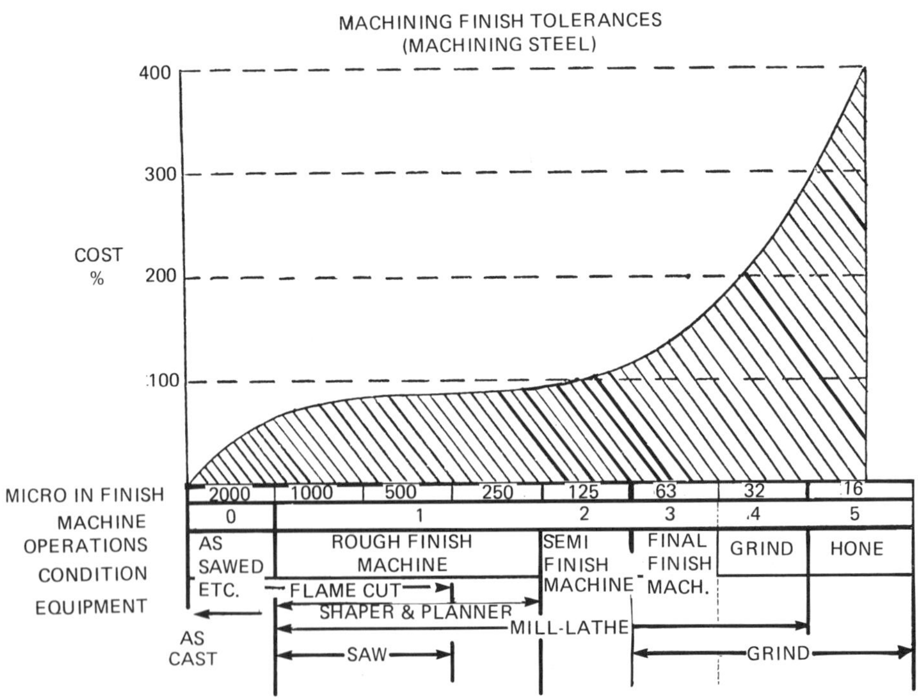

FIGURE 8

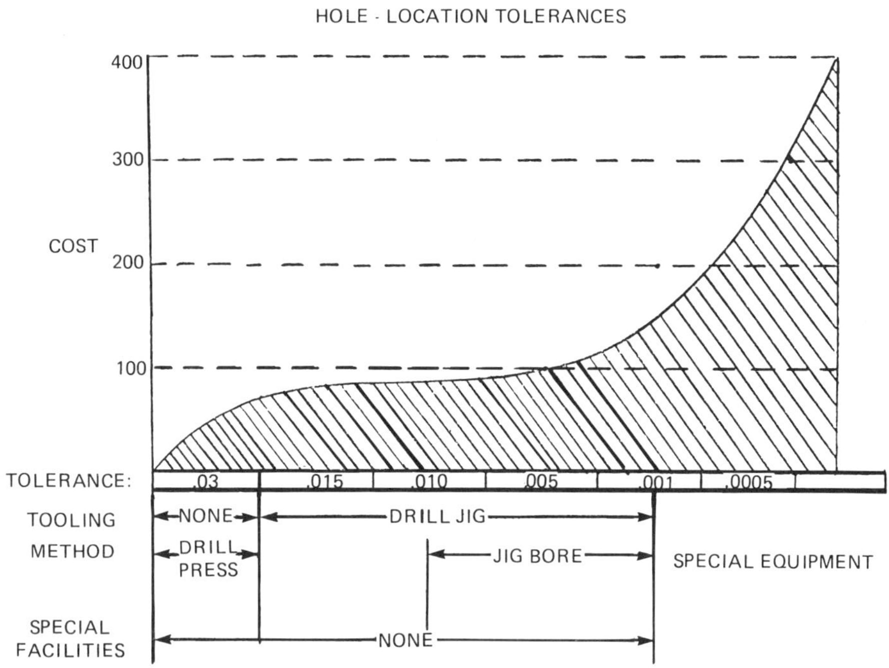

FIGURE 9

is quite apparent that as the surface finish tolerance is reduced, the number of process operations and relative costs increases. Likewise, Figure 9 reflects the increase in process operations and costs when true position hole tolerancing is tightened.

Figures 7, 8, and 9 are important tools for the design engineer. Once parametrically entered into the database, the engineer has the ability to retrieve important information in a timely manner.

Boundary conditions in the form of parametrics also create a sense of discipline in the engineering community. For the manufacturing engineer, beside revisions, there is nothing worse than having to create geometric entities that are outside existing capability.

The parametric can be positioned within the infrastructure by the database manager to prevent the release of parts if boundary conditions are violated. Notification flags could advise particular people in the release process that special approval is required.

PARAMETRIC PROGRAMS

A parametric program is a software program than can be generated to provide automatic generative capability for design, analysis, numerical control, and process planning where consistant and similar attributes can be assembled to enact a functional requirement. Elements of data are associated and strung together to create piece parts, process plans, or numerical tool paths.

The parametric, once created, is a powerful tool that pulls incremental information from the database to satisfy particular requirements.

A typical session using a parametric might result in the following:

The manufacturing engineer would log into the computer system and begin a set of interactive sessions. The objective is to create a process plan for a specific part classified into family number 1234.

The engineer would enter the piece part number and the system would respond.

1. What do you want to do?

   a. Design
   b. Analysis
   c. Process

The engineer selects "C"—process.

The next screen would provide the engineer with a second choice.

2. Search the database and retrieve the following information:

   a. Family information (similar to Figs. 7, 8, and 9)
   b. Family routing file
   c. Family process file
   d. Family tool package

Indeed the parametric could be sufficiently detailed to approach a true generative condition. It simply is a matter of time and the resources available to develop, condition, exercise, and implement.

The engineer can now make qualitative decisions based on the information presented.

ANALYSIS

Parametric programming can also be utilized to provide other process improvements. Portions of data which require analysis (stress, structural, heat, etc.) can be automatically routed to various applications (ANSYS, NASTRAN, STARDYNE, etc.).

PROCESS-GENERATED ATTRIBUTES

The ability to generate a family matrix of piece part attributes provides for many other areas of opportunity. The matrix and the attribute boundaries can be used to create a composite family part, depicting the worst-case entities. The process engineer can now develop specific processes to generate individual geometric entities.

A parametric program would assign to each geometric entity the generic process routing required to produce it. Additionally, tooling and numerical control cutter line files can be created and applied in the same manner.

A typical process format for a specific drilled hole tolerance might be:

Tolerance Band	Process
0.000–0.005	1. Centerdrill #_____ hole to __B__ depth using tool #_____
	2. Drill __X__ __Y__ hole to __A+__ depth using tool #_____
	3. Bore __X__ __Y__ hole to __A__ depth using tool #_____ and diameter __'C'__ leaving __'D'__ inches for finish.
	4. Ream X–Y hole to 'A' depth using tool #_____.
	5. Check finished dimension every 3rd piece using gage #_____.
0.005–0.010	1. Drill #3 hole to __'B+'__ depth using tool #_____.
	2. Ream #3 hole to __'B'__ depth using tool #_____.
	3. Check finished dimension every 20th piece using gage #_____.
Greater than 0.010	1. Drill #3 hole to __'A'__ depth using tool #_____.
	2. Check finished dimension every 50th piece using gage #_____.

Process formats could be created and entered into the database for each type of process function.

The engineer need only select one of the three (3) tolerance options along with the part number to be processed. The program will automatically select the correct process, assign geometric entity data, and generate a completed process for that entity.

There are many benefits for the creation of this form of a manufacturing database. By creating a family part of worst-case entities, proveout of the entire process suggests that all parts within that family can be manufactured utilizing the same process parameters (within minimal variation). This is a pure technical approach and does not consider piece part cost for creating those less severe entities.

From a business standpoint, is it cheaper to manufacture all piece parts within that family to a known set of conditions? One must also ask, "What is the price to achieve individual optimization and the related savings, if any?".

A second area of improvement would be the opportunity to reduce raw material variances. Bar stock, for instance, after examination of all turning parts, could be better incrementally defined to reduce the number of different diameter sizes. The parts could be analyzed and broken into 1/8 inch groupings instead of 1/32 inch. Test cases would show that the cost to purchase and carry 1/32 inch incremental diameter bar sizes far exceeds the cost to cut 1/8 inch of extra material.

FAMILY ROUTING

Once the family matrix, producibility tips, process-generated attributes, and boundary conditions are established, the family routing can be created.

The family routing is a generic flow of workcenters required to manufacture all parts within the family. The information required to form the routing can be collected from existing process sheets. This exercise is intended to standardize on specific workcenters and provide the opportunity to develop alternate routings for critical workcenters.

You can create a family routing matrix by listing all the family members in the vertical columns, and all the workcenter functions along the horizontal. Within each X, Y window, place a number that represents the sequential step in the overall routing. For workcenter combinations, list each center separately, but put the same sequence number in the X, Y block (Fig. 10).

Parts	Saw	Lathe/NC	Drill	Burr	Mill/NC	Lathe	Wash	Tumble	Plate	Inspect
A	1	2	3	4	5	–	6	7	8	9
B	–	1	–	2	–	–	3	–	–	4
C	–	–	2	3	–	1	4	5	–	6
D	–	3	6	2, 7	5	1	8	4	9	10
E	1	2	4	5	3	–	6	–	–	7
F	1	–	3	4	–	2	5	–	6	7

FIGURE 10 Family routing.

Family tool package	3/64 drill	1/16 drill	3/32 drill	#36 drill	#32 drill	1/8 drill	5/32 drill	11/64 drill	3/16 drill	#25 drill	1/8 ANPT drill	1/4 drill	9/32 drill	I drill	11/32 drill
	1	2	3	4	5	6	7	8	9	10	11	12	13	14	15
A	–	–	–	–	–	X	X	X	–	X	–	–	–	–	X
B	–	–	–	–	–	–	–	X	X	X	–	–	–	–	–
C	X	–	–	X	–	–	–	X	–	–	–	–	X	–	–
D	–	–	–	–	X	–	–	X	–	–	–	X	–	–	X
E	–	–	–	–	–	–	X	X	X	–	–	–	–	–	X
F	–	–	–	X	–	X	–	X	–	–	–	–	–	–	X
G	–	–	X	X	–	–	–	X	–	–	–	–	X	–	–
H	X	X	–	X	–	–	–	–	–	–	–	–	X	X	–
I	–	–	–	–	–	–	X	X	–	X	X	–	–	–	–
J	–	–	–	–	–	X	–	X	–	X	–	–	–	–	X
K	X	X	X	–	–	–	–	–	–	–	X	X	–	–	–
L	–	–	–	–	–	–	–	–	–	–	–	–	–	–	–
Total	3	2	2	4	1	3	3	9	1	5	2	2	3	1	6

FIGURE 11

25/64 drill	0.409 drill	21/64 drill	Center 118° drill	Center drill	10-2YNC2 TAP	6-32NC3 TAP	6-32NC2 TAP	7/16-20UNF-3 TAP	5/16-2YUNF-3 TAP	1/8 ANPT TAP	1" end mill	1/4" end mill	1 1/8" back spot face	BD510050-4PORT	BD510050-2PORT	5/8" end mill
16	17	18	19	20	21	22	23	24	25	26	27	28	29	30	31	32
X	–	–	X	–	X	–	–	X	–	–	X	X	–	X	–	X
X	X	–	X	–	X	–	–	X	–	–	X	X	X	X	–	–
X	–	–	X	–	–	X	–	X	–	–	X	X	X	X	–	–
X	–	–	X	–	–	X	–	X	–	–	X	X	X	X	–	X
X	–	–	X	–	X	–	–	X	–	–	X	X	–	X	–	X
X	–	–	X	–	–	–	X	X	–	–	X	X	–	X	–	X
X	–	–	X	–	–	–	X	X	–	–	X	X	X	X	–	–
–	–	–	X	–	–	–	X	–	X	–	X	X	X	–	X	–
X	–	X	X	–	X	–	–	X	–	X	X	X	–	X	–	–
X	–	–	X	–	X	–	–	X	–	–	X	X	X	X	–	–
–	–	X	–	–	–	–	–	–	–	X	–	–	–	–	X	X
–	–	–	–	–	–	–	–	–	–	–	–	–	–	–	–	–
9	1	2	10	1	5	2	3	9	1	2	10	10	6	9	2	5

Next, optimize the family routing by:

1. Determining the critical center(s)
2. Analyzing to establish if alternate processes are required and feasible
3. Reorganizing the matrix so that all parts follow the same sequence (except for special requirements which would be exceptions to the family process). Establish a critical sequence for the family routing to minimize different types of centers
4. Determining if selected equipment is capable of manufacturing these parts to the strategic plan forecast (workcenter loading, effectivity, productivity)

WORK CENTER TOOL PACKAGES

With the family part and routing matrices completed, the next step before the creation of a family process is to establish workcenter tool packages.

The objective would be to standardize and optimize tool requirements to produce geometrical entities routed to specific workcenters. The task is straightforward as follows:

For each critical workcenter, create a matrix that lists each family part member down the vertical, and across the horizontal list all the tools presently required to produce the part geometry. Group the tools by function and then size. At the bottom of the matrix, sum the total number of times that each tool is used.

Perform the same analysis for each family. Summarize the results and create a composite workcenter matrix for all families routed (Fig. 11). Total the results for each tool at the bottom of the composite matrix.

Analyze the results to standardize and optimize on the numbers, styles, and sizes of tools. Ask the following:

Do we really need all these different hole sizes?

Can we standardize on certain tools to create different geometrical configurations such as using a 3/8 inch endmill to create entities from 1/2 inch to 1 1/2 inch diameter without sacrificing quality and performance?

By optimizing the workcenter tool package, you can essentially provide for the following improvements:

1. Minimize the many variations of tools in inventory.
2. Standardize the types of tools required to generate entities within the workcenter.
3. Reduction in setup and teardown requirements between jobs of a given workcenter.
4. Throughput cycle of new parts introduction will be reduced by utilizing existing data.

With the creation of tool packages and a unique number assigned to each part, tool costs can become part of the piece part standard cost and not overhead. Reorder points for tools can be better controlled by determining total piece part requirements, assigning a tool wear factor, and establishing a pieces/tool number ratio. I fundamentally believe that it is better to discard tools at predetermined points and keep the process within a 2 sigma tolerance spread, then wear the tool to an out-of-tolerance condition. Consistancy and predeterminable results are critical to the ability to process new parts against proven techniques.

FAMILY PROCESSES

The creation of a family process that contains the attributes of all members, and the successful proveout of a composite family model will ensure that new members can be successfully manufactured the first time.

The family process is generic for the specific members and is composed of common parameters for the creation of any geometric entity.

The manufacturing database contains specific piece part information. In the form of producibility tips, geometric entities have been defined and standardized. The data contained in Figure 7 can now be utilized to define the family model.

Each geometric entity, attribute, and specification, within the family has been analyzed, and acceptable boundary conditions established. The engineer can assign to each boundary a specific process that will ensure conformance. The objective is not to optimize the definition of each process step, but rather to establish an operational, proven start point that will allow the operator to "tweek," if desired.

The family process contains all the steps necessary to manufacture any member. The composite family model is a real part that establishes acceptable process definition for each geometric entity.

The parametric program will select the appropriate process routings, definitions, tool packages, and generate a specific member-detailed process plan.

Through database management techniques, part revision notification can be associated with specific geometric entities in the family process and the piece part process. By associating the revision to the specific geometric entity, only that sheet within the process needs to be updated.

The database can be devised to flag the originator of the process that a change is pending and requires action, or, the parametric can be written to automatically make the change.

Electronic approval and notification to MRP that revision to the process, tools, N/C program, etc., have been performed will allow production to continue.

PIECE PART PROCESS GENERATION

The ability to generate a piece part process is based upon the conceptual model developed within the manufacturing database. The engineer should be required to provide minimal input to access data to generate a process.

Previously, we discussed the creation of certain data elements directed at generating processes. Classification and coding schemes were used to describe piece parts to not only retrieve data, but facilitate the creation of family attributes—routings, tool packages, and processes.

Through the use of parametrics and database retrieval mechanisms, the generation of piece part processes can be realized.

The engineer would enter the part number and request a process. The parametric would allow the database to be accessed, and retrieve part-specific information. Copies of files containing routing, process, tooling, costs, etc., would be available to the engineer. What took hours to research and retrieve historical part information, now takes seconds.

The parametric would pull from the family process, only those definitions that pertain to the part, and sequentially format according to the family routing.

Next, the producibility tip file would be accessed and specific geometric information retrieved and inserted into the process. Likewise, tooling information and N/C cutter files could be automatically generated.

The engineer need only review the documents, approve, and then release to production.

The ability to generate piece part processes in a timely, accurate manner that will ensure manufacturability is dependant on the ability of the engineer to access the necessary information.

The creation of a manufacturing database to access information in a timely fashion to generate processes, numerical control cutter line files, and tool geometry is fundamental to reducing costs and product lead times.

10
THE MANAGEMENT CHALLENGE: INTEGRATING CAD/CAM WITH CIM

For 20 years management from the president on down has been barraged with hundreds of acronyms, such as MIS, CAD, MRP, MRPII, CAM, CAE, and CIM. Some of us look back wistfully to when "getting out the iron" meant a simple phone call to the best expeditor. Why have we become so complex in our actions—our thinking—our planning? The facts are that complexity has increased because integration of our knowledge in the form of data has not occurred; this knowledge remains fragmented. The organization of our people remains fragmented also and reflects the same condition. While it is easy to make this statement, its implication and the importance of change are not so obvious. Let us divide the problem and the challenge into five sections:

1. What is integration?
2. The impact on organization:
   Structure
   Emotions
   Knowledge
   Data and flow
3. "Old geezers"—"young tigers"
4. A new look at old technologies
5. The future

WHAT IS INTEGRATION?

We probably all remember the days when we could request from a master mechanic or production control supervisor information about some part number, and this person would tell us what it was, its material, where it was stored, how to machine it, the black book dimensions used to "adjust" the engineering design, and more.

Through the force of memory, long employment, and tenacity, people made the manual system operate. In a way this person was an integrated system for he or she had stored and remembered a host of diverse data generated by several different departments.

Just a few years ago a famous pump manufacturer discovered that the assembly floor could no longer assemble pumps. It took a good deal of soul searching to discover that their skilled assemblers who had the "know-how" had retired. No written instructions existed to convert engineering design to a practical product.

Products are no longer as simple as Henry Ford's black Model-T. We have added hundreds of varieties, electronic controls, closer tolerances, NC tool manufacturing, multiplant operations, and many more variables to the equation.

Material requirements planning (MRP) and its associated data requirements and integration of knowledge began to integrate manufacturing and indicate not only the amount of data needed for the computer system to operate correctly, but the disciplines required for an effective result. However, this downstream activity almost completely ignored the engineering department, whose personnel were considered to design things, with no direct involvement in the flow of data. Thus many portions of engineering documentation were mechanized by the data processing department or the manufacturing department without engineering involvement.

Introduction of CAD, CAE, and CAD/CAM have illustrated how a preponderance of data used by all departments is initiated in engineering. Thus we have discovered that they must join the team and help us to integrate data and the corresponding flow of information.

Sharing data, not reinitiating it in each department, is integration. The need to share data has caused a severe strain on the present organization.

THE IMPACT ON ORGANIZATIONAL STRUCTURE

In the "old days" the sheer size of the manufacturing job required segmentation of authority and work effort. Engineering and manufacturing organized themselves vertically. Unfortunately systems are organized horizontally between and through departments. Since each department has been responsible for the accuracy of its own data and actions, they are naturally wary of using information generated by someone else. Figure 1 shows how the vertical and horizontal relationships create a troublesome paradox.

Preparation of data for computer systems was initiated with great fanfare 20 or more years ago. Unfortunately, the electromechanical devices and small-scale computers available at the time

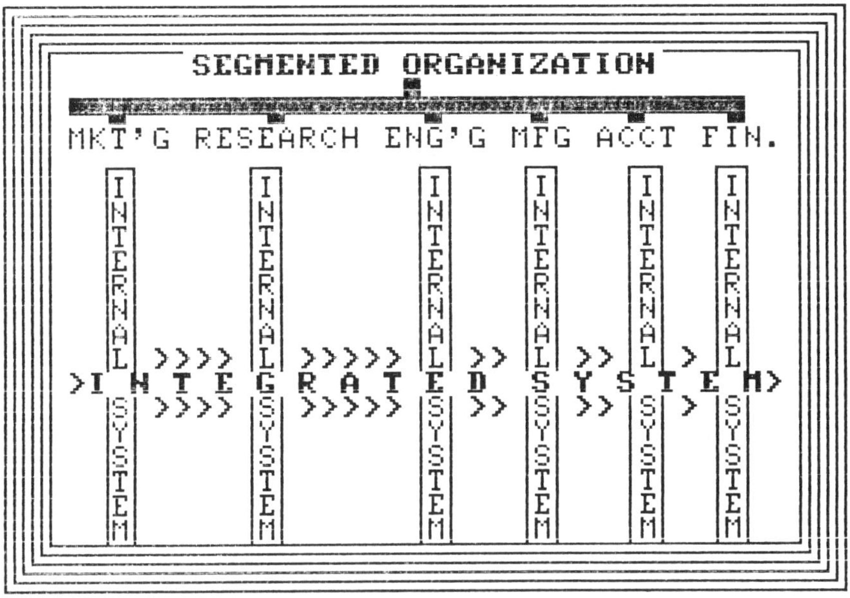

FIGURE 1 The vertical management reporting requirements often hinder the vertical relationships needed for a CIM system.

combined with segmented functions created a series of computer systems which, often, did not talk to one another and had much redundant data and overlap of function. Any activity to combine or merge these systems met with significant resistance, for such combinations implied reliquishing responsibility and authority as well as dependence upon others for performance of their job.

Recently Steve Rosenthal of Boston University queried 64 experts in the field of factory automation. This group was almost unanimous in concluding the most difficult problems in achieving computer integrated manufacturing (CIM) are primarily managerial rather than technical. Chuck Stata of Wang argues the manufacturing engineering director cannot implement CIM; it is a matter of the board room.*

With the significant increase in the power of computers and the ability to store large amounts of data, it is not difficult to imagine how identical data can be used more effectively by persons in a host of different departments. The possibility can be illustrated by

---

*Savage Associates, *CIM: The Organizational Issues, Fifth Generation Technology and Second Generation Organizations* (Wellesley, Mass., 1987).

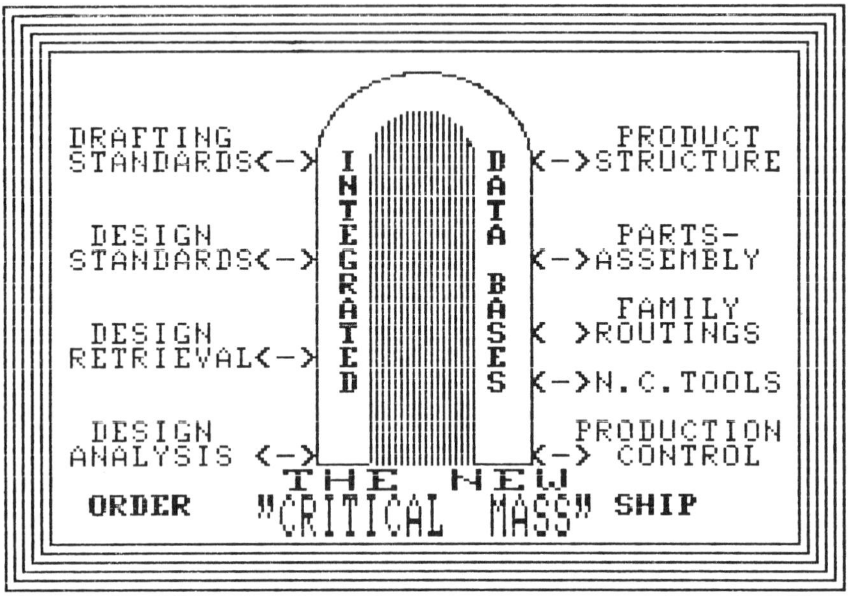

FIGURE 2 An integrated database allows economical, logical, and rapid use of identical data be several Engineering Manufacturing disciplines.

placing persons from such departments around a table, as shown in Figure 2. This critical mass of people:

1. Determines if a design requirement for a new order is available through a retrieval system.
2. If none is available, uses design analysis to create a new design to meet customer requirements.
3. Applies design standards based on company requirements.
4. Applies correct drafting standards to the new part and assembly designs.
5. Prepares a bill of material for the new product.
6. Prepares the geometry for the parts and assemblies of the product.
7. Utilizes a family routing for similar parts to prepare a specific routing for the new designs.
8. Prepares the NC tool path for the new geometry.
9. Enters the data for the parts and assemblies into the material requirements planning (MRP) and the production scheduling system.

This roundtable of people is a critical mass of activities performed today in separate rooms with separate data, and sequentially over many weeks.

By adding one ingredient, a database, all of the persons involved may now perform their tasks in parallel rather than sequentially, and with the same, not repeated, data. In this atmosphere, organizational barriers can't exist as they are now.

Dr. Joseph Harrington, Jr., a consultant with Arthur D. Little, states:

> There are hundreds, if not thousands, of compartments in the organizational structure of a typical large manufacturing company. While it is customary to find the lines of authority in an organization following a pyramidal structure (frequently known as the military structure), it is not uncommon to find that many managers have technical authority over the activities of groups reporting to another manager. This fragmentation of management is far from desirable even though it seems inevitable if very large organizations are to have an effective form of management. . . .
>
> For a century and a half American industry has been teaching its middle managers that it is their responsibility to maximize the productivity, efficiency, and output of their individual sectors. . . . But optimizing the subdivision does not guarantee that the total organization will operate optimally. On the contrary, while management structures can be fragmented as much as one wishes, the manufacturing function carried out under that management still remains a monolithic unit. If such a structure is to be optimized, it must be a totality and not piece-by-piece. Furthermore, these independent domains are inherent barriers to communication. While information theoretically should flow freely from point of origin to point of use anywhere in an organization, all too frequently it is forced to follow the line structure of the organization up through the hierarchy to a common point and then down again to the destination.
>
> There are many reasons for such restrictions, the principal ones being the human traits of self-protection and desire to control all the elements of one's own area.
>
> Business school professors have been cautioning against this system for years. Its defects are obvious to the impartial observer and libraries are full of documents pointing out the hazards encountered. In a paper by Wickham Skinner, Professor of the Harvard Graduate School of Business Administration, to the Technical Council of the Numerical Control

> Society in April 1970, he stated, "THE AMERICAN FACTORY SYSTEM IS OBSOLETE."
>
> The last two decades have seen the introduction of new technologies that not only revolutionize the manufacturing process but virtually demand the revolution of the management process that go with them. . . . As can be seen, the web of information exchange reaches into every part of the manufacturing organization and electronically circumvents the old communication problems and all their inherent barriers. . . . Obviously, the new system system will not be dominated by a single person such as the craftsmen of colonial days, but the same responsibilities will be fulfilled by an *INTEGRATED GROUP OF PERSONS*.*

How then shall *we* meet the challenge new technology presents to our organizations? Can we face them objectively, critically, and creatively? Our skill in solving this problem will be a major factor in the success of our future manufacturing efficiency, perhaps in our very survival.

THE IMPACT ON ORGANIZATION—EMOTIONS

We have discussed at some length the implied necessity for a new organizational structure. The impact on all persons involved can be highly emotional. Many managers will feel they have lost their territory, even their status. It may at times be difficult for personnel to know the exact function under which they operate and report.

Certainly, today one of the most significant problems of implementing a simple CAD/CAM system is organization. To whom should computer operations report? Engineering? Data processing? To whom should the designers and/or drafters on terminals report? To the Director of CIM? To Engineering? To whom should the parts programmers report? To the Director of CIM? To Manufacturing? To Engineering? Many other unresolved problems of organization have and will create some fear, anxiety, anger, and frustration.

THE IMPACT ON ORGANIZATION—KNOWLEDGE

There seems to be no question that technical knowledge is outpacing our ability to utilize it. Even as this book is being read, both the

---

*Joseph Harrington, Jr., *Computer Integrated Manufacturing* (Huntington, N. Y.: Robert E. Krieger Publishing Co., 1979).

author's and the reader's knowledge concerning CIM is becoming obsolete. Robert B. Reich, of the Kennedy School of Government at Harvard, clarifies this point:

> A flexible manufacturing system requires an organization designed for change and adaptability: high volume, standardized production requires an organization geared to stability.
>
> A flexible manufacturing system is rooted in discovering and solving new problems: high volume standardized production basically involves routinizing the solutions to old problems.*

The abundance of new knowledge will continually alter our concepts of manufacturing. New knowledge will change our design methods, parameters, and product specifications. For example, NASA has proposed that steel with special new characteristics be rolled in space. How will steel mill manufacturers face this challenge?

How will we educate our drafters and our engineers? How will top and middle management become accustomed to manage new and powerful techniques?

THE IMPACT ON ORGANIZATION—DATA AND FLOW

For the past 20 years data has been "flowing" more or less efficiently and effectively through a series of data processing business systems. Unfortunately, as mentioned previously, engineering documentation has often been provided by other than the engineering department. With the implementation of a significant new addition to this information and flow, part and assembly geometry, a new version and philosophy of data flow must be pictured. No longer can engineering departments sit by idly as others transform their data into computer information.

The engineering department is now part of the day-to-day flow of information. In fact, they are the originator of much of it.

Also, as we have tried to graphically illustrate, use of a database will create an atmosphere where on-line systems used interactively will be mandatory. We should remember that the implementation of a CAD/CAM system and CIM, including the graphics and design analysis portion, will be successful only when all users understand and happily embrace the new technologies.

How will we create enthusiasm in an engineering department to obtain their cooperation? How will other departments accept data

*Robert B. Reich, "The Next American Frontier," *Atlantic Monthly* (March, 1983).

from engineering without reinventing it? Will manufacturing engineering be satisfied with accuracy of geometry and other data created in engineering?

"OLD GEEZERS"—"YOUNG TIGERS"

"Young tigers" and "old geezers" describe different types of people to each of us. We should use these descriptions not to describe a chronological age, but a way of thinking, a philosophy used in addressing the day-to-day work place. The implementation of new technologies has created a new problem.

It is possible for a company to believe that a young enthusiast, a young tiger, with much knowledge about the new technologies be placed in a position of responsibility and authority to implement them. Unfortunately, this person may make little use of past knowledge, problems, company history, and other significant background material.

On the other hand, to make certain that this past background is not destroyed, an experienced person within the company may be assigned the responsibility and authority. Unfortunately, this person may be uncomfortable with the new technology—perhaps even antagonistic to its use.

In either case, the company may not be well served. How can we merge the best of both? How can we take a mixed group of old tigers and young geezers so they will integrate the best of the old and the best of the new? This may well be the most important consideration and challenge of all.

A NEW LOOK AT OLD TECHNOLOGIES

For the past several decades members of management have tended to be uninvolved in the solution of "minor" systems problems. This lack of participation created incomplete or parochial solutions. Now this lack of participation is showing up in the need to modernize present systems before investing time and money in more sophisticated and exacting new ones.

It is inevitable that members of management must now participate in solutions because most of the problems they represent are not departmental, but corporate in scope. These details are not glamorous. Often they cannot be evaluated clearly in economic terms. Nevertheless many of them have delayed modernization for a long time and must be addressed or an integrated CIM system may be impossible to implement. Let us discuss a few of them:

Part numbers
Piece part drawings
Bill of materials
Modular design
Design retrieval
Part families

Part Numbers

Almost 50% of companies in the United States have job-oriented, not part-number-oriented, systems, and the part number systems in a proportional number of companies are out of date with obsolete built-in significance, too long, and/or a difficult mixture of alphanumeric symbols.

Job number systems were impressed upon manufacturing to make certain costs of an order were collected. Since job number accuracy was accentuated, they soon replaced part numbers in significance within the system. Thus it is possible to find identical designs with identical part numbers being manufactured separately under different job numbers. Two cases may occur. Job numbers within the same customer order will be different, hiding duplicate parts, and job numbers between orders will be different, hiding duplicate parts. Integration of requirements, manufacturing planning, family routings, and design retrieval cannot increase productivity if this artificial segmentation is continued. Thus the accounting department as well as engineering and manufacturing must convert to a system which makes the part number the means to retrieve data and to combine requirements.

Many part numbers are too long because they been stuffed with significant data which is useful to only a few, is often redundant, and which may be obtained and maintained more easily today in other ways. (The writer has seen a part number system requiring 45 digits!) Streamlining activities from the creation of a new design in a CAD/CAM system to its manufacture on an NC machine tool requires systems which can be implemented rapidly and used as a tool not a millstone. In addition, far greater access to engineering documentation from design geometry to engineering drawings and parts lists requires a simple means to obtain the data. Recognizing that a part number should be the "key" to all other data concerning the design, doesn't it make sense that that key, the part number, be short and easy to use?

It will not be easy to change a part number system to make it reliable and to become more important than the job number, but productivity demands it. Few conversions have been successful without management participation, including awareness, understanding, and agreement.

Piece Part Drawings

A large segment of today's manufacturing companies also have multipart drawings and assemblies. These companies should convert to piece part drawings. Eleven reasons for such a conversion exists:

1. Clarity of fabrication details
2. Economical part design
3. Documentation changes
4. Part reusability
5. Design retrieval systems
6. Elimination of duplication
7. Raw material requirements
8. Principles of manufacturing
   a. Routing requirements
   b. Tooling requirements
   c. Group technology
9. Family drawing preparation
10. Computer-aided drafting
11. Purchased part documentation

Personnel within the engineering departments of companies who do not now have piece part drawings tend to resist this change, marshalling emotional forces rather than logic. Such statements as "it will create more paper," "it will take more time," "it will cost more," are heard. The corresponding benefits as outlined in the 11 reasons for the conversion are rarely heard. Members of management must participate in the resolution of this problem to avoid endless discussions and long delays.

Bills of Material

It seems that job number systems initiated by accounting departments were, at one time, a worthy alternate, or replacement for, a bill-of-material system. It must be remembered that the discipline of job number accounting met the requirements of an accurate costing system. However, a mechanized bill of material performs all the functions for which job costing was designed and meets many additional integrated system requirements which a job number system cannot hope to achieve.

A bill of material is an accurate and rapid means to record part and assembly relationships in an engineering design. In addition, it will eventually be the database method to store and retrieve geometric data for part and assembly designs. It has already become a mandatory system in the implementation of material requirement planning (MRP) and other mechanized engineering documentation systems. Thus it is imperative that a bill of material be installed and/or enhanced within a manufacturing company.

Modular Design

A significant advance in product documentation occurs when modular design techniques are used. The principle of modular design is to extract all design variables from the product and arrange them into separate bills of material known as attachments. These variable parts may also be moved to higher levels in the old bill of material, or split into segmented assemblies. The remainder, or "constant" portion, of the product is then divided into major assemblies, often called modules. It should be emphasized that the physical design must be converted, if necessary, to match this division. Several important results are obtained with this procedure:

1. Modules and attachments are used on a master schedule to generate MRP.
2. Modules and attachments are combined based on a customer order, thus eliminating a significant number of bills of material for slight variations to a product. It might be best explained by stating that a product is documented by part numbers which are expanded in quantity arithmetically utilizing modular design rather than increasing geometrically as required when each variation is documented with a specific bill of material.
3. Less inventory is required to maintain many product variations since the modules and attachments may be stocked in preparation for an order rather than stocking all possible combinations.
4. More standardization of products is possible because each module is designed to mate with other modules and attachments. Minor variations do not require complete new documentation, only that which shows changes.
5. Production can be increased in that the modules can now be identified and produced together where previously identical assemblies and parts may have been hidden under different bill-of-material part numbers and under several variations within them.

Design Retrieval

The most significant tool for successful implementation and integration of CIM and CAD/CAM is a classification and design retrieval system. This is often thought of as an engineering tool only, but management must understand the principles of these systems in order to see the corporate implication and importance of them.

Classification systems within a manufacturing company can be divided into four segments, raw material, purchase parts, designed parts, and assemblies and products. Let us look for a moment at the designed part segment.

Using a classification system, designed parts are divided into "families" with similar geometric shape, form, or functional relationships. One simple family, for example, may be those designs which are metallic, round, multiple diameter, with diameters successively decreasing, and with a throughgoing center hole. This family of parts will contain 15 to 20 parts if the system is properly designed. Later we will describe how families of parts are important for productivity improvement and product standardization.

A design retrieval system will have many uses and economic benefits. Some of them are as follows:

1. Design retrieval statistics have shown that from 4 to 20% of new designs already exist if they can be retrieved. Since cost for creating a new design may vary from $12,000 for a complex forging to $200 for a simple fabricated part, elimination of design redundancy generates a significant savings. (Many companies who have implemented a design retrieval system have paid for it in less than a year.)

2. Uniformity of data is encouraged by use of a design retrieval system. This also allows a more detailed description for purchased parts, avoiding one-vendor restrictions.

3. When designs are classified using permanent form, shape, or functional features, an important part family database can be established. This data will be key to the integration of CIM.

Engineers (and others) tend to resist the implementation of a design retrieval system. Thus management must assist in developing a climate conducive for such a system.

Part Families

A direct and important result of a classification system is part family data. Let us consider. Suppose we were asked to look at 80,000 designs and establish drafting standards, routings, and tooling for them. Wouldn't it be a lot easier, practical, and logical to consider them not as individual designs analyzed randomly, but rather as 1000 part families where similar parts with similar features are gathered? Isn't it more prudent, quicker, and economical to tackle standardization using 1000 families rather than 80,000 random parts? A properly designed retrieval system will allow this to take place.

Some of the family data which can be generated is as follows:

Design standards
Drafting standards

Producibility tips
Design analysis parameters
Family processes
Parts catalogs
Tooling catalog
NC cutter tool path
Parametric drawings and geometry

Let us look at how we might use this data and some of the benefits:

1. A parametric drawing can be used in conjunction with the classification system and a parts catalog. When a new design is required, a look at the appropriate part family and catalog section will make certain no existing design meets the new criteria. Once confirmed, the parametric family drawing is invoked to enter the dimensions and other criteria for the new design, and the computer prepares it automatically in the form of a drawing. (A parametric drawing should be considered a special program which literally draws the geometry of a part. The program has voids in it, however. With appropriate prompts, the user fills the voids by entering all the dimensional data necessary to create a specific design.)

2. The family of parts may be examined to create a family process and routing which optimizes the work centers, set-up time and requirements, and other data for the family. Of course, this first requires a study of the present routings and processes, which, when complete, will result in a more uniform and optimum set of routings. More, it will allow all new designs to be released quickly by preparation of a new routing and process through minor alterations to the family process and routing.

3. Once a family of parts is established, an index of perishable tools and jigs and fixtures can be prepared for each of them. This index may be used so that new designs will not generate new tooling until the family of tools index has been examined.

4. The classical use of part families is that of group technology or cellular manufacturing. Using family data such as design sizes, weights, raw materials and forecasted usage, machine cells can be developed.

5. A family tool path for those operations requiring NC tools will also expedite the preparation of a specific tool path for a new design. In addition, tool packages for an NC tool can be examined to make certain no additional features, such as nonstandard hole sizes, and taps, have extended the tool package requirements to create a tool change. This type of standardization is also accentuated in the use of producibility tips.

6. Producibility tips imply a joint agreement between engineering and manufacturing in the best means to standardize methods to make manufacturing more economical, higher in quality, and uniform.

Such items as standard drills and taps, bevels, bend radii, finishes, plating, and many more, standardized by family, create a fine benchmark for both engineering and manufacturing to avoid unnecessary costs and engineering changes. In addition, the producibility tips are not restrictive. Usually they are tabulated on the basis of cost with comparative cost increases also shown should they be changed. This allows designers alternate methods of design, but indicates to them "penalties" or costs of alternate methods.

If we review the "details" which require management attention, it is apparent that the majority of them are instituted in engineering. This recognition emphasizes again the need to bring the engineering activities into the mainstream of data flow in business and CAD/CAM systems. If CIM is to be realized both as a technique and as a truly integrated approach to manufacturing, however, all precisely, honestly, and with management knowledge and approval. The challenge to management is there for all to see.

THE FUTURE

It is important to understand that "future shock" will remain with us. We must learn the lessons of flexibility, continuing education, new management techniques, and, above all, expand and maintain our sense of humor. I believe the latter is truly important. At just the point where we are confident we understand some new technology, our knowledge becomes obsolete, perhaps worthless. The only certain knowledge we have today is that constant change is constant. Let us develop our management techniques and actively participate to take advantage of these challenging new ideas and inventions!

Appendix A
GLOSSARY OF TERMS

ENGINEERING

Bills of Material

The bill of material should be considered a document as well as a database, generated from the joint item master, product structure databases. At least five different documents are generated, including an indented, summarized and level-by-level bill of material, a costed bill of material, and a where-used, or used-on tabulation. (For consideration of a bill of material as a stored database see *Generic Bills of Material*.)

The data generated for this document is used in many business systems, such as manufacturing resource planning, parts lists, the master schedule, the marketing pricing manual, and assembly picking lists. Engineering changes many be reflected in the bill of material documentation in the form of changes and their corresponding effectivity. Control of the documentation must be based on configuration control rules.

While, normally, the lowest level of a bill of material system is the raw material required to make the designed part, a special consideration to avoid excessive paperwork is instituted by placing this data in the item master of the designed part. By inference rather than by product structure relationships, this data is extracted from the item master when the bill of material is exploded and is made a part of the explosion. This technique eliminates a sizable amount of product structures.

Computer-Aided Design (CAD)

Computer-aided design is a discipline used to capture the geometry of design in a manner by which both engineering and manufacturing can use it to create physical parts and assemblies accurately and rapidly.

Computer-Aided Engineering (CAE)

Computer-aided engineering uses a set of algorithms and formulas to design products, parts, and assemblies more rationally, effectively, and economically.

Configuration Management Rules

To initiate and maintain the identity of new designs, it is necessary to know, as design changes become necessary, when the part number or the revision level of a design should be changed. Rules of interchangeability assist in this decision and include the requirements necessary to know how such design changes affect each level of the bill of material from the point where the change is made up through each succeeding level.

The rules of configuration management control the contents of a product and may be reflected in the way the product is documented for customer requirements. (Also see *Effectivity* and *Bill of Material*.)

Design Analysis Programs

Many companies perform individual calculations for designs such as gears, tube bending, and solenoids. Often these are stored and maintained as separate entities. It is important to provide a means to store these mathematical algorithms in a coordinated database not only for ease of retrieval and maintenance, but also to consider them as integrated to other portions of a CAD/CAM system. For example, the results of a gear design calculation should not only be design data, but a drawing prepared using the dimensional data created during the calculation.

Design simulations and kinematic studies should be considered as working with other portions of a CAD/CAM database also. It may also be necessary to consider storing the results of these calculations in a database as proof or confirmation of the design for future reference.

Design Release Notice

When the engineering department completes a new design, it must be released to manufacturing and other departments of the company so it may be purchased or manufactured. The release notice is a typical means to achieve this communication requirement. In addition, the release notice contains the engineering data required for the item master database. Thus the release notice is a communication link to notify departments of a company that a new design has been initiated, and also obtain the necessary data increments for the item master database.

A release notice is required for each new design created, whether it is a designed part, a purchased part, an assembly, a kit of parts, or a product or module.

Design Retrieval

A design retrieval system is used to collect parts into families for analysis, retrieval of identical or closely similar designs, and as a basis for storing family data. Data such as parametric drawings, family processes and routings, tooling, fixtures, cutter paths, producibility tips, design standards, and drafting standards is tied together by the retrieval system.

Two forms of retrieval systems can be integrated as one unit, a monocode which retrieves designs based on similar shape, form, or functional characteristics, and a polycode which retrieves similar attributes, such as milling, drilling, boring, and grinding.

The retrieval system should be considered a main tool to consolidate data for a CAD/CAM system to fully utilize that which has been generated previously and to use such data for analysis, standardization, elimination of duplication, group technology, and flexible manufacturing.

Design Simulation Models

While a design analysis database is available, and simulation techniques have also been made available, special simulations may be required by marketing to certify compliance to customer specifications, for example. A database of these specific simulation models should be maintained to eliminate unknowledgeable users from searching for such data. These simulations may be customized for the customer, or made a general system for a specific set of customer requirements.

Design Standards

In order to fully utilize modular design concepts, the interrelationships between modules, attachments, and options need definition. These definitions may be considered one form of design standards. In addition, each product type has design standards of performance, weight, size limitations, and so on needing documentation. Limitations of designs may take the form of design analysis calculations or parametric rules. In any case, this type of documentation needs to be related to the drafting and design analysis databases.

The results will be found in the ability to combine modules into a finished product with confidence they will assemble as a unit, simulations of the features of a product, limitations of the capabilities and capacities of a product, and fulfilling the specifications of an order.

Retention of this data eliminates the total dependency on a particular designer's "know-how" and makes the data available for all.

Detacher Parts List

Prior to the mechanization of a bill of material, material lists were compiled directly on the assembly drawing. These have now been replaced with parts lists controlled and generated by a mechanized system. These detached parts lists form the core of information about a product since, when combined, they create a bill of material. Since it is important for accuracy of other systems to have the proper components related one to another in an assembly, use of predefined product structures becomes an important consideration. An accurate bill-of-material system has the following ingredients equal in all respects, a general product structure, a corresponding assembly drawing for a specific design, and a parts list. Coordination between data on the drawing and the parts list is maintained utilizing find numbers, also called bubble numbers. Each part on the assembly drawing is marked with them, and the same number is associated with the correct part number on the assembly parts list.

Dimensioning Standards

Many of the implied drafting standards used when documenting a design manually change when a CAD/CAM system is used. Dimensioning baselines, strings of dimensions, and tolerancing take on new meaning when tied directly to three-dimensional geometry which is utilized by others. No longer is the drawing interpreted by others. If the geometry is used directly for tool paths and other automatic manufacturing equipment, different ways of dimensioning a design imply different geometries. In addition, parametric drawings may be interpreted incorrectly depending upon the dimensions which are changed to make a new design.

For example, two holes on a part can be dimensioned in at least three ways and create three different designs. The holes may be individually dimensioned from one edge of the part, individually from the centerline of the part, or one hole can be dimensioned from these references and the second hole dimensioned from the first hole. Care must then be taken to set up instructions, rules, standards, and explanations ato minimize errors in geometry caused by poor drafting and dimensioning practices.

Drafting Standards

In a manner similar to that describing the formation of producibility tips, drafting standards should be considered part of the CAD/CAM database. Drafting standards specifically required to make a family of parts can be prepared. These standards should include dimensioning reference planes or surfaces and be customized for each family. Within these drafting standards, for example, distance of scribe lines and arrow heads from the actual design, welding symbols, and others will be maintained. Special care is required for part families where the parts are prepared on NC tools to make certain the geometrical dimensioning is consistent with machining practices, and uniform so new geometry definitions for the next new part in the family is not required. This consistency will allow a family tool path to be prepared and stored as part of the database.

Electronic Mail

Implementation of a CAD/CAM system in conjunction with a business system can be greatly enhanced not only by creating data in an on-line environment, but also by communicating it on line. Using electronic mail as a means to convey information about new design releases and changes makes it possible to convey such information to involved departments in a parallel manner rather than sequentially. This will speed up reaction time.

Thus the initiation of a release notice, a change notice, and a product structure should be considered as an on-line system.

Engineering Change Effectivity

When a design is changed in engineering, the documentation for it such as the drawing and detached parts lists are altered to reflect the change. However, it is often impractical or impossible to implement the change in the manufacturing process until some later date. The difference between these two events has been defined as effectivity of change.

Most modern bill of material systems accommodate effectivity of change by maintaining product structures of both the old and new configuration, including "in" and "out" parent-component relationships and the date when this change is to be incorporated. These changes are also reflected in the MRP system, which eliminates requirements for the old design and creates demands for the new.

Engineering Change Requests

Design change requests may be initiated by almost anyone in engineering, manufacturing, accounting, or by customers and other external functions. These requests, since they normally are created informally, following many formats, must be scanned, evaluated, interpreted, and generally considered for inclusion in a design alteration. The data in a request must be converted to a more concise statement of the change being considered.

Engineering Changes

A formal engineering change is used to document the actual design alteration being recommended. Documentation for a design change will normally appear on an engineering change notice and include the part numbers of all designs affected, the new configuration documentation requirements, the physical changes necessary. Related changes required such as disposition of any and all obsoleted or altered old designs, and all other documentation or physical entities affected such as tooling will be itemized.

Feature-Based Design for Parts and Assemblies

Every part may be considered composed of incremental pieces of geometry. These increments may be defined in one of two ways, by function, or by shape. This basically simple premise defines the most powerful tool for an Integrated CIM system. Let us look at these two increments:

Functional Geometry

Many designs can be described quite clearly by naming their function. Crankshafts, gears, pulleys, flywheels, yokes, sprockets, and more call up a rather specific set of pictures. Once we have established one of these basic designs, of course, each of them will also have specific design features based on form shape or function as follows.

Feature Geometry

Feature geometry includes the most simple design elements. Channel, angle, I-beam, bolt circle, cube, hole, thread, joggle, re-entry cut, round and/or diameter, polygon, bend, centerhole, slot, chamfer, and more, again call up a rather specific set of geometry.

Of course, when all geometric features are defined as solid models, the length of the list could give a user trouble in finding just the right set needed quickly when preparing a specific design. Fortunately, there is a relatively simple solution for this problem.

Piece Parts

Geometry features can be stored and retrieved in much the same way parts and assemblies are stored in a bill-of-material system. Consider each geometric feature as an entity which can be stored individually as an item master. Let us next assume a classification system has been developed which allows the retrieval of designs with similar shape or function. The retrieval system segregates all piece part designs into families of parts. Each of these families can be represented with a family drawing. The family drawing is, in reality, an assembly of geometry features. For each family of parts, then, a product structure in the form of a parts list can be used to associate, one to another, all the geometric features needed for a specific family, and can be used when a new design is prepared. We now have a means for rapid retrieval of geometric features for part designs. It should also be noted that, should an additional new design within a family require a new design feature, it can be added to the parts list and corresponding product structure. It should also be obvious that each geometric feature need be stored only once, since the family geometry parts lists become pointers to them as required.

Assemblies

Each assembly is a collection of part designs. Consequently an assembly parts list is a series of collected geometric features. Each assembly may also have additional geometric feature requirements, such as additional holes, welding, and machining. Thus each assembly may have its individually geometric parts list. If assemblies are classified in a manner similar to piece part designs, a family assembly drawing with a corresponding parts list of additional geometric features can be used to maintain geometric associations.

Bill-of-Material for Geometry

It would appear that a bill-of-material system tied by part numbers and product structures to all parts and assemblies, combined with additional product structures and their corresponding parts lists for geometric features, will make a powerful storage and retrieval system, avoiding redundancy and making it possible to create new solid models of designs rapidly and efficiently.

Finite Element Meshes

As a special case of design analysis, consideration should be given to creating meshes by part families for use in finite element analysis. The logic of this type of predesigned data needs careful consideration and investigation. It may be that such general meshes apply only to limited and/or specific families of designs, including assemblies. However, their generation may eliminate reinventing means for analyzing designs. Use of a family mesh presupposes part families and a design retrieval system.

Generic Bills of Material

In special cases a complete bill of material may be formed into a database. Of particular importance to a job shop, this database can be used to generate manufacturing requirements, lead times, and purchased part requirements before the complete product has been designed. The generic bill of material consists of a product structure similar to the product ordered but with a separate part number system. As the product is designed, these part numbers are replaced with actual design part numbers, converting production and purchasing requirements into actual orders. The generic bill of material may be used as a means of scheduling engineering design, since it is maintained by delivery date and lead times. A lead time chart generated from this bill of material will list, by lead time sequence, the parts and assemblies required first, second, and so on.

Item Master

Through common usage and terminology, the item master is considered to be the database required for business systems. Data found in this file is generated by several sources such as design engineering, manufacturing engineering, inventory control, and accounting. The item master data is used in conjunction with the product structure database in the bill-of-material system.

When properly designed, the item master database will have a record for every part number issued, both old and new, both active and obsolete. The data is a cornerstone of engineering documentation.

Kinematics for Finished Designs

In a manner similar to that described under *Design Simulation Models*, so should a database for kinematics be prepared.

Management Information System (MIS)

A management information system is designed to prepare and present timely data to operating personnel in their performance of daily activities.

Microfilm-Microfiche Interface

It would be nice to believe that all designs can be found on a CAD/CAM system. Unfortunately this probably will not be the case for many years. Thousands of current and adequate designs will still reside in drawings files. A microfilm system is a means to coordinate the old and the new by documenting both those drawings prepared on a CAD/CAM system and those manually prepared previously.

In addition, a classification system which classifies similar parts will be greatly enhanced if a microfiche system is used to store and maintain designs by family. Retrieval of similar parts is thereby enhanced and facilitated.

Parametric Drawings

Either through the use of a formal retrieval system or some other method, a database of parametric drawings for similar designs can be used to prepare part or assembly designs rapidly. It should be emphasized that the majority of all designs required are fundamentally simple parts. (It has been estimated that 40% of all drawings can be prepared using the geometry of a cube and a hole.) Use of the parametric drawing philosophy should not be limited to design and drafting activities only. Rather, the parametric drawings should be closely associated to the required process plan and routing, and the NC tool path requirements. Parametric drawings should include regulation drafting and design standards, producibility tips, and other geometry-influenced documentation standards.

Parametric drawings can be considered an extension of geometric features for a design and may well be an assemblage of such features when used to make a new design.

Part and Assembly Geometry

Two methods may be considered in developing and storing new designs.

Regenerating Designs

If parts and assemblies are to be created using feature-based geometry, it may be logical to presume only dimensional data, in text form, will be maintained for each design. Merging the geometric features with the variable dimensional data and other special text data and notes only when the design is needed could eliminate a tremendous amount of storage requirements. This feature must be compared to the possible disadvantage of the time required to reproduce the design, or "repaint" it on the terminal. At the piece part level, this principle may be quite successful, at least for simple parts, which compose probably 80% of all design requirements.

Finished Design Storage

Even if the entire design of individual piece parts must be maintained in storage, there is no reason to believe they must also be stored redundantly as part of every assembly geometry where they are used. If a product has five levels in the bill-of-material structure it could be found necessary to store the same piece part design five times at each successive assembly level. Rather, it seems more practical and economical to use the principle of a product structure and the corresponding parts list for an assembly to bring together the geometries of all the parts needed for the assembly.

A significant database which will eliminate redundant storage of identical geometry at each assembly level is that of the assembly relationship of one part to another in that assembly. The product structure can be considered the linkage for performing this activity. As each assembly is created, the components of the assembly may be located by part number, and the assembly regenerated when needed.

Assembly of Parts Transforms

When the piece part geometry is associated to their corresponding assemblies, additional data is required to assemble them correctly; this can be called assembly of parts geometry transforms. Three types of transforms may be considered. The most simple is the location of one part to another in an assembled condition. The second type of transform is that which shows the parts oriented on reference lines in exploded positions oriented to the proper assembly sequence and location. This type of transform will create exploded views of the assembly for spare part manuals, for example. A third set of transforms is that which considers the location of parts in bins or other receptacles prior to assembly and the movements of a robot required to transport the parts from these locations to the final assembly.

Storage of these transforms also eliminates the need to repeat the geometry found in one assembly when that assembly is in turn used in an assembly at a higher level. As assemblies are built from one level to another, the transforms rather than the geometry may be stored and utilized, together with the product structure data, when an assembly drawing is prepared.

Part Number Log

A part number log, normally a manual engineering department system, is used to manage part number issue, not only to ensure that a part number is not used more than once, but also to keep track of the person to which numbers are issued.

To maintain high productivity of a CAD/CAM system, the issuing of part numbers should be automatic. This procedure, however, presumes some orderly sequence in the part number system. If significance is built into the number, manual intervention may be necessary. While not the most important improvement in an engineering documentation system, automatic issue of a part number eliminates one of the many petty annoyances in engineering.

Parts Catalog

Closely associated with a retrieval system is a cataloging system which includes all parts which have been classified. In addition to the cataloging of parts by family, the ability to include critical characteristics of each part in each family must be accommodated. These characteristics may be dimensions, specific features, material chemistry, and others.

The ability to rapidly retrieve geometry of either a specific part geometry, the family general geometry, and other associated data as that found in the item master should be considered.

Producibility Tips

A potentially significant feature of a CAD/CAM database will be the ability to store jointly approved methods for economically manufacturing parts and assemblies. Called producibility tips, this data may be global for all designs, or specific for a family of parts. Examples of these manufacturing methods include standard drilled hole sizes, chamfers, finishes, tap sizes, and tolerances.

The significance of maintaining this data is that the designer will build into his design the most economical method for manufacture before it is released, thus shortening the time between engineering design and manufacturing. A significantly increased uniformity of manufacturing planning, tooling, and NC tool paths will result.

Product-Module Specifications

To integrate marketing activity into the CAD/CAM system, it is necessary to be able to identify product configurations and their corresponding bills of material and geometry in such a manner that marketing personnel can view these configurations and estimate design changes, costs, and lead times. In those companies making use of modular design techniques, the product modules, attachments, and options must be identified. In addition to such documentation, operating charcteristics, designs parameters, and simulation ability should be associated to this data.

Product Structure

The item master database contains no parent-component part relationships. (However, see *Bills of Materials* for an exception for raw material requirements.) The product structure database is specifically designed to maintain the parent-component relationship of one design to another in the form of a structure which represents a level-by-level set of parts lists and their corresponding physical assemblies.

In addition to this parent-component relationship, other data specifically related to the relationship is maintained here as quantity of each part required, find numbers (corresponding to the find numbers on the assembly drawing), and effectivity.

Traceability Structure

As a special case of manufacturing requirements, lot tracing of parts may be required. Under these conditions each part in a design may require a serial number for such tracing. Control in this manner is a paradox since identical parts also require different serial numbers. Traceability control can be maintained in the Product Structure by considering the lot number(s) as a level below the piece part or assembly for which they have been assigned. Then, in combination with the inventory system which will assign different bin locations for different lot numbers, and the bill-of-material picking list used to remove parts from inventory, the parts thus picked and their corresponding lot number will be a function of the inventory system and posted on the picking list prior to its release to manufacturing.

MANUFACTURING

ABC Inventory Analysis

An important management tool which assists in the analysis of both inventory and manufacturing requirements is an ABC inventory analysis. Before material requirements planning (MRP) was implemented, stock parts in inventory were analyzed by taking historical usage of each part and multiplying the quantity by its individual dollar value. The total value for all parts was then summarized. The percentage of dollar usage for e ch individual part compared to the total value was then calculated, and listed from high to low dollar usage. The values thus generated follow Pareto's Law since approximately 80% of all dollar usage is equivalent to 5% of designs, 15% of all dollars equivalent to 30% of designs, and the final 5% of the dollars equivalent to 65% of the designs. (These percentages vary from one company to another.)

Today, MRP, combined with the master schedule, can be used to create the same analysis, but on forecasted requirements not past usage, making the analysis far more powerful and useful.

CAM Computer-Aided Manufacturing

Computer-aided manufacturing is a process to convert design entities into physical parts and assemblies, using the latest manufacturing techniques, such as generative process planning, NC tools, and flexible manufacturing or group technology.

CIM Computer-Integrated Manufacturing

Computer-integrated manufacturing is the integration of all these support systems with a significant emphasis on an Integrated data-base to avoid redundancy, increase productivity, and create a synergistic pool of information.

Family Routings

By a means similar to that used to prepare parametric drawings, a family process can be generated. It is important to recognize the components of a process and the types which can be prepared.

Process

A process may be defined as a detailed "how" to perform the machining or fabrication operation on the design. This detailed set of instructions includes the means to set up the machine, the tools required, the feeds and speeds of the cutting cycle, and so on. Embedded in the process is the routing.

Routing

A routing can be considered a shortened version of the process. All detailed descriptions of the "how" are eliminated, and a shortened version of the process created. Information to be found on the routing includes work centers, operations, run times, set-up times, and tool part numbers. The routing is often issued as a portion of a shop packet when the design is actually manufactured. Two types of family processes utilizing design retrieval methods may be considered.

Part Geometry

When preparing a family process based on a design retrieval system utilizing similar geometry, the close similarities of one design to another make the actual operations identical. Some variations may occur in the actual machining operations as feeds and speeds due to material variations. In addition, some geometry not found normally in the family may require special operations.

Part Machining Features

If the part family has been based on individual manufacturing techniques, such as milling, drilling, or grinding, the parts may be quite different in shape and configuration and also not have the same sequence of operations. This family of parts will require more customization to make a specific process for a new design.

Group Technology Cells

As a special case of the manufacturing process, parts may be manufactured more economically using machine cells. Group technology is defined as the process whereby different designs, similar in form, shape, function, or manufacturing attribute, may be grouped and manufactured in a single batch. This practice avoids the many setups required should each different part be manufactured separately. In addition, grouping machines into a cell to perform various operations avoids excessive transit from one operation to another. A new name for this technique is flexible manufacturing.

Inventory Control

An important tool used with material requirements planning (MRP) is the inventory control function. Before MRP was first used, some, but not all, of the parts used in a product were considered inventory, and normally designated as stock items. The vast majority of parts and assemblies were considered work in process, and controlled by dollar value, not necessarily by quantity. MRP has changed this concept in that all parts and assemblies are considered to be in inventory and controlled by quantity as well as by dollars. Four segments of inventory are usually considered, raw material, finished parts and assemblies, work in process, and finished products. Purchased parts are considered to be in inventory as well as "on order." These records are used in conjunction with the master schedule and the bill-of-material explosions for the products to reduce gross requirements to net requirements.

Jigs and Fixtures

Special tools required to fabricate or machine parts are called jigs and fixtures. Fixtures are normally considered to be holding devices for parts which become weldments. Jigs are usually considered to be devices to restrain and position parts accurately while being machined.

These tools are usually designed for a specific operation on a specific part and can be quite expensive. More effort is being expended to classify such tools so they may be retrieved and altered slightly to be used successfully on more than one design. Numerically controlled machine tools require less of these special tools and thus comprise one of the economic benefits cited when purchasing them.

Manufacturing Geometry

A significant philosophy which must be considered in developing new designs is that of "additive" geometry for manufacturing. Design engineers prepare geometry to express the specific configurations they wish to achieve. To manufacture the design, additional material must usually be added to the finished design dimensions to fabricate or machine the part. Thus the original design dimensions are not changed, but additions made to them.

Using this philosophy, an "envelope" of additional dimensions may be added to the original design to create the necessary rough design. Thus the design database integrity may be maintained.

Manufacturing Resource Planning (MRPII)

Manufacturing resource planning is an extension of MRP which adds to the technique of material logistics by scheduling production machines and labor requirements as an integrated system.

Master Schedule

A master schedule is normally generated as the result of a product requirement forecast. Each product, possibly divided into a series of modules, has individual part numbers which are then used to explode their corresponding bill of material to generate part and assembly requirements by time periods. This activity is called material requirements planning.

Advances in this procedure now include pegged requirements, net change, manufacturing resource planning (MRPII), and just in time (JIT) planning. Pegged requirements allow each part and assembly requirement to be recorded, showing for what next higher level in the bill of material the requirement was generated. Net change allows a selective set of products to be re-exploded and the changes to the original requirements documented. (The original method was called regenerative material requirements planning, which required a completely new bill-of-material explosion and a corresponding completely new record of all requirements.)

Manufacturing resource planning is simply the integration of associated department systems, such as production planning, shop floor control, and vendor scheduling. JIT planning is simply a philosophy which applies the rules for material requirements planning even more rigorously and include programs to eliminate excessive lot sizes, quality rejects, machine breakdowns, and excessive transit times between operations.

The result of the master schedule system and material requirements planning is purchased and manufactured part and assembly requirements by time periods. Time periods can be as short as an hour or a day, and as long as a week or a month, and may be extended as far in the future as five years.

Material Requirements Planning (MRP)

Material requirements planning is a technique to schedule material into the manufacturing process at the proper time and in the proper quantities.

Numerically Controlled (NC) Tool Paths

In the past, each new design to be processed on an NC machine tool required an individual set of tool path geometry to be generated. This process is time consuming and liable to errors, often requiring excessive test piece machining. If, as a byproduct of part families, a family tool path is prepared, variations to parts within the family can be documented rapidly as new tool paths. If the geometry of the part family and the process for the family are integrated, these NC tool paths should become more accurate, be generated more rapidly, and conform to both engineering design standards and manufacturing requirements.

Perishable Tools

Perishable tools have often been considered expense items. Drills, taps, cutters, and more are used up during the machining process. More care is now being taken to inventory these items to make certain lack of their availability does not slow down the production cycle.

Planning Database

By combining the master schedule, material requirements planning (MRP), an ABC inventory analysis, and families of parts a significant planning database of information can be prepared and analyzed. Since it is important to improve design and manufacturing in the areas most needed and most used, the planning database can be the focal point for such actions.

Since all parts have been analyzed in the ABC inventory analysis, they can then be segregated into part families. Summarizing these parts by family allows the generation of an ABC analysis by family also. Thus the dollar values of parts, routing standardization, production planning, and other analyses can be given priority by family. If, for example, we assume a manufacturing company with 80,000 active parts classified into 1,000 families, it is far more efficient to analyze 1,000 families of collected parts than 80,000 individual items.

Product Costs

As parts and assemblies are manufactured, material and labor costs are collected by part number. Picking lists are utilized to collect parts together for issue to the shop floor for assembly, and these picking lists are extracts from the engineering bill of material. Once the product has been assembled, a cost roll-up can be used to determine its cost. Several different costs may be utilized in the cost roll-up. Standard cost, last actual cost, and average cost are some of the alternate methods used. The bill of material is the key system required to perform this activity.

Production Planning

Once the material requirements planning (MRP) system has generated manufactured part and assembly requirements, a production plan is prepared. Since each part and assembly has a process plan and a routing which reflects the work centers through which they must pass and the time required in each work center, a production plan and the corresponding work center loads can be generated.

Two types of shop loading may be generated, infinite, and finite. Infinite loading does not take into account any limitations of work centers. As a result, some or all work centers might be overloaded. Usually, compensation for an overload condition is made through increased work shifts, overtime, or subcontracting. When using finite loading techniques, the computer algorithms are designed to move loads from one work center to another either through alternate routings or by forward or backward scheduling. In a truly optimum rescheduling algorithm, string loading is required since, if one part of an assembly must be back loaded or forward loaded to level the shop load, all parts in the assembly should be changed in a corresponding manner to avoid manufacturing surplus inventory. Finite loading is a complex problem and rarely utilized effectively today.

Process Polycodes

A polycode is normally considered to be a specific attribute which is found on many designs regardless of its form, shape, or function. For example, attributes such as knurling, boring, grinding, material chemistry, and bolt circles might be considered some of these attributes.

Polycodes developed for these attributes can be used by themselves to group parts with similar characteristics, or they may be used to augment a design classification system for a family of parts with similar shapes, form, or function. In either case identification of these features may be used to determine whether parts may be manufactured together in a group. Used in conjunction with similar part geometry creates a powerful tool for group technology techniques and cellular manufacturing.

Quality Assurance

Emphasis has been placed on standardization of concepts, including design geometry, tolerances, producibility tips, families of parts, and more. Even though designs are different, standardization of concepts creates an atmosphere for more logical inspection techniques and quality measurements. Feature-based geometry becomes a significant tool for quality assurance methods and improves the possibility for electronic inspection techniques also. More reasonable tolerances, designs drawn to scale, and dimensions prepared with the manufacturing requirements considered make uniformity of inspection and the corresponding accuracy of results much more feasible and probable.

Raw Material

One of the more complex principles to understand in manufacturing is the conversion of a finished design into a corresponding raw material requirement. Four different conversion must be considered—purchased parts, forgings, castings, and raw materials.

Purchased Parts

A purchased part, whether it is an assembly, such as an electric motor, or a piece part, such as a nut or bolt, is considered a single piece. Quantities are measured in units of "each" or "pieces."

Forgings

All finished designs made from forgings have part numbers different from the purchased forging part number. In this manner the rough forging may be used, with variations, for other finished designs. Thus each finished forging requires a parts list or product structure showing this relationship, and the unit of measure is "each" or "piece."

Castings

All finished designs made from castings also have part numbers different from the purchased casting part number. Thus, as with forgings, each finished casting requires a parts list or product structure showing this relationship, and the unit of measure is "each" or "piece."

Raw Materials

The most complex relationship between a finished design and the material required is when raw materials are used. For example, a 3 in. diameter by 9 in. long finished design may require raw material 3 1/4 in. in diameter to machine to the finished dimension and 9 1/4 in. rough length for a cut-off allowance. The part can be cut from raw material purchased to a specific length, purchased in random lengths, purchased in multiple lengths, or to an exact cut length.

If a length other than cut length is specified, a parts list for the finished part must be prepared, and the unit of measure will be inches or an other corresponding linear measurement equal to that specified by the designer. If an exact cut length is designated, the raw material can be purchased using the finished design part number and specifying "operating 10," or the cut length can be ordered using a raw material part number and a unit of measure of "each" or "piece." In the former case no parts list is required. In the latter case a parts list is required.

When plate is purchased for a finished design, a parts list is required and the unit of measure is usually square inches, or other unit of measure equivalent to the dimensional units specified by the designer.

The units of measure required in this procedure are extremely important to maintain accuracy. For example, steel may be purchased in pounds, recorded for usage in feet, inches, or square feet or square inches, and disbursed in the same manner to production.

It should be recognized that the lowest level of a bill of material must end in one of these final configurations.

Shop Floor Control

During the planning stages of material requirements planning (MRP), three types of orders may be generated in the system, planned, firm, and released. Planned orders may be changed many times without affecting external paperwork or actions. Firm orders are used to reflect changes to requirements based on the reality of shop restrictions and other variables not considered in MRP. Released orders are entered into the shop floor control systems.

Once orders are released to manufacturing for parts and assemblies, shop packets are prepared, including the proper process or routing, drawings, and other materials. Labor is collected by part number and operation. Once a part or assembly is completed, it is reported and the finished design sent to inventory for storage.

Standard Processes

Once processes and routings have been developed for a family of designs, it is possible to standardize them for each family. If all geometric variations are presumed included, two means to develop specific processes for a new design may be considered.

Generative Process Planning

Generative process planning presumes that all machining features are identical between the standard process and the new design. If this presumption is true, the few variables, such as machinability, can be entered and the entire routing for the new design, including sequence of operations, will be created automatically. There seems to be no question that certain types of part families lend themselves to this proposition. However, and unfortunately, these cases are comparatively rare and do not solve the global problem of the simple but large varieties of slight differences.

Variant Process Planning

Variant process planning presumes families of parts with shape, forms, and/or functions, similar enough to be able to create a general process which covers a majority of the machining or fabrication requirements to make a new design for the family. Under this principle, a family routing is used as the base to make the necessary changes and/or additions to the process to make the new design. If this principle is utilized, it will almost certainly be more economical, and probably create a process more rapidly, meeting the general requirements more practically. In any case, it should be emphasized that generative process planning or variant process planning requires a significant amount of effort to standardize operations within a family of parts prior to applying the principles.

Standard Time Data

Standard time data is based on the principle of time and motion studies which have been used to develop standard movements of workers as they perform their work. Movements of a machine operator can be broken down into detailed elements, such as reach, grasp, place, locate, drill, and remove.

These standard elements have a corresponding time associated with them, including the variables of weight, complexity of the part, and fixtures used. Using this data, the total machining time for a design can be estimated and simulated quite accurately. Standard time data is thus used to create the set-up and run time data for a process and routing of a new design.

Transit Time

Each process for a design must be planned considering the time between operations. Since, normally, parts are machined using several machine tools, transit time becomes a significant portion of process time. Normally transit time is divided into three sections, wait time, move time, and queue time. Wait time is that which occurs during the machining operation as one part at a time is machined until a batch is completed. Move time is that which occurs from the time a batch is completed until it is transferred to the next operation. Queue time is that which occurs as the batch waits at the next operation until the machine operator can begin working on it.

Transit times are important in production scheduling and shop loading. However, they have often been created artificially using inaccurate parameters, and are often not as accurate as they should be.

Work Center Parameters

Each machine tool, whether it be a simple drill or a complex numerically controlled work center, has operating characteristics and limitations. These variables include operations which can be performed, part size limitations, tool travel, feeds and speeds, accuracy, and tooling requirements. Knowledge of these parameters is necessary when the process planner determines the most economical process and routing for a new design or for a set of designs.

The parameters of a work center often are the governing limitations for routing a new design and can become bottlenecks in the production should too many designs be routed over the same machine tool.

Appendix B
CIM CRITERIA

SCOPE

As a CAD/CAM system is implemented, the necessary ingredients for a successful utilization are required. Through several implementations, these have been carefully noted. While not necessarily all inclusive (a great deal depends upon the state of the existing systems in a manufacturing plant), they are some of the more important ones, with particular emphasis on configuration management and engineering documentation generally.

If these items are not considered, the possibility of a successful system is certainly reduced.

This appendix refers the reader to other sources for in-depth treatment, but a general roadmap of requirements has been presented.

THE BENCHMARKS FOR SUCCESS

Initiation and control of engineering documentation depend upon the following ingredients:

- Part number system
- Revision level control
- Piece part drawings
- Rules of interchangeability
- Product structures
- Drawing content definition
- Bill-of-material system
- Engineering change control
- Design retrieval
- Configuration management
- Engineering policies

Engineering standards
Engineering procedures
CAD/CAM requirements
Tool design with CAD/CAM

The intent of this appendix is to define each of these segments of engineering documentation sufficiently to gain an understanding of their importance, their interrelationship, and the written documentation needed to define a system utilizing them. Since modern systems tend to utilize a computer, the emphasis in this report will be on computer system implementation. However, all systems or items listed can be implemented manually should it be required and desired.

Since tool design is also a significant user of a CIM systems, consideration will be given to use in their design, also.

PART NUMBER SYSTEM

If a proper part number system and corresponding rules of use are poorly initiated, all other systems may fail. The reason for this blunt statement is built on history. Often drawing numbers have replaced part numbers. Often these are two different systems. Sometimes the customer order number is used as a replacement of the part number. In any case, there are often redundant documentation numbering systems, which tend to confuse the user and document the same design under a variety of different systems, making documentation changes difficult, redundant, and dangerous. Thus the following requirements for a part number meet the criteria for a clear, concise, and singular identity system.

A part number should be considered the singular identity for a design. Each design, whether purchased or manufactured, piece part or assembly, product or raw material, must have an identity, or part number.

A part number should be considered the singular identity for a design. Each design, whether purchased or manufactured, piece part or assembly, product or raw material, must have an identity, or part number.

A design may be documented on one or on many pieces of paper or on several electronic computer files, or in several geometric data files. In all cases the key to the data is the part number.

One common misconception is that the drawing and the part number can be different. If a drawing is considered to be just one of many potential portions of the design documentation schema, however, this illogical system will not be implemented, or converted if in use. Clarity of drawing content as well as other reasoning behind design documentation requirements will also help clarify this point.

Since the part number is used by all persons within a manufacturing company, not just engineering, it should be easy to use and designed to maintain accuracy. Thus the following rules should be considered in the design of a part number system:

1. The part number should be numeric only.
2. The part number should be no longer than eight digits and issued sequentially.
3. No meaning should be applied to or within the part number, should have no suffixes or prefixes, and should never vary in length.
4. Every part number must represent a design and the contents or definition of the design documented on some physical and/or electronic record.
5. The revision level of a design is not part of the part number.
6. Rules of interchangeability will be followed in changing the part number of a design and/or its associated revision level.
7. Means to obsolete a design and the corresponding part number will be prepared.

For additional references to these rules, see:

Engineering Documentation for CAD/CAM Systems

p. 13 Part Number Systems General
p. 13 History
p. 41 Revision level example
p. 45 Accuracy and Length
p. 141 Rules

CAD/CAM Systems, Planning and Implementation

p. 59 Part Numbers
p. 61 Rules

REVISION LEVEL CONTROL

Since the revision level is not considered a part of the part number, what is its use? Several systems rely on a good revision level control system:

Interchangeable design changes
The release procedure
Documentation quality and content
Communication of changes

Design Changes

It is important to recognize that the revision level changes only if an interchangeable change to the design occurs. In some companies where the revision level is considered part of the part number, a revision level change is the equivalent of a part number change and thus the release of a new design. Of course this creates unnecessary new designs with all the attendant costs of new processes, tools, routing, and more.

A new design could be released under three different revision level control systems. These levels will become even more significant as a CAD/CAM system is tied to the documentation of a new design since a design within the computer must be recognized through several steps of documentation credibility.

The Release System

If the revision level is designed so it can be "00," or numerical, or alpha, the "00" level will be considered a design in process and one which has not been released from engineering. This condition means that it is for engineering use only and no prototype or production items may be manufactured or purchased. In addition, the design is not ready for anything other than further engineering documentation. Engineering changes may be made to this type of design with no documentation changes, revision level changes, or consideration of the rules of interchangeability. This may be called Class C documentation.

A numerical revision level denotes a prereleased design which has some but not all the documentation necessary. This could be called a design with Class B documentation. However, sufficient documentation is available to purchase raw material for the design, or order a limited supply of the design for prototype production. Rules of interchangeability shall apply to this documentation as well as the engineering change procedures and other documentation control requirements.

Documentation Quality and Content

An alpha revision level signifies Class A, or complete, design documentation. While the rules of interchangeability and documentation control apply to both Class A and Class B documentation, a more formal approval of an engineering change may be initiated for Class A documentation, including an engineering change review board.

Communication of Changes

The revision level change, used as a communication medium, is the trigger which announces an interchangeable change has occurred.

Thus a design revision level will be found on an engineering drawing, design specifications, purchased part drawings, purchase orders, routings and processes, engineering change notices, item master and product structures within a computer system, part geometry in a CAD/CAM systems, assembly parts lists, bills of materials, and picking lists, among others.

When a revision level change is made to a design, the documents which change depend on whether the design is a piece part or an assembly. When the components of an assembly are changed, this often triggers a revision level change on the next higher level assembly and parts list. Rules of configuration control tied to product structuring guide this change requirement.

For additional references to these rules, see:

Engineering Documentation for CAD/CAM Systems

p. 150 Revision Level
p. 152 Interchangeability
p. 169 Release Conditions
p. 172 Documentation Levels

CAD/CAM Systems, Planning and Implementation

p. 37 Layers for Revisions
p. 146 Parts List Documentation

PIECE PART DRAWINGS

One of the most important concepts of a configuration management and engineering documentation system is that of individual piece part documentation. Often companies have documented a series of piece part designs on the same drawing as that of the assembly within which the piece part designs are used. Often multipart drawings are made which contain a variety of different parts simply to consolidate designs. Often an assembly drawing is prepared which has sufficient detailed dimensions that the parts comprising the assembly can be made by extracting the part dimensions. Many companies use tabulated drawings that are comprised of a picture with alpha dimensions and a table of parts each with their own list of alpha conversions to actual dimensions. There are nine reasons why these practices reduce good design control and increase design cost. These reasons are as follows:

Clarity of fabrication details
Economical part design
Documentation changes
Part reusability
Design retrieval

Elimination of duplication
Raw material requirements
Principles of manufacturing
  Routing requirements
  Tooling requirements
  Group technology
Family drawing preparation
Computer-aided design
Purchased parts

If configuration control is a significant factor in engineering documentation, then single part documentation becomes mandatory. Perhaps a simple example will illustrate. If two different designs and their assembly appear on one drawing, the drawing number and item number become the part numbers for the parts. If an engineering change is made to one of the parts, the only part number which can show such a changeis the combination drawing-item number. (If, at this time, each item on the drawing has a separate and individual part number which is not associated to the drawing number, a cross-index of part number and location on a drawing number must be maintained. Other ramifications of such a system will not be described here.)

An interchangeable change requires only a revision level change. However, the documentation is such that a revision level appears only on the drawing of the combined parts and assembly. A change to the drawing revision level will require item master changes in the computer of all parts and the assembly even though only one part has changed. If a table of revision levels is maintained on the drawing, a revision level change on the part will require a revision level on the drawing also to inform people that the drawing has changed also. This requires maintaining two revision levels for control of both the part and the drawing upon which the part is documented. If a mechanized parts list is maintained for such a drawing, the changed part, requiring only a revision level change, forces a change to the assembly drawing upon which the design appears, forcing a revision level change on the assembly parts list so both documents will track. This is an unnecessary set of confusing, costly, unnecessary, and misleading documentation changes.

If, on the other hand, a piece part on such a drawing must be redesigned, the part drawing on the multipart drawing must be changed. The part number should now be changed since the two designs are not interchangeable. However, they cannot change since the old design part number is irrevocably tied to the drawing number upon which the design appears. Thus the change to the design is, in fact, equal to a revision level change. However, the change to the drawing immediately erases the configuration of the former part. Even though notes might be kept of the former configuration of the part, how will the integrity of the former design be maintained? How will later users judge the interchangeability of old and new designs? What is the part number of the part made obsolete?

This dissertation may sound tedious and contrived, but it is necessary to simulate such systems to dramatize the problems with them. Alternative solutions are also expensive, requiring the initiation of new assembly drawings for minor changes. Such changes require significant approval if the ultimate user is the federal government. It is far better, and more economical to maintain singular piece part design control.

Last, but not least, a good and mutually acceptable product structure used by both engineering and manufacturing to build a product and create a bill of material is best served with piece part drawings.

For additional references to these rules, see:

Engineering Documentation for CAD/CAM Systems

p. 47 Piece Part Drawings
p. 140 Rules
p. 141 Part and Assembly Drawings
p. 172 Database

CAD/CAM Systems, Planning and Implementation

p. 53 Similar Parts
p. 82 Detached Parts List
p. 101 Parts Similarity
p. 164 An Application
p. 186 Families of Parts

RULES OF INTERCHANGEABILITY

Since several references have been made to interchangeability and the rules for changing part numbers and/or revision levels of designs, it is appropriate to insert this requirement here. For this report only a brief discussion will be prepared, with several references to general configuration management rules.

The rules of interchangeability are key factors in configuration management systems. In addition to the rules, a decision tree is also available, which assists the designer or drafter to determine, for his or her specific example, whether a part number or revision level must change. A specific reference to such a decision tree is shown below.

The rules of interchangeability are as follows:

Interchangeable: When two or more parts possess such fundamental and physical characteristics as to be equivalent in performance and reliability and capable of being exchanged one for the other without alteration of the parts themselves or any adjoining parts except for normal adjustments, and do not require selection for fit or performance, the parts are interchangeable. A change to a previously designed part which still allows it to be used according to this rule shall have a revision level change only.

Noninterchangeable: When a part is to be changed and the changed part does not physically fit or will not function as a replacement for the unchanged part in all applications of its previous revisions, and all unchanged parts will not physically fit or function as a replacement for the proposed change, the parts are considered noninterchangeable, and a new part number and corresponding drawing are issued for the new design.

These two rules govern documentation control. It is important to keep them in mind at all times as other facets of engineering documentation and control are discussed.

For additional references to these rules, see:

Engineering Documentation for CAD/CAM Systems

p. 152 Rules
p. 157 Examples

CAD/CAM Systems, Planning and Implementation

p. 147 Parts List Changes
p. 151 Decision Tree

PRODUCT STRUCTURES

The combination of two or more parts in a dismantleable or nondismantleable group of designs is called an assembly. A special case covered by this definition is a kit of parts which are not attached one to another. An assembly is but one rung of a ladder in a complete product structure. The definition for a product structure follows:

A product structure is defined as a graphical relationship showing the hierarchial relationships and combinations of parts and assemblies in a product.

The product structure is normally used to picture how both engineering and manufacturing wish to assemble a product. In a simple product, the structure may be only two levels, an assembly and the component parts. Some products, simple tools, for example, may require no structure. (However, in many cases a finished product, while not requiring assemblies, may require several raw materials. Thus the "piece part" is in reality a series of raw materials brought together into a product structure represented by a parts list.)

The product structure is normally converted into a series of parts lists which may also be defined as single level bills of material. The single level parts list also represents an assembly drawing and is called a detached parts list.

Once the parts lists are combined, particularly in a mechanized system, they may be joined together into an indented, or structural, bill of material by the computer. Another form of the bill of material is the summarized bill of material. A manufacturing form of the parts list is the picking list.

One of the key ingredients of a product structure is effectivity. When a design change is required on an assembly, often one part or subassembly design is replaced with another. However, this change is often not required immediately, and the economics of such a change dictate that old designs be used until they no longer exist. The date on which the old designs will be replaced with new is defined as the effective date. The product structure may be used to reflect both the implementation of the new design and the removal of the old design through the use of "in" and "out" designations in the product structure combined with the date of effectivity and an engineering change number, if the tracking of the reason for change is required.

Once a product structure is created for a product, variations to such a structure may be examined. Four methods of identifying and documenting these variations are known, or use of attachments, moving the variations to a higher level, dividing parts in one assembly into two thus adding a new level, and using the principle of a standard bill of material to which parts required by the customer are added, and those not needed are subtracted.

Once good product structuring procedures are documented, modular design techniques can be used. In general, modular design means documenting a product through the use of a series of disconnected modules. Each module may have two or more variations. When a customer order is received, the order number becomes the top level part number, and the parts list created is that of a series of modules, attachments, and special options. By using this technique, the bills of material required to document a product and variations are arithmetical, not geometrical in number.

For additional references to these rules, see:

Engineering Documentation for CAD/CAM Systems

p. 162 Bill of Material Ingredients
p. 175 Effectivity Date
p. 185 Product Structuring
p. 210 Modular Design

CAD/CAM Systems, Planning and Implementation

p. 82 Detached Parts List
p. 147 Parts List
p. 157 Sales Order

DRAWING CONTENT

Various types of parts often require different kinds of data on their drawing. Since all single part designs and all assemblies each have separate documentation, these variations can be more specifically defined. In general, however, the following criterial are recognized generally for adequate drawing content of designs:

1. The first most general rule is that the drawing for a design shall have sufficient data that no reference to other drawings is required. (Exceptions to this rule are reference through engineering specifications and other standard short cut documentation means.)

2. All dimensions required to make the part or assembly will be reflected on the drawing. The corollary to this rule is that part dimensions, unnecessary in the assembly process, will not appear on an assembly drawing.

3. Part numbers for piece parts and subassemblies will not appear on an assembly drawing. Find, or bubble numbers will be used to correlate piece parts of an assembly with their part numbers shown on a detached parts list.

4. The revision level of a design will be maintained not only on its drawing, but on all other documentation represented by the same part number. In the case where several sheets with the same part number are required, all such sheets will be identified by the same part number and the same revision level even though a change made to the design is reflected on one sheet only.

5. The detached parts list and the corresponding assembly drawing will have identical revision level numbers.

6. If material specification numbers are used as standard engineering documentation, they may be used in place of a complete raw material description for the design. However, any change to a raw material used for a design will require a new part number.

7. The dimensions shown on an engineering design are the "what" of the design, not the "how." Thus engineering, not manufacturing, dimension will appear on the drawing. No manufacturing techniques, practices, or instructions will appear on the engineering drawing.

8. A purchased part drawing will be prepared. This drawing does not necessarily mean a "picture," but, rather, a specification of the part. Where pictures and/or specifications are used to describe a purchased part, various qualified vendors and their corresponding vendor part number may appear on the the drawing. If such parts are not interchangeable between vendors, different part numbers must be assigned them.

9. A kit of parts requires a parts list but does not require a drawing.

10. A piece part cannot be documented on a bill of material with such documentation the sole source of the design content.

11. A part replaced by another can be so documented in the item master of a bill-of-material system, allowing any reference to the old part number be transferred to the new.

12. No design may be obsoleted while it is maintained as a portion of a product structure.

For additional references to these rules, see:

Engineering Documentation for CAD/CAM Systems

p. 62 Documentation Changes
p. 76 Purchased Parts
p. 127 Family Drawings
p. 141 Piece Part Drawings
p. 172 Raw Material

CAD/CAM Systems, Planning and Implementation

p. 91 Piece Part Drawings
p. 91 Drafting Standards
p. 111 Drawing Content Examples

BILL-OF-MATERIAL SYSTEM

The bill-of-material system is derived from two segments of the same database. These segments are the item master data and the product structure data, as noted previously. The bill-of-material system represents a series of methods to extract data from these files.

Several of the results of this system are listed below:

Item master data in several formats
Detached parts list

Picking list
Costed assembly
Single level used-on
Multiple level used-on
Spare part designation list
Spare part catalog
Indented (structural) bill of material
Summarized bill of material
Engineering scheduling data
Costed bill of material

The mechanized bill-of-material system eliminates the need to create a material list for an assembly drawing. In fact, the implementation of a mechanized bill-of-material system makes it mandatory to eliminate such a duplication in order to avoid errors of redundancy.

The item master database has several manufacturing data increments added to it to perform inventory control, production scheduling, and other manufacturing systems activities. The product structure may be tied directly to an assembly process so that parts of the assembly may be issued in some assembly-oriented and lead-time-oriented sequence.

In addition, scrap percentage and other manufacturing variables may be attached to the product structure database to facilitate sufficient parts or order enough raw materials to meet production demand and forecasted scrap. Thus the database used in a bill of material becomes a jointly used system and is the "spine" of all other engineering and manufacturing systems.

For additional references to these rules, see:

Engineering Documentation for CAD/CAM Systems

p. 170 Bill of Material Outputs
p. 173 Engineering Schedule
p. 178 Implementation
p. 265 Database Example
p. 266 Output Examples

CAD/CAM Systems, Planning and Implementation

p. 145 Definition
p. 91 Illustration

ENGINEERING CHANGE CONTROL

An engineering change procedure is the most important system required for control of documentation and the corresponding designs.

No product is stagnant. Since design changes must, and do, occur, the means to control and document them are significant. Thus an engineering change procedure is necessary.

The rules of interchangeability are key to design changes. Not only may a design change, but the part number change may have to be reflected in subsequent levels of the bill-of-material documentation. In addition, the effectivity of the change, as mentioned previously, will require not only documentation, but also agreement between engineering and manufacturing.

Often engineering changes occur as a result of personnel in departments other than engineering finding some discrepancy in the design or some problem with the manufacture of the part as is. Thus an engineering change request may be required to initiate a formal change. As noted previously, it is possible to have three levels of engineering documentation, each allowing a different engineering change discipline. Regardless, the communication of the change is vital. In the past, an engineering notice, a manual system requiring paper documentation for a change, has been utilized. Often the signatures of several persons are required before the change can be initiated and implemented. Since these approvals are often sequential, an inordinate amount of time is spent unnecessarily. Thus more systems are relying on more modern electronic mail procedures to implement the system. Thus approvals and knowledge of changes will be in parallel, not in sequence, saving considerable time and red tape.

Once a database is on line, an engineering change may be researched rapidly, and effectivity dates, for example, computed rapidly for exhausting old designs before using new.

For additional references to these rules, see:

Engineering Documentation for CAD/CAM Systems

p. 154 Engineering Changes
p. 155 Change Notice Types
p. 177 Engineering Change Board
p. 268 Change Order Example

CAD/CAM Systems, Planning and Implementation

p. 7 The Largest Database
p. 152 Engineering Changes

DESIGN RETRIEVAL

The avoidance of creating unnecessary design is becoming more important to manufacturing companies than ever before. The cost for initiating one new design has grown asymtotically during the past 10 years. A simple design may cost $200; a complex design $24,000. Thus design retrieval is becoming a popular system. While generally thought to be a system for retrieving designs used in products, many companies have found a design retrieval system in tool design, combined with piece part documentation and modular design has significantly cut tool design costs.

Several types of design retrieval systems have been developed over the years. The most common has been making the part number a significant code which then includes data about the design. Unfortunately, this data is usually quite parochial, including data such as product type, drawing size, material, and other such data. Overlap of data and/or definition of the data occurs also, and these systems do not create an economical, rational, logical, of effective design retrieval system.

Name coding has also been attempted. Unfortunately, a noun description must be augmented with several adjectives to segment designs into groups which can be distinguished one from another. The adjectives as well as the nouns can become so complex and designs can be defined in so many ways by so many people, that standardization becomes difficult or impossible.

Picture coding is another example of an attempt to classify parts together by similar characteristics. While this seems a better method than name coding, the proliferation of pictures and the subtle difference in such pictures makes it difficult to determine the code number and corresponding picture which should be selected.

While other manual or homegrown systems have been developed by many manufacturing companies, the most effective systems today appear to be those which have been designed and implemented by professional U.S. consulting firms. Three of them are described below.

Organization for Industrial Research, Inc. (OIR)

240 Bear Hill Road
Waltham, Massachusetts 02154
617-890-4030
Alexander Houtzeel, President
(Formerly TNO, Amsterdam, Holland)

This consulting firm, formerly TNO, has begun an aggressive campaign to sell their classification system. OIR has excellent marketing brochures, and a great deal of credibility in the marketplace. The former president, Mr. Houtzeel, is a personable, outgoing, credible salesman of their product.

OIR specializes in group technology techniques, utilizing a multiple code to utilize the system.

MI-CLASS is the name of the classification system. MI-PLAN is the name of the process plan generator. MI-GRAPHICS is the name of an interface of computer graphics classification, coding, and group technology.

OIR has strong ties to Computervision, and both are jointly marketing CAD/CAM and classification systems.

OIR confesses to be weak in design retrieval but is strong in manufacturing processes and developing multiple codes for manufacturing.

OIR professes that it is unnecessary to preplan a classification system, but, rather, simply begin the classification process, and it will operate efficiently.

OIR is strong in manufacturing techniques and has developed some unique interfaces between a multiple code and the resulting machine tool routing requirements.

Brisch, Birn and Partners, Inc.

1656 S. E. Terrace
Fort Lauderdale, Florida 33316
303-525-3166
Bill Hyde, President
(Formerly, 1960, Brisch Ltd., England)

The Brisch firm was founded in England in the early 1940s. It soon gained fame in England, on the continent, in Russia, and in Africa. The first system installed in the United States was De-Vilbiss, and the second, Caterpillar Tractor Company, in 1956, and 1957.

The Brisch classification system is both a decision tree and a multiple code system. At one time, the Brisch firm was foremost in group technology skills also, but has not promoted the technique for some years.

The firm is small with not over 14 people. However, they have had several classification assignments, including New York Air Brake, Watertown New York, Morgan Construction Company, Worcester, Mass., Boeing Aircraft, and others.

The system is recognized as the "Cadillac" of the industry. The Brisch firm is not noted for their marketing ability.

Until recently, the Brisch system was completely manual, but the implementation of the TAGS cataloging system has greatly improved their system and its ability to be used by a host of people. It is rapid and user friendly.

The Brisch firm has been instrumental in promoting classification by holding a series of seminars, asking users to present their findings in the implementation of the system. These seminars have been successful.

Implementation of a Brisch system starts with a survey on the client's site, followed by a formal report stating costs, savings, and scheduled completion time.

The company places one or more of their personnel on site as the classificaton system is designed, training the client's personnel in the process.

Computer Aided Manufacturing Laboratory

Brigham Young University
433 CB
Provo, Utah 84602
DCLASS

The DCLASS system is being promoted by the University of Utah. The system originated from a take-off of the Boeing Aircraft BUCCS system. The BUCCS system was designed and implemented by the Brisch, Birn and Partners firm.

The philosophy behind DCLASS is almost precisely that of the Brisch firm. However, DCLASS is claimed to be a universal classification system. The university provides a series of predesigned decision trees for the use of their clients.

Much of the classification logic and retrieval has been mechanized and is available on computers as small as the Apple to a VAX, IBM, UNIVAC, and other computers. Emphasis is placed on the mechanization of the system rather than the quality of the classification procedures. In brochures prepared by the university, there are implications that users must customize the system for their personal use.

DCLASS is also tied to process planning logic, decision tree formats, and time standards, all computer generated.

No consideration is given, nor recognized, that implementing a DCLASS system will cause a possible imbalance in the number of parts within a family where too large a group may appear, or even none at all.

As engineering documentation becomes more integrated, and with part numbers which have no significance, the database of both text and geometry must have some means for rapid and logical retrieval. A professional classification system appears to be the answer. Part family data in a CAD/CAM system will become the most significant database in engineering in the future.

For additional references to these rules, see:

Engineering Documentation for CAD/CAM Systems

p. 79 Design Retrieval
p. 127 Family Drawings for CAD/CAM

p. 257 Design Retrieval Example
p. 276 Family Requirements Planning

CAD/CAM Systems, Planning and Implementation

p. 105 Design Retrieval Principles
p. 133 Design Retrieval

CONFIGURATION MANAGEMENT

Configuration management may be considered the effective implementation and use of all the ingredients of an engineering documentation system described previously. The more formal definition for configuration management follows:

> Configuration management is the systematic approach to identifying, controlling, and accounting for the status of the parts and assemblies required in a contract and/or design from the point of its initial definition throughout its entire life.

Configuration management for a new product may be divided into four sections, a functional definition, a hardware definition, the manufacturing definition, and the field definition. As a portion of this definition, a product number and its proper configuration control are required. Thus the product configuration baseline is based on the following items previously described:

Product designation number
Released drawings
Released specifications
Released manuals
Standard product and options
Spare parts
Bill of material
Family tree

A special case of configuration management is called traceability, and/or lot control. Lot control, usually a requirement of the government, is based on the premise that a lot of raw material or parts may be identified with a unique number. This lot number is then associated with every design which uses the raw material or parts.

Traceability is a more concise set of controls in that each design or segment of raw material must have an individual identity number (other than the part number). This unique number must be recorded in such a manner that the design, regardless of where it is used, will be trackable. To some extent this traceability procedure seems a paradox. On one hand, a manufacturing company is striving to identify identical parts with an identical part number to manufacture them economically in one batch. On the other hand, traceability requires different numbers on identical parts to identify their differences. The system required to control this documentation is rather expensive, requiring a significant documentation effort.

For additional references to these rules, see:

Engineering Documentation for CAD/CAM Systems

p. 143 Configuration Management
p. 146 Product Baseline
p. 147 Product Definition

CAD/CAM Systems, Planning and Implementation

p. 149 Policies, Standards, Procedures

POLICIES

Policies are an excellent way to document the intended activities of a company. Well-written policies signal the intent of all departments and their willingness to work as a team under rules agreed to by all. Thus, as a means for communication, a policy is extremely important for the implementation of a configuration management system. Most policies, when approved, require an engineering standard and/or an engineering procedure to carry out the intent.

A formal definition of a policy follows:

> A policy is a statement of requirements and objectives; guiding principles governing action.

STANDARDS

An engineering standard is defined as the rules of the game, the "what" of an activity. Often the means of accomplishing what is described is left to the procedural aspect. Thus the methods and the procedure are not usually considered part of such a standard.

PROCEDURES

A procedure is considered a written document which assists users to follow a path of data flow, fill in forms, perform daily activities, or itemize instructions for standard actions. The written procedure is often confusing and redundant. Thus care must be taken to make them clear and simple, with alternate methods clearly defined.

A more formal definition of a procedure follows:

> A procedure is a written definition of methods and practices required to direct and guide the activities in support of an established system.

For additional references to these rules, see:

Engineering Documentation for CAD/CAM Systems

p. 147 Policy Standards Procedure Timing
p. 148 Policy Standards Procedure Matrix

CAD/CAM REQUIREMENTS

With the implementation of a CAD/CAM system, the requirements of configuration management described previously must be executed even more carefully. No longer will manual systems or visual systems control drawing content. No longer can engineering changes take place on a drafting board, because the design drawing data is now stored on the computer. Thus the need for a large database, a database manager, and configuration control of such data becomes a significant requirement.

None of the requirements mentioned previously will change. The procedures on their implementation may. In addition, many procedures presently performed sequentially may now be replaced with a system which allows the procedure to operate in parallel.

For additional references to these rules, see:

Engineering Documentation for CAD/CAM Systems

p. 5 CAD/CAM Integrated Data Flow
p. 74 Drafting Database
p. 98 CAD/CAM Family Database
p. 99 Family Drawings
p. 255 Detail Design Analysis
p. 295 Quality Control

CAD/CAM Systems, Planning and Implementation

p. 3 Ingredients of a CAD/CAM System
p. 7 The Database
p. 54 Producibility Tips
p. 193 Process Planning
p. 234 Engineering-Manufacturing Database

TOOL DESIGN WITH CAD/CAM

Probably the most concise description of tool design requirements is to consider them a product and simply repeat all the rules mentioned previously to document and control each of them. Thus configuration control for tooling requires part numbers, piece part drawings, and retrieval system, configuration control rules and procedures, and all the other ingredients of a control system.

For additional references to these rules, see:

Engineering Documentation for CAD/CAM Systems

p. 292 Tooling

CAD/CAM Systems, Planning and Implementation

p. 3 Ingredients of a CAD/CAM System
p. 252 Tooling

SUMMARY

All the ingredients of a workable configuration management system have now been described. Obviously policies, standards, and procedures are required to define the activities necessary to integrate these ingredients to have them operate as cogent and economical system. A few companies have made such a system operate. Many more will.

INDEX